Space Law and Policy

Series Editor

Maria A. Pozza, Gravity Lawyers, Christchurch, New Zealand

Space activities across the Asia-Pacific, specifically in Australia and New Zealand, are increasing and diversifying. Over the last decade, countries in this region have seen significant growth within their national space industries and economies which in turn benefit their wider societies due to the downstream application and use of, space-related technologies, products, and services. This phenomena presents not only a great opportunity for these countries, but also represents a situation necessitating considered and fit-for-purpose space law and policy. Robust yet flexible regulatory regimes are needed if countries in the Asia-Pacific, such as Australia and New Zealand, are to benefit from the strategic and commercial advantages associated with a presence in outer space and functioning domestic space sector. These regulatory regimes must respond to, and address, numerous challenges ranging from adherence to international legal obligations to issues of geo-political space security and military concerns through to questions about how best to regulate and support commercial space endeavours. These myriad factors will have an impact upon how countries develop, implement, and refine their space laws and policies. This series offers a unique perspective from this region and intends to act as a source of knowledge and guidance for space actors across government, the private sector and academia in the pursuit and furtherance of their respective space activities and operations. The contributors to this series are all experts in their field who share a common interest in strengthening space governance via the promotion and development of space law and policy.

Sandeepa Bhat B · Adithya Variath
Editors

Legal Responses to Space Resource Utilisation

Springer

Editors
Sandeepa Bhat B
The West Bengal National University
of Juridical Sciences
Kolkata, West Bengal, India

Adithya Variath
Jindal Global Law School
O.P. Jindal Global University
Sonipat, Haryana, India

ISSN 2662-9054 ISSN 2662-9062 (electronic)
Space Law and Policy
ISBN 978-981-95-8863-3 ISBN 978-981-95-8864-0 (eBook)
https://doi.org/10.1007/978-981-95-8864-0

This Springer imprint is published by the registered company Springer Nature Singapore Pte Ltd.
The registered company address is: 152 Beach Road, #21-01/04 Gateway East, Singapore 189721, Singapore

Preface

The journey of this book traces back to our earlier edited volume on the Artemis Accords, which itself emerged from the deliberations at the *Global Conference on Resource Mining in Outer Space: Artemis Accords and Beyond*, an event we curated in 2023. That project brought together an extraordinary range of scholars and practitioners to grapple with the legal and political dimensions of space resource activities and the geopolitics around Artemis Accords. It revealed not only the urgency of the subject but also the diversity of perspectives that could enrich the debate. This present book is a natural continuation of that effort. However, it also marks an important shift in our approach. While the first book, *Artemis Accords and Resource Mining in Outer Space* (Springer, 2025), concentrated primarily on the Artemis Accords and its implications, this second project moves beyond and offers theoretical frameworks, normative debates, and national approaches to the governance of space resources.

The expanding presence of States and private actors in outer space is forcing the international community to revisit fundamental principles of space law. Questions concerning appropriation, benefit-sharing, and cooperation have gained unprecedented prominence. While the United Nations is still deliberating on many of these issues in situating the resource utilisation within broader patterns of international law and diplomacy, this book attempts to capture the evolving landscape of governance of outer space resources.

As we read this book, the field of space mining is undergoing a remarkable transformation. Unprecedented advancements are being made in asteroid mission technologies and the exploration of lunar resources, with private companies announcing ambitious asteroid missions and pursuing the use of lunar regolith for in-situ resource utilisation. These multi-million dollar investments are flowing into critical infrastructure, ranging from next-generation lunar landers to advanced propulsion systems to integration of artificial intelligence in space technologies. These developments underscore the timeliness and urgency of answering the questions explored in this book.

This book is organised into two parts. The first part addresses theoretical and normative frameworks, critically interrogating the foundations of space law and governance in light of emerging resource challenges. The second part examines diverse national approaches, spanning the United States, China, Australia, the United Kingdom, Japan, New Zealand, and the United Arab Emirates, illustrating how domestic legal developments and strategic choices intersect with international norms. Together, these contributions provide a comprehensive and comparative view of the global discourse on space resource utilisation.

As editors, our intention has been to develop a project where contrasting perspectives can coexist. We have not sought uniformity of opinion but rather diversity and multiplicity of analysis, so that readers can appreciate the richness and complexity of the debate. We believe that such academic disagreements and agreements foster a deeper understanding of the legal, geopolitical, and ethical issues surrounding the exploitation and utilisation of celestial resources.

We remain grateful to the contributors for their intellectual generosity and commitment, and to Springer for once again providing a scholarly home for this project. We are also grateful to Mamun Mishra for her support and belief in this project. We also thank Cynthia Kroonen and Raghavi Jayagopal for coordinating this book project. This book stands as a collective endeavour, enriched by the insight and dedication of all involved. This book is as much a reflection of our contributors' dedication as it is of our own. We hope that it will contribute meaningfully to ongoing discussions on resource utilisation and space governance, inspire future research, and guide policymakers, industry stakeholders, and academics and scholars in shaping a sustainable framework for space resource utilisation.

Kolkata, India Prof. (Dr.) Sandeepa Bhat B
Sonipat, India Adithya Variath

Contents

Editors and Contributors

About the Editors

Sandeepa Bhat B is working as a Professor of Law, Dean—Post Graduate Studies, and the Director of the Centre for Aviation and Space Laws at the National University of Juridical Sciences, Kolkata. He has the teaching and research experience of over twenty-two years after completing his master's degree with first rank and double gold medals in 2003. His seven Major Research Projects are sponsored by the World Bank, ISRO, the WB Judicial Academy, the Ministry of Justice, the Ministry of Environment, Forest and Climate Change, the Ministry of Commerce, and the Government of West Bengal. He holds the unique feat of being the first Indian to be inducted as the UNIDROIT Correspondent for India. He has been a member of four distinguished international bodies, including the American Society of International Law. Dr. Bhat has published thirteen books, nine of which are on aviation and space laws. In addition, he has published more than seventy articles in the journals of international and national repute. He also has the distinction of being a member of the Indian Space Research Organization's Expert Group for drafting the National Space Act for India.

Adithya Variath is an Assistant Professor and Associate Dean at the Jindal Global Law School, O.P. Jindal Global University, India. Adithya was an Elizabeth Moir Scholar at the Oxford India Centre for Sustainable Development at the Oxford Department of International Development, University of Oxford. Before moving to Oxford, he served as an Assistant Professor of Law and founding Coordinator of the Centre for Research in Air and Space Laws at Maharashtra National Law University Mumbai. He is the recipient of the 2025 Sachdeva Award for Space Law. Adithya's research focuses on space and aviation laws, emerging technologies, and global institutional design. He is the recipient of the 2025 Sachdeva Award for Space Law.

Contributors

Sandeepa Bhat B Centre for Aviation and Space Laws, The WB National University of Juridical Sciences, Kolkata, India

Anna Marie Brennan University of Waikato, Hamilton, New Zealand

Art Cotterell Faculty of Law and Justice, University of New South Wales, Sydney, NSW, Australia

Lucia Kozák Csajková Prague University of Economics and Business, Prague, Czech Republic

Melissa de Zwart Andy Thomas Centre for Space Resources, ARC Centre of Excellence in Plants for Space, Adelaide University, Adelaide, South Australia, Australia

Chelsea Francek University of Waikato, Hamilton, New Zealand

Karl Grossman State University of New York at Old Westbury, Old Westbury, NY, USA

Michelle L. D. Hanlon Center for Air and Space Law, The University of Mississippi School of Law, Oxford, MS, USA

Stacey Henderson Jeff Bleich Centre for Democracy & Disruptive Technologies, Flinders University, Adelaide, Australia

Joel Lisk Jeff Bleich Centre for Democracy & Disruptive Technologies, Flinders University, Adelaide, Australia

Miloslav Machoň The Ambis University, Prague, Czech Republic

Leonardo Alberto López Marcos Department of Public International Law, Complutense University of Madrid, Madrid, Spain

J. Lauren Napier Northumbria University, Newcastle, UK

Christopher J. Newman Northumbria University, Newcastle, UK

Nancy Riordan University of Massachusetts, Boston, MA, United States

G. S. Sachdeva Chandigarh University, Mohali, Punjab, India; NALSAR University of Law, Shameer Pet, Hyderabad, India

Scott Schneider Special Counsel, International Aerospace Law & Policy Group, Brisbane, Australia

Charles W. Stotler University of Mississippi School of Law, Oxford, USA

Yumiko Takatori Takatori Law Office, Tokyo, Japan

Fabio Tronchetti Northumbria Law School, Northumbria University, Newcastle upon Tyne, UK

Adithya Variath Jindal Global Law School, O.P. Jindal Global University, Sonipat, India

Theoretical and Normative Frameworks for Space Resource Utilisation

Changing Legal Dimensions of Mining Space Resources

Sandeepa Bhat B

Abstract With the unfolding of the space age and subsequent technological developments, we have defied the rule of gravity to enter outer space and to land on celestial bodies. Manifold commercial uses of outer space ensured that the States are not the only players in outer space. Investments from private entities in commercial space activities have increased extensively in the twenty-first century. Meanwhile, the technological breakthroughs resulted in devising plans for the commercial utilisation of space resources. Both State and nonstate actors are now aspiring to benefit from space resource mining. The most significant question arising out of these developments is whether the current developments are in tune with States' commitments under international space law. To answer this question, one needs to go through the fundamental norms of space governance developed under the Outer Space Treaty 1967, which stands as the Magna Carta of international space law. The common rights regime reflected by the principles like benefit and interests of all, the province of all mankind, freedom of access, freedom of exploration, use and scientific investigation, and national non-appropriation is key for our understanding in this context. Current developments in the form of different national space laws at the domestic levels and Artemis Accords at the international level need to be tested through the lens of international space law, which shall be the focus of this chapter.

Keywords Space resources · Space mining · Outer Space Treaty · Commercial space activities · International space law

1 Introduction

When the advent of space took place, most of the countries were being liberated from their colonial powers. As newly independent nations, they were struggling to overcome economic and political tensions. The economic disparity between them

S. Bhat B (✉)
Centre for Aviation and Space Laws, The WB National University of Juridical Sciences, Kolkata, India
e-mail: sandeep@nujs.edu

S. Bhat B and A. Variath (eds.), *Legal Responses to Space Resource Utilisation*, Space Law and Policy, https://doi.org/10.1007/978-981-95-8864-0_1

and other countries was evident.[1] Those countries had a common understanding that overcoming economic problems would also help them in effectively dealing with political concerns. Hence, the developmental needs became a matter of prime importance to them. This was also reflected in the international trade negotiations under the General Agreement on Tariffs and Trade (GATT) 1947.[2] Incorporation of Part IV on "Trade and Development" in 1965 as a part of the GATT is the result of global awareness to respect the developmental needs of marginalised States.[3]

As Carl Q. Christol has rightly remarked, the entry into outer space was a ray of hope for newly independent States to overcome their economic troubles.[4] Since outer space and the resources it offers are bounties of nature, collective rights of all in their use were considered as a foundational principle applicable in carrying on space activities. Human activities in outer space were viewed as a magnificent opportunity to correct colonial injustices. This was also considered to be one of the reasons for States not objecting to the orbiting of satellites over their territories.[5] The negotiations leading to the drafting of the Outer Space Treaty[6] unveiled several principles of the common rights regime, which were codified in the Outer Space Treaty. The drafters of the Moon Agreement[7] intended to provide a logical conclusion for respecting common rights in celestial resources by incorporating the concept of common heritage of mankind (CHM).

The concept of CHM undoubtedly stands as the most advanced concept in the common rights regime. With its focus on individual non-appropriation, international management system, equitable sharing of benefits, sustainable use, and peaceful uses, the CHM regime was perceived as the torchbearer of protecting the rights of all without any discrimination.[8] Unfortunately, space-faring nations consider this to be against their self-interest. As a consequence, the Moon Agreement suffered from a very limited number of ratifications. Even among the parties to the Moon Agreement, very few are space-faring nations. This effectively takes away the sanctity of CHM in its application to the moon and other celestial bodies. An argument of

---

[1] Luis Angeles, 'Income Inequality and Colonialism' (2007) 51(5) *European Economic Review* 1155, 1155–76.

[2] *General Agreement on Tariffs and Trade 1994* (15 April 1994) Marrakesh Agreement Establishing the World Trade Organization, Annex 1A, 1867 UNTS 187 (GATT).

[3] Protocol to introduce Part IV to GATT (adopted 1965, entered into force 27 June 1966) 'Part IV: Trade and Development' https://www.wto.org/english/res_e/booksp_e/gatt_ai_e/part4_e.pdf. Accessed 5 May 2025.

[4] Carl Q Christol, 'Important Concepts for the International Law of Outer Space' in *Proceedings of the Fortieth Colloquium on the Law of Outer Space* (1998) 73, 74.

[5] Stephen Gorove, 'Major Legal Issues Arising from the Use of the Geostationary Orbit' (1984) 5 *Michigan Journal of International Law* 3, 4.

[6] *Treaty on Principles Governing the Activities of States in the Exploration and Use of Outer Space, including the Moon and Other Celestial Bodies* (adopted 27 January 1967, entered into force 10 October 1967) 610 UNTS 205, 18 UST 2410, TIAS No 6347, 6 ILM 386 (Outer Space Treaty).

[7] *Agreement Governing the Activities of States on the Moon and Other Celestial Bodies* (adopted 5 December 1979, entered into force 11 July 1984) 1363 UNTS 3 (Moon Agreement).

[8] Arvid Pardo, 'An International Regime for the Deep Seabed: Developing Law or Developing Anarchy?' (1969) 5(2) *Texas International Law Forum* 204.

considering CHM as a part of customary international law in governing the moon and other celestial bodies also fails in the absence of established *opinio juris* and state practices.[9]

However, the failure of CHM to have strong roots in space law shall not be construed as the end of the common rights regime in space law. The widely accepted Outer Space Treaty still comprises several common rights principles, which are mandatory for observance in carrying on any activity in outer space and celestial bodies. Unfortunately, commercial space activities in this New Space era fail to respect common rights principles enshrined under the Outer Space Treaty. The legal developments in several space-faring nations indicate the acceptance of unwarranted deviation from established common rights principles. One of the significant aspects evident in such legal developments is permitting the mining of space resources. This has far-reaching consequences on the space treaties of the twentieth century, which were carefully negotiated instruments to cater to the needs of both the developed and the developing parts of the world.

2 Recent Legal Developments on Space Resource Mining

In 2015, the United States became the first country to pass a law authorising space resource mining. It introduced a chapter on "space resource commercial exploitation and utilisation" in the Commercial Space Launch Competitiveness Act 2015.[10] With a disclaimer on the assertion of sovereignty and ownership over celestial bodies, the United States law permits the commercial exploration and commercial recovery of space resources by United States citizens. It highlights the need for encouraging commercial exploration and recovery of space resources by United States citizens, and discourages government barriers in this process. It confers property rights to United States citizens over mined resources by granting a right of possession, ownership, transportation, use and sale of those resources.[11] Though this law led to much controversy, the United States remained aloof from it. Relying on some internal deliberations and brushing aside a valuable dissenting opinion expressed by Prof. Joanne Gabrynowicz, the United States asserted its position to be in conformity with its obligations under the Outer Space Treaty.[12]

While the United States' position was controversial, Luxembourg followed it in its national legislation passed in 2017. The Luxembourg Law of July 20th 2017 on the Exploration and Use of Space Resources is a more detailed piece of law to authorise

[9] David S Myers, 'Is there a "Common Heritage of Mankind?"' in *Proceedings of the Thirty-third Colloquium on the Law of Outer Space* (1991) 335, 336.

[10] *United States Commercial Space Launch Competitiveness Act 2015*, Public Law 114–90, 25 November 2015.

[11] *United States Commercial Space Launch Competitiveness Act 2015*, Public Law 114–90, 25 November 2015, Sec. 51303.

[12] See Mr. Smith of Texas, Space Resource Exploration and Utilization Act of 2015, House of Representatives, 114th Congress Report, 1st Session, 114–153.

private entities to utilise space resources commercially. In 2019, the United Arab Emirates enacted its Federal Law on the Regulation of the Space Sector. In Section 18 of this law, the Council of Ministers and its delegates were empowered to develop norms relating to space resource exploration, exploitation and use. In pursuance of this power, Cabinet Resolution number 19 was passed in 2023 regarding the space resource utilisation.[13] It empowers the UAE Space Agency to grant permits for the commercial recovery of space resources. The right to own, buy, sell, trade, transfer, store, use or dispose of extracted space resources is conferred to the operator of space resource activities. In 2021, Japan enacted the Act on the Promotion of Business Activities for the Exploration and Development of Space Resources. It also mentions the grant of licences for space resource activities and states under Article 5, "A person who conducts business activities related to the exploration and development of space resources shall acquire the ownership of space resources that have been mined…".

In 2023, India passed the Indian Space Policy 2023 (2023 Policy)[14] to promote commercial space activities. Non-governmental entities are encouraged to undertake end-to-end activities in the domain of outer space. In this process, the 2023 Policy also promotes the commercial recovery of space resources. It copies the 2015 United States legislation regarding ownership of mined resources by stating that the non-governmental entities involved in mining space resources "shall be entitled to possess, own, transport, use, and sell any such asteroid resource or space resource obtained…" Though one may consider this as non-binding and consequently a trivial development in the absence of a legislative framework, the Government of India has proceeded forward with empowering the Indian National Space Promotion and Authorisation Centre (IN-SPACe)[15] to enforce the 2023 Policy. The IN-SPACe has also developed Norms, Guidelines and Procedures for Implementation of Indian Space Policy-2023 in respect of Authorization of Space Activities 2024 (NGP). How an executive policy without the backing of a law, especially in a Parliamentary form of democracy like India, will have a binding legal effect is a matter to be considered by the Indian judiciary in due course.

While the above were the domestic legal developments in different countries, the United States attempted an international instrument on space resource mining by developing the Artemis Accords in 2020.[16] Describing it as a non-binding political commitment, the United States sought cooperation from other countries to explore and use outer space and its resources. The Artemis Accords contains ten core principles that focus on the mining and utilisation of space resources. Declaring it to be in accordance with the Outer Space Treaty, Section 10 of the Artemis Accords

---

[13] Cabinet Resolution No (19) of 2023 Concerning the Space Resources Regulations https://uaelegislation.gov.ae/en/legislations/1634/download#:~:text=Property%20rights%20over%20Space%20Resources&text=The%20Operator%20also%20has%20the,as%20authorised%20by%20the%20Agency. Accessed 5 May 2025.

[14] *Indian Space Policy 2023* https://www.isro.gov.in/media_isro/pdf/IndianSpacePolicy2023.pdf. Accessed 5 May 2025.

[15] Indian Space Policy 2023, Clause 5.

[16] *Artemis Accords* (signed 13 October 2020) https://www.nasa.gov/wp-content/uploads/2022/11/Artemis-Accords-signed-13Oct2020.pdf. Accessed 5 May 2025.

permits the extraction and utilisation of space resources. Section 11 supplements this by conferring the rights to State Parties to declare safety zones and prevent the entry of others into these zones to avoid possible interference with their activities. In addition, Section 9 allows the State Parties to take appropriate measures for preserving their outer space heritage.

Even though the Artemis Accords is a non-binding instrument and has no standing in international law, the number of signatories to it is increasing alarmingly. With Bangladesh becoming the latest signatory on 8 April 2025, 54 countries have become part of it.[17] Russia and China have rejected this US-centric instrument. As a counter to the Artemis Accords, Russia and China have jointly proposed the International Lunar Research Station Project (ILRS Project). The ILRS Project proposes to establish a permanent base on the moon by 2035, and to power this permanent base, it has an in-built plan of developing nuclear reactors to supply energy in harsh lunar conditions.[18] As it stands in April 2025, 17 countries and more than 50 research institutions have become part of the ILRS Project.[19]

The two parallel developments, the Artemis Accords and the ILRS Project, are posing a serious threat to the maintenance of the balance of power in space exploration. In addition, these recent legal developments, both at the national and international levels, threaten to disregard the established common rights principles of the Outer Space Treaty. In light of this, probing into the principles of the Outer Space Treaty and understanding them in the right spirit of common rights attains significance. The subsequent parts of this chapter attempt to perform this task.

3 Common Rights Principles of the Outer Space Treaty

3.1 Benefit and Interests Principle

Article I of the Outer Space Treaty encompasses the need to ensure that space activities benefit and cater to the interests of all States. It stipulates that "The exploration and use of outer space, including the moon and other celestial bodies, shall be carried out for the benefit and in the interests of all countries, irrespective of their degree of economic or scientific development…" Unfortunately, the benefit and interests principle of the Outer Space Treaty has not received due recognition in practical terms.

---

[17] NASA, 'The Artemis Accords' https://www.nasa.gov/artemis-accords/. Accessed 5 May 2025.

[18] Eirwen Williams, '"US in Total Shock": China and Russia's 2028 Plan to Build a Nuclear Reactor on the Moon Stuns the Global Space Community' *Sustainability Times* (26 April 2025) https://www.sustainability-times.com/low-carbon-energy/us-in-total-shock-china-and-russias-2028-plan-to-build-a-nuclear-reactor-on-the-moon-stuns-the-global-space-community/. Accessed 28 April 2025.

[19] 'International Lunar Research Station Attracts More Partners: CNSA' *The State Council, The People's Republic of China* (24 April 2025). https://english.www.gov.cn/news/202504/24/content_WS680976d7c6d0868f4e8f2045.html accessed 28 April 2025.

There have been arguments against its binding nature over the years by advocating an argument of convenience to push it to the realm of morality.[20] However, as part of the opening paragraph of a binding international treaty, the principle of the benefit and interests of all States cannot be considered a mere moral obligation.

The collective assertion of developing countries during the 1990s to practically implement the principle of benefit and interests is evidence of breaking the myth of its non-binding nature.[21] The absence of any direct benefit percolating to the developing countries from the space activities was the major bone of contention in this collective assertion. The need for international cooperation and codification of the rights and responsibilities of States in this regard was considered to be of paramount significance. However, the space-faring States argued before the Legal Sub-committee of the United Nations Committee on the Peaceful Uses of Outer Space (UNCOPUOS) that there is already a presence of high-level successful cooperation in space exploration.[22] The contribution of space activities to scientific developments is undeniably in accordance with the common interests of all. Even the indirect benefits of technological developments via space exploration, like direct television broadcasting, remote sensing, meteorology, telecommunication, etc., that the developing countries have received were argued by the space-faring States to be in consonance with the principle of benefit and interests of all States.[23]

The assertion of developing countries resulted in the United Nations General Assembly adopting the Declaration on International Cooperation in the Exploration and Use of Outer Space for the Benefit and in the Interests of All States, Taking into Particular Account the Needs of Developing Countries 1996 (1996 Declaration). Unfortunately, the much-needed clarification on the rights and responsibilities of States emanating from the principle of benefit and interests of all States remained illusory despite the 1996 Declaration. In addition, the non-binding nature of UN General Assembly resolutions made the 1996 Declaration ceremonial. Though the 1996 Declaration failed to go beyond Article I of the Outer Space Treaty, we should not forget the strong roots of the benefit and interests principle under the binding treaty norm.

The twentieth-century space activities involving scientific investigations and some commercial activities could be regarded as in the common interest, and understandably, some indirect benefits also percolated to all from such activities. However, the compliance with the principle of benefit and interests of all countries without the cost–benefit analysis cannot be ascertained. The developing countries that indirectly

---

[20] Kurt Anderson Baca, 'Property Rights in Outer Space' (1993) 58(4) *Journal of Air Law and Commerce* 1041, 1063.

[21] Nandasiri Jasentuliyana, *International Space Law and the United Nations* (Kluwer Law International 1999) 47.

[22] Edwin W Paxson III, 'Sharing the Benefits of Outer Space Exploration: Space Law and Economic Development' (1993) 14(3) *Michigan Journal of International Law* 487; see also Peter Jankowitsch, 'The Role of the United Nations in Outer Space Law Development: Past Achievements and New Challenges' (1998) 26(2) *Journal of Space Law* 101, 106.

[23] Elena Carpanelli and Brendan Cohen, 'A Legal Assessment of the 1996 Declaration on Space Benefits on the Occasion of its Fifteenth Anniversary' (2012) 38(1) *Journal of Space Law* 1, 3–6.

benefited from the commercial space activities of space-faring nations did not receive those benefits free of cost. They paid to receive space-related services from the developed countries as well as their commercial entities, and thereby, remained customers of services offered for sale. This has boosted the financial inflow to space-faring nations and drained the economy of developing States.

In comparison to the twentieth-century space activities, the twenty first century is witnessing a higher level of space commercialisation. In the zeal to promote commercial space activities, the foundational principle of the benefit and interests of all countries is being discarded. Permitting space mining and the conferment of individual ownership rights over the mined resources can never serve the interests of all countries. In the absence of any mechanism to equitably share the profit derived from space mining, benefits can never percolate to all. Hence, authorising private commercial mining of space resources constitutes a blatant violation of the Outer Space Treaty's obligation.

3.2 Province of All Mankind

During the drafting of the Outer Space Treaty, the two Roman law concepts that were readily available for application to space activities were *res nullius* and *res communis*. *Res nullius* is something that does not belong to anyone; however, the first person to get hold of it can get property rights over it.[24] On the contrary, *res communis* belongs to all in common, and no one can claim individual rights over it. It imbibes the right to use common resources without an obligation to share the benefits.[25]

The application of either *res nullius* or *res communis* in the regime governing outer space would have resulted in depriving the rights of developing countries. A *res nullius* regime would have resulted in both the United States and the Soviet Union colonising outer space and celestial bodies. The leaning towards the *res communis* regime could not have provided a different result since both States, being the only two in space at that time, could have exploited space resources without sharing any benefit with the international community. Hence, the drafters of the Outer Space Treaty used the expression "province of all mankind", which is considered to be an advanced version of the *res communis* regime.

Christol views the province of all mankind as a starting point of legal reasoning in space law-making.[26] Even though it does not incorporate the elements of mandatory sharing of benefits and the international management system like CHM, the province

[24] Said Mahmoudi, *The Law of Deep Sea-Bed Mining: A Study of the Progressive Development of International Law Concerning the Management of the Polymetallic Nodules of the Deep Sea-Bed* (Almqvist & Wiksell International 1987) 86.

[25] Sudhir K Chopra, 'Antarctica as a Commons Regime: A Conceptual Framework for Cooperation and Coexistence' in Christopher C Joyner and Sudhir K Chopra (eds), *The Antarctic Legal Regime* (Martinus Nijhoff 1988) 163, 167.

[26] Carl Q Christol, 'Important Concepts for the International Law of Outer Space' in *Proceedings of the Fortieth Colloquium on the Law of Outer Space* (1998) 73.

of all mankind is viewed as an improvement over the *res communis* regime. When viewed from the perspective of space resources, the province of all mankind should at least involve a voluntary process of sharing benefits. This is again reflected in Christol's argument when he says that the benefits derived from space resources must serve the whole of mankind.[27]

In the absence of a clear definition under the Outer Space Treaty, the meaning and background of the province of all mankind have been distorted in the writings of several Western writers. There are arguments that the province of all mankind principle has no property overtones. It is confined to jurisdiction and responsibility over a territory, and the aspects of property rights are covered only within the ambit of CHM.[28] Such an interpretation is made to draw a watertight compartment between the province of all mankind and CHM, which is imaginary and was never foreseen in the process of drafting the space treaties. Moreover, the phraseology of Article I of the Outer Space Treaty does not support this argument of confining the application of the province of all mankind principle to the territory. Since Article I does not mention "outer space or celestial bodies" as the province of all mankind, instead, "exploration and use" are declared as the province of all mankind; non-application of this principle to space resources is not a tenable argument.

3.3 Right of Free Access

In addition to the principles of benefit and interests and the province of all mankind, Article I of the Outer Space Treaty declares the freedom of exploration, use and scientific investigation to be available to all States in outer space and celestial bodies. Understandably, it is a corollary principle to the absence of State sovereignty in outer space. However, the freedom ensured here is not unregulated.[29] The freedom of exploration and use shall not be misconstrued as the freedom of exploitation. Other fundamental principles, including the province of all mankind, the benefit and interests of all countries, compliance with international law, peaceful uses, international cooperation and environmental protection, which are mentioned in other provisions of the Outer Space Treaty, also need to be respected. Since every state enjoys this freedom, one's exercise of freedom shall not interfere with others' freedom, which is also reflected under Article IX of the Outer Space Treaty in terms of preventing harmful interference.

---

[27] Carl Q Christol, 'Important Concepts for the International Law of Outer Space' in *Proceedings of the Fortieth Colloquium on the Law of Outer Space* (1998) 78.

[28] B Maiorsky, 'A Few Reflections on the Meaning and the Interpretation of "Province of All Mankind" and "Common Heritage of Mankind" Notions' in *Proceedings of the Twenty-ninth Colloquium on the Law of Outer Space* (1987) 58, 59; see also J E S Fawcett, *Outer Space: New Challenges to Law and Policy* (Clarendon Press 1984) 4.

[29] Narendra Singh, 'Legal Status of Outer Space' in RC Hingorani (ed), *International Law Through United Nations* (NM Tripathi Private Limited 1972) 125, 127.

In order to respect the principle of freedom for all in outer space, Article I also declares everyone's right to have access to all areas of celestial bodies. The establishment of mining bases and the declaration of safety zones by any state go against the freedom of access available to all. Given the huge investments and the need for several mining equipment and installations in celestial mining, the mining bases cannot be temporary in nature. States and private companies involved in mining activities will require a longer period of time to recover their investments. Abrupt shifting of equipment and installation from one mining site to another to guarantee the freedom of access to others will not be practically feasible. In addition, the exercise of exclusionary rights in the pretext of safety zones will effectively take away others' freedom of access.[30] Even if the State sovereignty is not claimed over safety zones, the exercise of jurisdiction and application of State laws therein will have equally negative connotations on other States' rights. The right to free access to all parts of celestial bodies will have a natural death when different States and their authorised entities create several safety zones on celestial bodies.

3.4 International Cooperation

The principle of international cooperation runs like a golden thread throughout all five UN space treaties.[31] Based on the success of State cooperation during the International Geophysical Year, the principle of cooperation between States took deep roots in international space law.[32] Since developing countries can enjoy their freedom in outer space only with the assistance of space-faring nations, cooperation between States in the field of space activities is of utmost significance. The Outer Space Treaty explicitly stipulates the need for cooperation under Articles I, III, IX, X, and XI. These provisions highlight the significance of international cooperation in scientific investigation, maintenance of international peace and security, respecting corresponding interests of all states, providing opportunity for others to observe space flights, and publicly disclosing the nature, conduct, locations, and results of space activities. Every such area of cooperation has the potential to play a significant role in ensuring the developing countries' entry into the domain of outer space. The

[30] Rossana Deplano, 'The Artemis Accords: Evolution or Revolution in International Space Law?' (2021) 70 *International and Comparative Law Quarterly* 799, 807.

[31] Apart from the Outer Space Treaty and the Moon Agreement, three other treaties are entered in the domain of outer space. They are: Agreement on the Rescue of Astronauts, the Return of Astronauts and the Return of Objects Launched into Outer Space, 22 April 1968, 672 UNTS 119, 19 UST 7570, TIAS No 6599, 7 ILM 151 (entered into force 3 December 1968) [Rescue Agreement]; Convention on International Liability for Damage Caused by Space Objects, 29 March 1972, 961 UNTS 187, 24 UST 2389, 10 ILM 965 (1971) (entered into force on 1 September 1972) [Liability Convention]; and Convention on Registration of Objects Launched into Outer Space, 6 June 1975, 28 UST 695, 1023 UNTS 15 (entered into force on 15 September 1976) [Registration Convention].

[32] Bailey DeSimone, 'How the Antarctic Treaty of 1959 Influenced the Outer Space Treaty of 1967' *Library of Congress* (28 January 2022). https://blogs.loc.gov/law/2022/01/how-the-antarctic-treaty-of-1959-influenced-the-outer-space-treaty-of-1967/ accessed 1 May 2025.

Moon Agreement builds upon the spirit of cooperation enshrined in the Outer Space Treaty by imbuing stronger provisions on State cooperation, including mandatory sharing of benefits derived from resource exploitation.

Under the Rescue Agreement, the entire process of rescue and return of the personnel, and recovery and handing over of space objects to the launching authority, is based on the cooperation between States. It also solicits the cooperation of the launching authority in taking effective steps to eliminate the possibility of causing harm by any hazardous or deleterious components of its space objects.[33] The Liability Convention, despite involving the issue of fixation of liability, is based on State cooperation. The settlement of compensation for damage caused by space activities is primarily through diplomatic negotiations.[34] If it fails, the next step involves the establishment of a Claims Commission,[35] which also functions on the basis of State cooperation. The Claims Commission is a three-member body consisting of one each appointed by the claimant State and the launching State, respectively. The third member, who shall be the Chairman of the Commission, is appointed jointly by the parties. Even the decision of the Claims Commission becomes binding on the parties only if the parties agree in advance. Hence, the entire process of dispute resolution under the Liability Convention is based on the cooperation between States. Finally, the dual registration process under the Registration Convention, in the absence of any provision stipulating the consequences of failure to register space objects, is completely dependent on the cooperation of State Parties to it.

One may criticise the principle of international cooperation in international space law due to the absence of clarity on its specific nuances.[36] However, one cannot deny the fact that repetitive stipulation of international cooperation in the UN space treaties adequately represents the common rights of all States in outer space and celestial bodies. The cooperation stipulated under the Artemis Accords is not the same as envisaged in the UN space treaties. It is rooted in the US dictate without any consultation with other countries. The States' eagerness to join the Artemis Accords stems more from their fear of being left out rather than from the right spirit of cooperation and mutual assistance. Moreover, the limited plurilateralism of the Artemis Accords based on geopolitical considerations goes against the much-needed multilateral cooperation at the global level.

3.5 Non-appropriation Principle

Non-application of the concept of *res nullius* as well as the sovereignty of States in outer space is well established by the principle of national non-appropriation

[33] Rescue Agreement, Art. 5(4).

[34] Liability Convention, Art. IX.

[35] Liability Convention, Art. XIV.

[36] N Jasentuliyana, 'Article I of the Outer Space Treaty Revisited' (1989) 17(2) *Journal of Space Law* 129, 140.

mentioned under Article II of the Outer Space Treaty,[37] especially when it is read with the concept of State responsibility stipulated under Article VI of the Outer Space Treaty. Article II prohibits national appropriation of outer space and celestial bodies "by claim of sovereignty, by means of use or occupation, or by any other means". Going by the principle of common rights, it rejects the application of the first-come, first-served principle that ruled terrestrial property rights on the earth. However, this provision has been subjected to several misinterpretations.

Those who support the mining of space resources and granting of ownership rights argue that the resource exploitation is not prohibited under Article II since it does not amount to national appropriation.[38] This argument is not based on the correct understanding of the provision. It must be noted that national appropriation is prohibited not just when it is attempted by way of claim of sovereignty, use or occupation, but also "by any other means". The phrase "by any other means" shall essentially include the exploitation of space resources amounting to national appropriation. Segregating land territory and the natural resources that are part of the land territory for the purpose of interpreting this provision is not based on any sound legal premise.

Some writers have even tried to extend the analogy from the law of the sea by basing their arguments on the concept of the global commons. Fishing rights on the high seas that are enjoyed by individual States as well as subject to unilateral licences to private entities, are indicated as examples to be used while interpreting Article II of the Outer Space Treaty.[39] Such an argument can only stem from the incorrect understanding of the concept of the global commons. Though there are different elements and regions that form part of the global commons, we need to understand that they stand at different levels. The high seas regime under the United Nations Convention on the Law of the Sea (UNCLOS) is distinct from that of the regime developed for application to outer space and celestial bodies. Freedom of fishing in the high seas is an explicitly recognised right under Article 87 (Part VII) of the UNCLOS. However, the Outer Space Treaty does not recognise any such right to exploit space resources. Instead, it stipulates the non-appropriation principle. Part VII of the UNCLOS, which governs the high seas, has no mention of the non-appropriation principle. A proper analogy from the law of the sea in this context can only stem from Part XI of the UNCLOS, which stipulates the non-appropriation principle.[40] Part XI of the UNCLOS deals with the norms governing the Area and its resources, which are also global commons. While applying the concept of CHM to the Area and its resources, Part XI of the UNCLOS does not permit either individual State exploitation or unilateral licences to mine resources.

---

[37] Grier C Raclin, 'From Ice to Ether: The Adoption of a Regime to Govern Resource Exploitation in Outer Space' (1986) 7(4) *Northwestern Journal of International Law and Business* 727, 733.

[38] Theodore E Wolcott, 'Reaching for the Moon' in *Proceedings of the Twenty-third Colloquium on the Law of Outer Space* (1981) 87–88.

[39] Frans G von der Dunk, 'Asteroid Mining: International and National Legal Aspects' (2017) 26 *Michigan State International Law Review* 83, 93.

[40] *United Nations Convention on the Law of the Sea* (UNCLOS), art 137.

Another incorrect interpretation of Article II of the Outer Space Treaty is witnessed with respect to the mining of space resources by individuals and non-governmental entities. It is argued that only States are prohibited from appropriating outer space and celestial bodies, as Article II prohibits only national appropriation.[41] References are also made to the drafting history of the Outer Space Treaty to argue that the initial drafts of the Treaty included a prohibition on both national and individual appropriation. However, since the final text of the Outer Space Treaty did not include a separate provision on prohibiting individual appropriation, it is argued to be permitted. This argument has also been the basis of domestic laws that have mushroomed to grant resource mining rights in outer space to private commercial entities.

At the outset, these arguments are unconvincing and absurd. Once a state is prohibited from appropriation, how its individuals can obtain rights to appropriate remains an unanswered question in every literature supporting private space mining. It is pertinent to note here that Article VI of the Outer Space Treaty imposes a direct responsibility on States with respect to space activities of non-governmental entities.[42] It also stipulates the requirement of authorisation and continuing supervision of all space activities by the appropriate State Party to the Treaty. Failure to authorise and supervise private space activities attributes international responsibility to States. Hence, each State is duty-bound to establish a proper system of authorisation and supervision in compliance with its international treaty obligations. When the State itself is not authorised to appropriate outer space and celestial bodies or their resources, it cannot authorise its individuals or commercial entities to do so. If they attempt to do so, it goes against the basic legal tenets that one who does not possess a right cannot grant it to others (*nemo dat quod non habet*).

Hence, the non-appropriation principle under Article II of the Outer Space Treaty is not an easy escape clause for those supporting space resource mining, either by a select few countries or by their private commercial entities. However, this shall not be misconstrued to mean that the Outer Space Treaty completely prohibits space resource mining. The drafting history and the incorporated provisions clearly indicate the need to respect the common rights of all in using space resources. What is prohibited is only the exploitation of resources by a select few to cater for their self-interest without sharing the benefits that are due to others. Developing a well-negotiated international regime by involving all States to utilise space resources for the benefit of all is not subject to any prohibition under the UN space treaties.

---

[41] Wayne N White Jr, 'Real Property Rights in Outer Space' in *Proceedings of the Fortieth Colloquium on the Law of Outer Space* (1998) 370, 378.

[42] Vladimír Kopal, 'Treaty on Principles Governing the Activities of States in the Exploration and Use of Outer Space, Including the Moon and Other Celestial Bodies' *United Nations Audiovisual Library of International Law* (2008) 4 https://legal.un.org/avl/pdf/ha/tos/tos_e.pdf. Accessed 6 May 2025.

4 Search for an International Response

The self-centred domestic space laws mentioned above and the Artemis Accords required international legal responses to address the debates unfolded by them. The two noteworthy developments in this regard are the establishment of the Hague Space Resources Governance Working Group (Hague Working Group) and the Working Group on Legal Aspects of Space Resource Activities (UNCOPUOS Working Group). The former was established by the Hague Institute for Global Justice in 2014, and the latter was the brainchild of the Legal Sub-Committee of the UNCOPUOS. While the discussion on space resources utilisation was going on at the UNCOPUOS level for a considerable period of time, the formal establishment of the UNCOPUOS Working Group took place in 2021. Proposals from Austria, Belgium, China, Czech Republic, Finland, Germany, Greece, Poland, Portugal, Romania, Russia, Slovakia and Spain have highlighted the need for establishing the Working Group regarding space resource activities. Austria and China's proposals were detailed and are widely considered to be instrumental in the development of the UNCOPUOS Working Group's mandate.[43]

4.1 The Hague Building Blocks

The Hague Working Group was made up of members from different domains to put forward views from different perspectives. It had thirty-five members with a mixed representation from international organisations, States, academics, NGOs and industry. A consortium of academic institutions led by Leiden University, the Netherlands, contributed extensively to the deliberations of the Hague Working Group. Probing into the legal developments relating to space resource activities in the context of UN space treaties and advising on the future legal framework was the major mandate of the Hague Working Group. Acting upon the mandate, the Hague Working Group had a series of deliberations between 2016 and 2019. Ultimately, it developed and adopted the Building Blocks for the Development of an International Framework on Space Resource Activities (Hague Building Blocks) on 12 November 2019.[44] In 2020, the Hague Building Blocks were introduced in the UNCOPUOS, and subsequently, a Commentary to the Hague Building Blocks was developed.[45]

---

[43] Anne Campbell, 'Space Resource Discussions within the United Nations Committee on the Peaceful Uses of Outer Space Legal Subcommittee: The Past, Present and Future' *Institute for Earth and Space Exploration White Papers* 1 (2021) https://ir.lib.uwo.ca/cgi/viewcontent.cgi?article=1000&context=iesewp. Accessed 2 May 2025.

[44] International Institute of Air and Space Law, 'The Hague International Space Resources Governance Working Group' https://www.universiteitleiden.nl/en/law/institute-of-public-law/institute-of-air-space-law/the-hague-space-resources-governance-working-group#consortium-partners. Accessed 3 May 2025.

[45] Committee on the Peaceful Uses of Outer Space, Legal Subcommittee Sixty-First Session (28 March–8 April 2022) UN Doc A/AC.105/C.2/2022/CRP.23 (5 April 2022).

The Hague Working Group could not architect an international framework on space resource activities; however, it attempted to provide some guidelines in the form of building blocks. The Hague Building Blocks reiterate the provisions of the Outer Space Treaty regarding the benefit and interests of all countries, peaceful purposes, avoiding harmful interference, State responsibility, liability for damage, emergency assistance, the principle of due regard for the interests of all, avoidance of harmful environmental impact, etc. Some elements of the Moon Agreement are also selectively found under it. Adaptive governance to regulate space resource activities, sustainable development, securing the orderly and safe utilisation of space resources, rational, efficient and economic use of space resources, and taking into account the needs of developing countries are some of the key aspects derived from the Moon Agreement.[46]

The Hague Building Blocks supports the development of an international framework to recognise jurisdiction and control of States over any space-made products, grant priority rights to an operator, register priority rights, grant resource rights, share information, provide advance notification of space resource activities, etc. It stipulates the need for an international framework to set technical standards and prior review to ensure the safe conduct of space resource activities. The respect for the non-appropriation principle under Article II of the Outer Space Treaty for establishing safety zones in the process of space resource activities is highlighted by the Hague Building Blocks. It suggests international consultation as a mechanism to avoid conflicting safety zones as well as possible interference with the enjoyment of freedom of access by others. States and international organisations must undertake to monitor and redress the harmful impact resulting from space resource activities. Developing an international registration system and maintaining an international database for ensuring transparency are insisted under the Hague Building Blocks. In terms of dispute settlement, strengthening the process of consultation and promoting the use of arbitration are also pointed out.

Though the Hague Building Blocks is worthy of appreciation, it could not establish the much-needed international legal framework to address manifold issues in space resource utilisation. Instead of establishing a concrete framework, it ended up just suggesting some guidelines to be followed in the process of establishing a future international framework. This looks like a researcher suggesting some general reforms towards the end of the research. Accordingly, the Hague Working Group has not achieved much success in practical terms. In addition, the biggest compromise that the Hague Working Group made in the process of developing the Hague Building Blocks is on the sharing of benefits. It enlists only the indirect benefits, similar to those that were argued to be gained by the developing countries during the twentieth century, as a part of the benefit-sharing obligation. It explicitly mentions, "The international framework should not require compulsory monetary benefit-sharing".[47]

The benefit-sharing clause of the Hague Building Blocks clearly indicates the undue influence of the developed countries and their industries, which is detrimental

[46] See Moon Agreement, Art. 11.

[47] Hague Building Blocks, Art. 13.2.

to the interests of the developing countries. While the argument of indirect benefit to the developing countries in the exploration and use of outer space may appeal to a certain extent, such an argument sounds absurd in the commercial utilisation or exploitation of space resources. Benefit sharing in the commercial utilisation of common resources of outer space can only happen meaningfully with monetary sharing. A viable mechanism of allowing the investors to recover their cost of space resource utilisation, and subsequently, equitable sharing of profits derived over and above their investments, should have been worked out instead of lifting the obligation to share monetary benefit. Such a system is inevitable to eliminate the increasing troubles originating from economic inequalities.[48]

4.2 The UNCOPUOS Working Group

Though an agenda item titled "General exchange of views on potential legal models for activities in exploration, exploitation, and utilisation of space resources" appeared in the UNCOPUOS meeting during 2016, no substantial progress was made on it until the UNCOPUOS Working Group was established. Recognising the dangers of unilateral regimes and understanding the need for respecting multilateralism in space resource utilisation, the UNCOPUOS started fresh initiatives on the sidelined agenda in 2021.[49] The UNCOPUOS Working Group started its functioning in 2022 with a five-year mandate to report on a set of principles to govern space resource activities. It has held several high-level meetings and is currently involved in the process of information collection and stocktaking. In this process, it has invited submissions from States as well as non-State members. At the time of this writing, 22 States and 7 organisations and non-State entities have submitted their views.[50] The UNCOPUOS Working Group, being a formal international body, is expected to come up with more concrete results in having a well-balanced international legal framework governing space resource utilisation.

5 Conclusion

The mining and utilisation of space resources requires an approach based on equity. This has been recognised deeply under the UN space treaties. Unfortunately, the current legal developments at the domestic level of the countries highlighted above

[48] United Nations, 'Rising Inequality Affecting More Than Two-Thirds of the Globe, but It's Not Inevitable: New UN Report' *UN News* (21 January 2020) https://news.un.org/en/story/2020/01/1055681. Accessed 3 May 2025.

[49] Committee on the Peaceful Uses of Outer Space, Legal Subcommittee Sixtieth Session (31 May–11 June 2021) UN Doc A/AC.105/C.2/L.317 (12 April 2021).

[50] UNOOSA, 'Working Group on Legal Aspects of Space Resource Activities' https://www.unoosa.org/oosa/en/ourwork/copuos/lsc/space-resources/index.html. Accessed 3 May 2025.

fail to respect it. The Artemis Accords are causing further damage to the notion of equity in outer space law. Since the space resources are subject to the collective rights of all, their utilisation needs to respect everyone's rights. Other commercial space activities, which do not involve the collection or removal of space resources, may be justified with an argument of indirect benefits derived by developing countries. However, resource mining in outer space should yield direct benefits to the developing countries since it results in the depletion of natural resources belonging to all. Percolation of any indirect benefit to stakeholders does not make any sense in this context. Hence, space resource mining rights cannot be granted either to States or to individuals without incorporating the benefit-sharing principle.

Prof. (Dr.) Sandeepa Bhat B is working as a Professor of Law and the Director of the Centre for Aviation and Space Laws at the National University of Juridical Sciences (NUJS), Kolkata. He has the teaching and research experience of over twenty-two years after completing his master's degree with first rank and double gold medals in 2003. His seven Major Research Projects are sponsored by the World Bank, ISRO, the WB Judicial Academy, the Ministry of Justice, the Ministry of Environment, Forest and Climate Change, the Ministry of Commerce, and the Government of West Bengal. He holds the unique feat of being the first Indian to be inducted as the UNIDROIT Correspondent for India. He has been a member of four distinguished international bodies, including the American Society of International Law. Dr. Bhat has published nine books on aviation and space laws and four books on medical law. In addition, he has published more than seventy articles in the journals of international and national repute (substantially on aviation and space laws). He is the recipient of the inaugural Sachdeva Award for Space Law 2022 in recognition of his outstanding contribution to space law. He has presented over 200 papers in international and national conferences, including the coveted International Astronautical Congress, as well as in international conferences held at Austin, Cambridge, Changsha, Charles Town, Hamilton, Istanbul, Jakarta, Osun State, Paris, Seoul, Sharjah, Singapore and Southampton. He is a Listed Arbitrator at the Court of Arbitration for Space Enterprise, IASL, Moscow. He also has the distinction of being a member of the Indian Space Research Organization's Expert Group for drafting the National Space Act for India.

The Development of an International Legal Framework for Space Resource Activities

Stacey Henderson and Joel Lisk

Abstract The prospect of mining on the Moon, asteroids, and other celestial bodies appears to be closer than ever, yet the legal framework to govern this type of space activity is unclear. While several spacefaring States appear to be reaching a common understanding that space mining activities are not contrary to existing space law, including the principle of non-appropriation in the Outer Space Treaty, this is by no means settled. Over the last decade, there have been numerous efforts to make progress on a legal framework for space resource activities. This includes the Hague Building Blocks for the Development of an International Framework on Space Resource Activities, the UNCOPUOS Working Group on Legal Aspects of Space Resource Activities, the Artemis Accords, and domestic space law. This chapter examines these different efforts in light of existing international space law and asks whether they are solid building blocks or shaky foundations for the development of an international legal framework for space resource activities. The chapter also explores whether there are lessons to be learned for an international legal framework for space resource activities from how mining activity is regulated in other areas beyond national jurisdiction, such as Antarctica and the deep seabed. The chapter concludes that existing international space law provides a solid foundation for the development of an international legal framework for space resource activities, but that more work is needed to ensure that the interests of all nations are respected and protected, and that space remains the province of all humankind.

Keywords Space mining · International legal framework · Outer Space Treaty · Non-appropriation principle · Space resource activities · Global governance

---

S. Henderson (✉) · J. Lisk
Jeff Bleich Centre for Democracy & Disruptive Technologies, Flinders University, Adelaide, Australia
e-mail: stacey.henderson@flinders.edu.au

J. Lisk
e-mail: joel.lisk@flinders.edu.au

S. Bhat B and A. Variath (eds.), *Legal Responses to Space Resource Utilisation*, Space Law and Policy, https://doi.org/10.1007/978-981-95-8864-0_2

1 Introduction

The prospect of mining on the Moon, asteroids, and other celestial bodies appears to be closer than ever, yet the legal framework to govern this type of space activity remains unclear. While several spacefaring States appear to be reaching a common understanding that space mining activities are not contrary to existing space law, including the principle of non-appropriation in the Outer Space Treaty,[1] this is by no means settled at law. This chapter examines efforts over the last decade to make progress on a legal framework for space resource activities, including the Hague Building Blocks for the Development of an International Framework on Space Resource Activities,[2] the United Nations Committee on the Peaceful Uses of Outer Space (UNCOPUOS) Working Group on Legal Aspects of Space Resource Activities,[3] the Artemis Accords,[4] and domestic space law.[5] This chapter examines these different efforts in light of existing international space law applicable to space resource activities.[6]

The chapter enquires whether these efforts are solid building blocks or shaky foundations for the development of an international legal framework for space resource activities. The chapter also explores whether there are lessons to be learned for an international legal framework for space resource activities from how mining activity is regulated in other areas beyond national jurisdiction, such as Antarctica and the deep seabed. The chapter concludes that while existing international space law can provide a solid foundation for the development of an international legal framework for space resource activities, more work is needed to ensure that the interests of all States are respected and protected, that space remains the province of all humankind, and that an appropriate balance can be struck between protecting the space environment,

[1] Treaty on Principles Governing the Activities of States in the Exploration and Use of Outer Space, including the Moon and Other Celestial Bodies (adopted 27 January 1967, entered into force 10 October 1967) 610 UNTS 205, 18 UST 2410, TIAS No 6347, 6 ILM 386 [Outer Space Treaty], art II.

[2] UN Doc A/AC.105/C.2/L.315 (3 February 2020).

[3] 'United Nations Office for Outer Space Affairs, 'Working Group on Legal Aspects of Space Resource Activities' https://www.unoosa.org/oosa/en/ourwork/copuos/lsc/space-resources/index.html. Accessed 15 July 2024.

[4] 'National Aeronautics and Space Administration, *The Artemis Accords: Principles for Cooperation in the Civil Exploration and Use of the Moon, Mars, Comets, and Asteroids for Peaceful Purposes* (13 October 2020) https://www.nasa.gov/wp-content/uploads/2022/11/Artemis-Accords-signed-13Oct2020.pdf?emrc=653a00. Accessed 15 July 2024 [Artemis Accords].

[5] United States, *Space Resource Exploration and Utilization Act of 2015*, 51 USC 513; Luxembourg, *Loi du 20 juillet 2017 sur l'exploration et l'utilisation des ressources de l'espace*; United Arab Emirates, *Federal Law No (12) of 2019 on the Regulation of the Space Sector*; Japan, *Law Concerning the Promotion of Business Activities Related to the Exploration and Development of Space Resources 2021*.

[6] Outer Space Treaty; Agreement Governing the Activities of States on the Moon and Other Celestial Bodies (adopted 18 December 1979, entered into force 11 July 1984) 1363 UNTS 3 [Moon Agreement].

enabling in-situ space resource utilisation, and permitting commercial extraction of space resources.

2 Space Resource Exploitation and Use

Mining has been an essential aspect of humanity's technological advances over the ages. The discovery and extraction of metals, minerals, and other elements have allowed societies to grow in technological capacity and facilitated the invention of new technologies and approaches. As a global community, we have become reliant on the extraction of resources from below the surface of the Earth; this is especially so for the development of new technologies that are essential for the decarbonisation of the global economy, an effort that is extremely reliant on access to elusive rare earth elements.[7] With access to rare earth elements becoming harder and demand increasing, some have looked to outer space as providing the potential solution with its hypothesised endless volumes of various elements.[8] It has been suggested that some metallic asteroids have extremely high levels of elements such as platinum, iridium, palladium, and rhodium, amongst others.[9] These elements play essential roles in the development of modern technologies. For example, iridium is used in medical imaging, as a catalyst for complex chemical processes, and is an essential element required for the development of organic light-emitting diodes (OLEDs, a technology that has become an essential component of modern television, mobile and computer screens). Platinum, while being a metal commonly seen in jewellery, is also an essential catalyst used for the development of green hydrogen, a technology that is expected to play a key role in the decarbonisation of the global economy. It has been suggested that if space mining activities were to ever reach viability, there is the potential for space resource mining operations to change global resources markets and revolutionise the way humanity accesses and uses otherwise hard-to-obtain resources.

A more immediate area of interest with respect to space resources is extraction to facilitate exploration missions.[10] One of the most prominent limiters on space

[7] Mureji Fatunde, 'The race to produce rare earth elements' *MIT Technology Review* (5 January 2024) www.technologyreview.com/2024/01/05/1084791/rare-earth-materials-clean-energy/. accessed 10 August 2024; V Balaram, 'Rare earth elements: A review of applications, occurrence, exploration, analysis, recycling, and environmental impact' (2019) 10(4) *Geoscience Frontiers* 1285.

[8] Monisha Ravisetti, 'Rare asteroids near Earth may contain precious metals worth $11.65 trillion' *CNET* (4 October 2021) www.cnet.com/science/rare-asteroids-near-earth-may-become-targets-for-space-mining/. Accessed 10 August 2024.

[9] Jacob Lornic, 'The Dream of Space Mining is Still an Expensive Challenge' *Bloomberg* (7 December 2023) www.bloomberg.com/news/newsletters/2023-12-06/mining-rare-earths-and-metals-from-asteroids-is-an-expensive-challenge. Accessed 10 August 2024.

[10] Neel V Patel, 'Here's how we could mine the moon for rocket fuel' *MIT Technology Review* (19 May 2020) www.technologyreview.com/2020/05/19/1001857/how-moon-lunar-mining-water-ice-rocket-fuel/. Accessed 10 August 2024.

exploration missions has been the need to launch all required assets and consumables into orbit at the commencement of a mission. This includes thousands of tonnes of fuel. This requirement is a major limitation on space exploration missions. With States developing more ambitious missions to the Moon and Mars, the requirement to lift all required consumables to Earth orbit and beyond has been reconsidered in light of the discovery of water ice on the Moon as well as hypothesised frozen deposits of water on Mars.[11] While these deposits present immediate benefits with respect to accessing water, they may also serve as feedstock for the production of fuel for spacecraft and oxygen for human habitation. There have also been proposals to use lunar and Martian regolith in the development of facilities, which also has the potential to reduce the loads being launched from the surface of the Earth to these celestial bodies.[12]

With various States and corporate entities seeking to travel to the Moon and Mars, there is a genuine need to consider how the resources present on celestial bodies—whether they are precious metals or resources that will sustain human presence in outer space—are accessed by the States and corporate entities that seek to exploit them. On Earth, there is a network of laws, regulations and guidelines that govern the access, extraction, and use of resources.[13] These laws do more than simply allow mining or resource extraction to take place. In many States there are complex laws that set out the rights that landowners have to the resources below the land they occupy, laws that dictate the rights a corporate entity has to prospect or undertake studies to find resource deposits, and additional laws that set out the order in which entities can commence (or attempt to commence) extraction activities. This is then supported by further regulations on how resources are extracted and the conditions that apply to any extraction activities. Beyond the laws that specifically apply to resource extraction activities on the ground, there are also laws extracting entities must comply with, associated with environmental protection, work health and safety, taxation, export, and refinement of extracted resources.[14] While extensive, these networks of laws provide a legally 'certain' operating environment in which entities extract and exploit resources with clear boundaries and guardrails that dictate how these activities take place. These systems of laws can be highly burdensome, but the fact that they exist and provide a pathway to permissible resource extraction activities allows businesses to seek out investors and institutional supports that enable them to

[11] Huicun He and others, 'A solar wind-derived water reservoir on the Moon hosted by impact glass beads' (2023) 16 *Nature Geoscience* 294; J Pinson, 'Moon May Hold Billions of Tons of Subterranean Ice at Its Poles' *EOS* (20 November 2020) eos.org/research-spotlights/moon-may-hold-billions-of-tons-of-subterranean-ice-at-its-poles. Accessed 10 August 2024.

[12] Juan-Carlos Gines-Palomares and others, 'Laser melting manufacturing of large elements of lunar regolith stimulant for pacing on the Moon' (2023) 13 *Scientific Reports* 15,593; Altan Alpay Altun and others, 'Additive manufacturing of lunar regolith structures' (2021) 5 *Open Ceramics* 100058.

[13] Andrew Cannon, 'The Great Space Rush: Regulating Space Mining' (2020) 39 *Australian Resources and Energy Law Journal* 1.

[14] Andrew Cannon, 'The Great Space Rush: Regulating Space Mining' (2020) 39 *Australian Resources and Energy Law Journal* 1.

meet the high input costs associated with a resource extraction activity (such as costs associated with prospecting, applying for permits, purchasing land and equipment).

Turning to the space sector, there are no laws as extensive as those described above for space resource extraction activities. In some respects, this is due to the lack of space resource extraction activities had not occurred at the time of writing. Despite this position, there are still some laws that consider space resource activities.

3 Space Resource Activities and Existing Space Law

The Outer Space Treaty is the foundational treaty governing all activities in space. All major spacefaring States are party to the Outer Space Treaty; many of its provisions are considered reflective of customary international law and thus are also binding on non-State Parties.[15] The Outer Space Treaty neither explicitly permits nor prohibits space resource activities. However, such activities appear to be a permitted 'use' of space in accordance with Article I, particularly when conducted for scientific research or for in-situ use, including for human survival. The principle of non-appropriation contained in Article II of the Outer Space Treaty provides that 'space, including the Moon and other celestial bodies, is not subject to national appropriation by claim of sovereignty, by means of use or occupation, or by any other means'.[16] While it may be argued that the principle of non-appropriation prohibits appropriation of celestial bodies as a whole, and not appropriation of resources once extracted from the celestial body, this is not settled at law. In the absence of such an interpretation of Article II, the commercial viability of space resource activities appears low. However, space resources activities are not restricted to commercial activities and include in-situ use of extracted resources to produce consumables for space missions and human survival,[17] which would appear to be permissible uses of space under existing international space law. The Outer Space Treaty also requires that space activities be authorised and supervised,[18] and be carried out in a manner that gives 'due regard' to the interests of other States.[19] However, just what is required for States to satisfy the obligation to authorise and continuously supervise or what is meant by giving 'due regard' to the interests of other States in undertaking space resource activities in practice, is not set out in the Outer Space Treaty or the other space law treaties. Many States implement their obligation to authorise and continuously supervise

---

[15] Francis Lyall and Paul B Larsen, *Space Law: A Treatise* (2nd edn, Routledge 2018) 50.

[16] Treaty on Principles Governing the Activities of States in the Exploration and Use of Outer Space, including the Moon and Other Celestial Bodies (adopted 27 January 1967, entered into force 10 October 1967) 610 UNTS 205 [Outer Space Treaty], art II.

[17] Melissa de Zwart, Stacey Henderson and Michelle Neumann, 'Space resource activities and the evolution of international space law' (2023) 211 *Acta Astronautica* 155, 156.

[18] Outer Space Treaty, Art. VI.

[19] Outer Space Treaty, Art. IX.

space activities through licensing regimes in domestic law,[20] which may signal that domestic space law has a role to play in shaping any future legal framework for space resource activities, as well as in its implementation.[21]

The 17 State Parties to the Moon Agreement have additional obligations specifically related to space resource activities and the establishment of an international regime for the governance of space resources. The Moon Agreement provides that the 'Moon and its natural resources are the common heritage of mankind'.[22] In 2008, State Parties to the Moon Agreement released a joint statement noting that the principle of the 'common heritage of mankind' does not mean that space resource activities are prohibited.[23] Further, in echoing the principle of non-appropriation in the Outer Space Treaty, that

> 'Neither the surface nor the subsurface of the Moon, nor any part thereof or natural resources in place, shall become property of any State, international intergovernmental organization, national organization or non-governmental entity or of any natural person'.[24]

It is notable that the prohibition on ownership in the Moon Agreement applies to 'natural resources in place' and not natural resources once they have been extracted from a celestial body. Article 11(5) obliges States Parties to 'establish an international regime, including appropriate procedures, to govern to exploitation of the natural resources of the Moon as such exploitation is about to become feasible'.[25] Article 11(7) outlines the main purposes of the foreshadowed international regime as follows:

a. The orderly and safe development of the natural resources of the Moon;
b. The rational management of those resources;
c. The expansion of opportunities in the use of those resources;
d. An equitable sharing by all States Parties in the benefits derived from those resources, whereby the interests and needs of the developing countries, as well as the efforts of those countries which have contributed either directly or indirectly to the exploration of the Moon, shall be given special consideration.[26]

The requirement for equitable benefit sharing is often touted as the reason behind the low level of ratifications of the Moon Agreement. It is not clear from the Moon

---

[20] See eg *Space (Launches and Returns) Act 2018* (Australia); *Outer Space and High-altitude Activities Act 2017* (New Zealand); *Space Industry Act 2018* (United Kingdom); *Outer Space Act 1986* (United Kingdom); *Law of 17 September 2005 on the Activities of Launching, Flight Operations or Guidance of Space Objects* (Belgium); *Wet ruimtevaartactiviteiten 2007* (Netherlands); *Loi nº 2008–518 de 3 juin 2008 relative aux operations spatiales*, 3 June 2008 (France).

[21] Philip de Man, 'State practice, domestic legislation and the interpretation of fundamental principles of international space law' (2017) 42 *Space Policy* 92.

[22] Moon Agreement, Art. 11(1).

[23] Committee on the Peaceful Uses of Outer Space Legal Subcommittee, *Joint Statement on the Benefits of Adherence to the Agreement Governing the Activities of States on the Moon and Other Celestial Bodies of 1979 by States Parties to That Agreement*, 47th sess, UN Doc A/AC.105/C.2/2008/CRP.11.

[24] Moon Agreement, Art. 11(3).

[25] Moon Agreement, Art. 11(5).

[26] Moon Agreement, Art. 11(7).

Agreement whether the equitable benefit sharing required is monetary in nature, or could include the sharing of other benefits such as consumables produced or new knowledge gained. Regardless, the United States, which is not a party to the Moon Agreement, has made it clear that it 'shall object to any attempt by any other state or international organization to treat the Moon Agreement as reflecting or otherwise expressing customary international law'.[27] Further, it is the US view that 'utilisation of space-based resources—including commercial utilisation—is consistent with the four main United Nations space treaties'.[28] As such, the Moon Agreement and its more detailed provisions relating to space resources are of no relevance to non-State Parties and of little help in developing an international legal framework for space resource activities, having been shunned by the majority of States.

4 Efforts to Develop a Legal Framework for Space Resource Activities

Given the lack of clarity about the rules applicable to space resource activities in existing space law, over the last decade, there have been numerous efforts to make progress on the development of a legal framework for space resource activities. This section examines some of these efforts.

4.1 Hague Building Blocks for the Development of an International Framework on Space Resource Activities

The Hague International Space Resources Governance Working Group was established in 2016 as a 'multi-stakeholder dialogue' with the aim of developing 'draft building blocks' for an international framework for space resource activities.[29] The Building Blocks for the Development of an International Framework on Space Resource Activities ('Building Blocks')[30] were released in 2019 and consist of a set of recommended practices and guidelines regarding space resource activities. The Building Blocks lay the groundwork for further international discussions on

[27] 'Executive Order on Encouraging International Support for the Recovery and Use of Space Resources', Exec Order No 13,914, 85 Fed Reg 20,381 (6 April 2020) s 1.

[28] G Swiney, 'US Statement Agenda Item 14: potential legal models for activities in exploration, exploitation and utilization of space resources', 60th sess, COPUOS Legal Subcommittee (2021).

[29] Thomas Cheney, 'There's No Rush: Developing a Legal Framework for Space Resource Activities' (2019) 43 *Journal of Space Law* 106, 106.

[30] Hague International Space Resources Governance Working Group, *Building Blocks for the Development of an International Framework on Space Resource Activities* (2019) [Building Blocks].

the potential development of an international framework for space resource activities.[31] They are expressed as being guided 'by the principle of adaptive governance' noting it was 'neither necessary nor feasible to attempt to comprehensively address space resource activities in the building blocks: space resource activities should be incrementally addressed at the appropriate time on the basis of contemporary technology and practices'.[32] Despite not being a binding law, the Building Blocks provide some valuable insights into the necessary content of any future international legal framework for space resource activities.

The Building Blocks define space resource activity as:

> 'an activity conducted in outer space for the purpose of searching for space resources, the recovery of those resources and the extraction of raw mineral or volatile materials therefrom, including the construction and operation of associated extraction, recovery, processing and other transportation systems'.[33]

As Masson-Zwaan has observed, a major source of tension in debates about space resource activities and the regulation thereof relates to a 'misconception between commercial space mining activity and the general use of space resources'.[34] In defining space resource activity in this way, the Building Blocks ensure that all aspects of space resource activities are captured and regulated by a future legal framework.

The objective of an international legal framework for space resource activities is expressed to be to 'create an enabling environment for space resource activities that takes into account all interests and benefits all countries and humankind'.[35] The Building Blocks propose that an international legal framework provide that space for resource activities:

a. 'be consistent with international law'[36];
b. '[p]romote compatibility and predictability of domestic frameworks of States and internal frameworks of international organizations'[37];
c. '[c]ontribute to sustainable development'[38] and '[p]romote the use of sustainable technology'[39];
d. '[p]revent disputes arising out of space resource activities'[40];;

[31] Thomas Cheney, 'There's No Rush: Developing a Legal Framework for Space Resource Activities' (2019) 43 *Journal of Space Law* 106, 106.

[32] UN Doc A/AC.105/C.2/L.315 (3 February 2020) 1.

[33] Building Blocks, Para 1.1.

[34] Tanja Masson-Zwaan and Dimitra Stefoudi, 'Space Mining' (2019) *Issues in Science and Technology* 16, 16.

[35] Building Blocks, Para 1.1.

[36] *Id.*, Building Block 4.1.

[37] *Id.*, Building Block 4.2(b).

[38] *Id.*, Building Block 4.2(c).

[39] *Id.*, Building Block 4.2(g).

[40] *Id.*, Building Block 4.2(d).

e. '[p]romote the sustainable, rational, efficient and economic use of space resources'[41];
f. '[t]ake into particular account the needs of developing countries'[42];
g. 'be carried out for the benefit and in the interests of all countries and humankind irrespective of their degree of economic and scientific development'[43];
h. that '[a]ppropriate international consultations shall be undertaken in accordance with Article IX of [the Outer Space Treaty] if there is a reason to believe that any potentially harmful interference may be caused'[44]; and that
i. '[s]pace resources shall be used exclusively for peaceful purposes'.[45]

The *Building Blocks* propose the creation of priority rights whereby an operator would have the right to prospect for space resources in a designated area for a maximum period of time.[46] To facilitate the safe conduct of space resource activities, s 11.3 provides that:

> 'Taking into account the principle of non-appropriation under article II OST, the international framework should permit States and international organizations responsible for space resource activities to establish a safety zone, or other area-based safety measure, around an area identified for a space resource activity as necessary to assure safety and to avoid any harmful interference with that space resource activity. Such safety measure shall not impede the free access, in accordance with international law, to any area of outer space by personnel, vehicles and equipment of another operator. In accordance with the area-based safety measure, a State or international organization may restrict access for a limited period of time, provided that timely public notice has been given setting out the reasons for such restriction'.[47]

The *Building Blocks* also contain provisions aimed at the long-term sustainability of space resource activities[48] and contemplate proactive measures for avoiding and mitigating potentially harmful impacts of space resource activities.[49] Building Block 13 provides that space resource activities should be 'beneficial for all countries and humankind' and provides for benefit sharing.[50] The *Building Blocks* provide valuable guidance about the sorts of benefits that might be included in any benefit sharing obligation, explicitly stating that an international framework for space resource activities 'should not require compulsory monetary benefit-sharing.[51] The sorts of benefits foreshadowed by the *Building Blocks* include, but are not limited to, enabling facilitating, promoting and fostering the following:

---

[41] *Id.*, Building Block 4.2(f).
[42] *Id.*, Building Block 4.2(i).
[43] *Id.*, Building Block 4.3(b).
[44] *Id.*, Building Block 4.3(c).
[45] *Id.*, Building Block 4.3(a).
[46] *Id.*, Building Block 7.
[47] *Id.*, Building Block 11.3.
[48] *Id.*, Building Blocks 10 and 12.
[49] *Id.*, Building Block 11.
[50] *Id.*, Building Block 13.
[51] *Id.*, Building Block 13.2

a. The development of space science and technology and of its applications;
b. The development of relevant and appropriate capabilities in interested States;
c. Cooperation and contribution in education and training;
d. Access to and exchange of information;
e. Incentivization of joint ventures;
f. The exchange of expertise and technology among States on a mutually acceptable basis;
g. The establishment of an international fund.[52]

The *Building Blocks* provide a good foundation for the development of a future legal framework for space resource activities, albeit one that is very Eurocentric, while at the same time acknowledging the interests of developing States.

4.2 UNCOPUOS Working Group on Legal Aspects of Space Resource Activities

The Working Group on Legal Aspects of Space Resource Activities was established by the UNCOPUOS Legal Subcommittee in 2021. During debates about the creation of the Working Group, the United States observed that the use of space resources is critical to the long-term viability of space activities.[53] Russia supported the creation of the Working Group and noted that any activities 'related to the exploration, exploitation and utilisation of space resources must be based on a clear legal framework which is widely recognised on a global level'.[54] China and the Group of 77 argued that there was an urgent need to examine domestic space laws that allow 'for the exploitation of celestial bodies for economic purposes' in light of international space law 'to avoid contradictions or gaps within the legal framework…and in order to provide a clear understanding of the legal obligations of States'.[55]

The mandate of the Working Group is to collect information regarding the exploration, exploitation and utilisation of space resources, including the existing legal framework for such activities, and specifically to develop 'a set of initial recommended principles for such activities, taking into account the need to ensure that they are carried out in accordance with international law and in a safe, sustainable, rational and peaceful manner'.[56] The Working Group, led by Chair Mr. Andrzej Misztal

[52] *Id.*, Building Block 13.2.

[53] G Swiney, 'US Statement Agenda Item 14: Potential Legal Models for Activities in Exploration, Exploitation and Utilization of Space Resources' 60th sess, COPUOS Legal Subcommittee (2021).

[54] Russian Delegation, 'General Exchange of Views on Potential Legal Models for Activities in Exploration, Exploitation and Utilization of Space' 60th sess, COPUOS Legal Subcommittee (June 2021).

[55] Alejandro Solano Ortiz, 'G-77 and China Statement during the Sixtieth Session of the Legal Subcommittee of the United Nations Committee on the Peaceful Uses of Outer Space' 60th sess, COPUOS Legal Subcommittee (June 2021).

[56] United Nations, *Report of the Committee on the Peaceful Uses of Outer Space* 64th sess, UN Doc A/76/20, Annex III ('Mandate, terms of reference, and workplan and methods of work for the

(Poland) and Vice-Chair Professor Steven Freeland (Australia), has a five-year work-plan. States have been invited to submit responses to the mandate and purpose of the Working Group. Australia's response affirmed that Australia is 'committed to contributing to the development of norms that ensure the long-term safety, stability and sustainability of the outer space environment'.[57] Australia also observed that in its opinion, the threshold for the requirement of an international legal framework for space resource activities in the Moon Agreement had not yet been reached, stating that 'the world has not yet arrived at the point where the exploitation of space resources could be considered "feasible"'.[58]

At the time of writing, the Working Group has not yet produced any guidance for a future legal framework for space resource activities. This is not surprising; it is in its initial stages, and things tend to move slowly in international fora. However, as time progresses, the work of the Working Group may provide a more solid foundation for a future legal framework for space resource activities, one that is agreed to by the major space-faring States and represents the interests of all States.

4.3 Space Resource Activities Under the Artemis Accords

The United States' National Aeronautics and Space Administration ('NASA') announced the Artemis Accords in May 2020 as expressing a 'shared vision for principles, grounded in the Outer Space Treaty of 1967, to create a safe and transparent environment which facilitates exploration, science, and commercial activities for all of humanity to enjoy'.[59] At the time of writing, there are 43 signatories to the Artemis Accords, making the Accords more significant for the development of an international framework for space resource activities than the Moon Agreement, despite being a political agreement and not a legally binding one. However, Russia has criticised the Accords and NASA's Artemis program as being too US-centric.[60] China has also been excluded from participation, in part because of US domestic

working group established under the Legal Subcommittee agenda item entitled "General exchange of views on potential legal models for activities in the exploration, exploitation and utilization of space resources"').

[57] Australia, *Response to the invitation to provide information on the mandate and purpose of the Working Group on Legal Aspects of Space Resource Activities under the Legal Subcommittee of the Committee on the Peaceful Uses of Outer Space* (2023) https://www.unoosa.org/documents/pdf/copuos/lsc/space-resources/LSC2023/StatesResponses/Australias_Submission_to_Space_Resource_Working_Group.pdf. Accessed 22 July 2024, 2, 6.

[58] *Id.*, at 6.

[59] National Aeronautics and Space Administration, *The Artemis Accords: Principles for Cooperation in the Civil Exploration and Use of the Moon, Mars, Comets, and Asteroids for Peaceful Purposes* (13 October 2020) https://www.nasa.gov/wp-content/uploads/2022/11/Artemis-Accords-signed-13Oct2020.pdf?emrc=653a00. Accessed 15 July 2024.

[60] Jeff Foust, 'Russia Skeptical about Participating in Lunar Gateway' *SpaceNews* (12 April 2020) https://spacenews.com/russia-skeptical-about-participating-in-lunar-gateway/. Accessed 22 July 2024.

law.[61] Additionally, criticism has been levelled at the Artemis Accords for being structured as a series of bilateral agreements between NASA and State space agencies, rather than as a multi-lateral agreement, risking fragmentation, divergence, and favouritism.[62]

Principle 10 of the Artemis Accords relates to space resources. It states that 'the utilization of space resources can benefit humankind by providing critical support for safe and sustainable operations'[63] and affirms the view that 'the extraction of space resources does not inherently constitute national appropriation under Article II of the Outer Space Treaty'.[64] Principle 10.4 provides that:

> 'The Signatories intend to use their experience under the Accords to contribute to multilateral efforts to further develop international practices and rules applicable to the extraction and utilization of space resources, including through ongoing efforts at the COPUOS'.[65]

This may signal a more significant role for the UNCOPUOS Working Group on Space Resources in the future in the development of a legal framework for space resource activities.

The main way that space resource activities are facilitated in the Artemis Accords is through the use of safety zones around space resource activities,[66] which are expressed to be a practical implementation of the obligation of due regard and mitigation against harmful interference.[67]

The principles in the Artemis Accords relating to space resource activities seek to extend the agreement of States on what is permissible under existing international space law. Professor von der Dunk argues that the United States is using the Artemis Accords to 'gather consensus' of its interpretation of the *Outer Space Treaty* in relation to space resource activities.[68] There is a risk that any future legal framework for space resource activities based on the Artemis Accords principles relating to space resources will represent only the interests of powerful, US-aligned States and not be truly international.[69]

---

[61] Hannah Kohler, 'The Eagle and the Hare: US–Chinese Relations, the Wolf Amendment, and the Future of International Cooperation in Space' (2015) 103(4) *Georgetown Law Journal* 1135, 1135.

[62] Laura C Byrd, 'Soft Law in Space: A Legal Framework for Extraterrestrial Mining' (2022) 71(4) *Emory Law Journal* 801, 283.

[63] Artemis Accords, Principle 10.1.

[64] *Id.*, Principle 10.2; See also, Melissa de Zwart and Joel Lisk, 'The Artemis Accords and Subsequent Developments' in Sandeepa Bhat B, Dilip Ukey and Adithya Variath (eds), *International Space Law in the New Space Era: Principles and Challenges* (Oxford University Press 2024) 231, 244–45..

[65] *Id.*, Principle 10.4.

[66] *Id.*, Principle 11.7.

[67] *Id.*, Principle 11.

[68] Frans von der Dunk, '"For All Moonkind": Legal Issues of Human Settlements on the Moon' (Conference Paper 55891, International Astronautical Congress, October 2020).

[69] Rachel Neef, 'Artemis Accords: A New Path Forward for Space Lawmaking?' (2021) 42(2) *Adelaide Law Review* 569, 579.

4.4 Domestic Space Law and Space Resource Activities

Four States have enacted domestic laws relating to space resource activities. The United States' 2015 *U.S. Commercial Space Launch Competitiveness Act* was the first law to explicitly address ownership of extracted space resources. The operative provision of the 2015 law provides that:

> 'A United States citizen engaged in commercial recovery of an asteroid resource or a space resource under this chapter shall be entitled to any asteroid resource or space resource obtained, including to possess, own, transport, use, and sell the asteroid resource or space resource obtained in accordance with applicable law, including the international obligations of the United States'.[70]

In a report on a precursor to the 2015 law on space resources, the United States Congress' House of Representatives Committee on Science, Space, and Technology described the purpose of what would eventually become the above provision as 'enabling' a new industry and 'providing clarity for future entrepreneurs'.[71] The report of this Committee directly addressed concerns from space resource exploitation proponents that 'legal and regulatory uncertainties [were] impeding their development and threatened to disrupt their continued investment and eventual activities in outer space'.[72] The Committee largely dismissed any potential conflict with the international obligations of the United States, stating that there was nothing in the law which 'call[ed] for the United States to violate its existing international obligations' under the treaties applicable to outer space, that the law did not 'claim sovereignty over outer space or any celestial bodies', and that '[r]emoving, taking possession, and using in-situ celestial resources … is not to be construed as an act of national appropriation by claim of sovereignty, by means of use or occupation, or by any other means'.[73]

In 2020, then-President Trump signed Executive Order 13,914 on 'Encouraging International Support for the Recovery and Use of Space Resources'. This Executive Order explicitly stated that 'Americans should have the right to engage in commercial exploration, recovery, and use of resources in outer space, consistent with applicable law'.[74] The Russian State space agency, ROSCOSMOS, responded to the Executive Order by characterising it as an attempt 'to expropriate outer space' condemning what it perceived as 'aggressive plans to actually seize territories of other planets'.[75]

---

[70] 51 USC §51303 (2024); *U.S. Commercial Space Launch Competitiveness Act*, Pub L No 114–90, §402,129 Stat 704, 721 (2015).

[71] H.R. Rep No 114–153, 3 (2015).

[72] *Id.*

[73] *Id.*, 6 (2015).

[74] Exec Order No 13,914, 'Encouraging International Support for the Recovery and Use of Space Resources' (6 April 2020) 85 Fed Reg 20381, sec 1.

[75] 'Russian space agency says Trump paving way to seize other planets' Reuters (7 April 2020) https://www.reuters.com/article/us-usa-russia-space/russian-space-agency-says-trump-paving-way-to-seize-other-planets-idUSKBN21P1SY. Accessed 22 July 2024.

Despite the condemnation of the United States, Luxembourg was quick to follow, introducing its domestic law relating to space resource activities in 2017. Luxembourg's *Law on the Exploration and Utilisation of Space Resources* which explicitly stated that (in English translations of the French text) 'space resources are capable of being owned',[76] and that 'no person can explore or use space resources without holding a written mission authorisation from the minister or ministers in charge of the economy and space activities'.[77] Unlike the United States' law, the Luxembourg law set out a licensing framework to permit space resource exploitation activities. This approach sought to provide increased regulatory certainty to those seeking to utilise the framework to undertake space resource exploration and exploitation activities.[78]

The United Arab Emirates legislated to regulate space resources activities in 2019 with its Federal Law on the Regulation of the Space Sector.[79] The approach of this law was to include space resource exploration or extraction activities (whether for scientific, commercial or other purposes) in a list of regulated activities.[80] The law later provides that 'terms and conditions relating to Authorization for the extraction, exploitation and utilization of Space Resources, including their ownership, purchase, sale, trade, transportation, storage … shall be determined by a Decision issued by the Council of Ministers or whomever it delegates'.[81] This provision effectively allows for the government of the UAE to impose conditions on operators as it sees fit at a future time. This allows businesses to consider developing space resource exploitation plans from the UAE's national jurisdiction, but does not necessarily provide the certainty or clarity of operating conditions that the Luxembourg law provides.

In 2021, Japan introduced its own space resources law: the *Act on the Promotion of Business Activities for the Exploration and Development of Space Resources.*[82] This law provided that a person who obtains space resources in accordance with an approved business activity plan pertaining to their licence acquires ownership of those space resources.[83] Much like the approach from Luxembourg, Japan's 2021 law is a complete licensing regime capable of providing certainty to those businesses seeking to undertake space resource exploitation activities.

As each of the domestic laws relating to space resource activities is expressed to be consistent with that State's obligations under the Outer Space Treaty, this is relevant State practice for the interpretation of what activities are a permissible use of

---

[76] *Loi du 20 juillet 2017 sur l'exploration et l'utilisation de ressources de l'espace,* Mémorial A N° 674, 28 July 2017 (Luxembourg), Art. 1.

[77] *Id.*, Art. 2(1).

[78] Mahulena Hofmann, PJ Blount, Gabrielle Leterre, Antonino Salmeri and Laetitia Zarkan, *Space Legislation of Luxembourg* (Wolters Kluwer 2022).

[79] Federal Law No 12 of 2019 on the Regulation of the Space Sector (UAE).

[80] *Id.*, Art. 4(i)-(j).

[81] *Id.*, Art. 18(1).

[82] Act on the Promotion of Business Activities for the Exploration and Development of Space Resources (Act No 83 of 2021) (Japan).

[83] *Id.*, Art. 5.

space under existing space law.[84] In 2020, the Canadian-based Outer Space Institute issued an open letter expressing concern over States enacting domestic laws and engaging in bilateral negotiations in the absence of legal certainty and urging the United Nations to begin the process for developing a new treaty on space resource utilisation.[85]

5 Resource Activities in Other Areas Beyond National Jurisdiction

There may be lessons for any future legal framework for space resource activities to be drawn from the regulation of resource activities in other areas beyond national jurisdiction, specifically Antarctica and the deep seabed.

5.1 Regulation of Resource Activities in Antarctica

International exploration of Antarctica is governed by a set of treaties, collectively known as the Antarctic Treaty System, building on the Antarctic Treaty, which entered into force in 1961.[86] One of the main objectives of the Antarctic Treaty was to ensure that 'Antarctica remained a place where scientific research could be freely undertaken'.[87]

Resource activities other than for scientific research have been prohibited on Antarctica since 1991 under the Madrid Protocol on Environmental Protection.[88] This means that there is no prospect of commercial resource activity being conducted on Antarctica while the Madrid Protocol remains in force. There remain concerns, however, that some States may choose to act outside of the Antarctic Treaty System,[89] particularly when presented with the lure of mass profits to be derived from resources. It is possible that a similar restriction on resource activities may be adopted for a future

[84] Melissa de Zwart, Stacey Henderson and Michelle Neumann, 'Space resource activities and the evolution of international space law' (2023) 211 *Acta Astronautica* 155, 157.

[85] Outer Space Institute, 'Letter to Tijjani Muhammad-Bande, President of the United Nations General Assembly' (Aug 2020) http://www.outerspaceinstitute.ca/docs/InternationalOpenLetterOnSpaceMining.pdf. Accessed 22 July 2024; Laura C Byrd, 'Soft Law in Space: A Legal Framework for Extraterrestrial Mining' (2022) 71(4) *Emory LJ* 801, 806.

[86] Antarctic Treaty (opened for signature 1 December 1959, entered into force 23 June 1961) 402 UNTS 71.

[87] Donald R Rothwell, 'The Antarctic Treaty At Sixty Years: Past, Present and Future' (2021) 22 *Melbourne J Intl L* 1, 4..

[88] Protocol on Environmental Protection to the Antarctic Treaty (adopted 4 October 1991, entered into force 14 January 1998).

[89] Donald R Rothwell, 'The Antarctic Treaty At Sixty Years: Past, Present and Future' (2021) 22 *Melbourne J Intl L* 1, 10.

legal framework for space resources. If so, commercial space resource activities would be prohibited whereas space resource activities for scientific research would remain a permissible use of space. Indeed, such a moratorium on commercial space resource activities for a specified period of time may be one way to preserve the space environment and the knowledge to be gained from exploration of celestial bodies before large-scale destructive commercial space resource activities are conducted.

5.2 *Regulation of Resource Activities on the Deep-Sea Bed*

In 1966, then US President Johnson stated:

> '[U]nder no circumstances must we ever allow the prospects of rich harvest and mineral wealth to create a new form of colonial competition among the maritime nations. We must be careful to avoid a race to grab and to hold the lands under the high seas. We must ensure that the deep seas and the ocean bottoms are, and remain. the legacy of all human beings'.[90]

During debates at the United Nations in 1967, the delegate from Malta introduced the language of 'common heritage of mankind'.[91] He warned of 'the truly incalculable dangers for mankind as a whole were the seabed and ocean floor beyond present national jurisdiction to be progressively and competitively appropriated, exploited and used for military purposes by those who possess the required technology'.[92]

The *Law of the Sea Convention* entered into force in 1994 and is generally considered to be reflective of customary international law. Echoing the Outer Space Treaty, the *Law of the Sea Convention* creates rules for the international seabed and its mineral resources as the 'common heritage of mankind', with such resources not being subject to claims of sovereignty or appropriation.[93] A new international body, the International Seabed Authority, was established to 'provide for the equitable sharing of financial and other economic benefits' that derive from resource activities.[94] At the time of writing, the rules for resource activities developed under the International Seabed Authority require a licence to conduct resource activities on the deep-sea bed and cover the exploration, but not the exploitation, of polymetallic nodules, seafloor massive sulphides, and ferromanganese crusts.[95] The International

---

[90] Graham Nicholson, 'The Common Heritage of Mankind and Mining: An Analysis of the Law as to the High Seas, Outer Space, the Antarctic and World Heritage' (2002) 6 *NZ J Envtl L* 177, 184.

[91] *Id.*, 181.

[92] *Id.*

[93] United Nations Convention on the Law of the Sea (adopted 10 December 1982, entered into force 16 November 1994) 1833 UNTS 397, art, 136.

[94] United Nations Convention on the Law of the Sea (adopted 10 December 1982, entered into force 16 November 1994) 1833 UNTS 397, art s.137(1) and 140(2).

[95] Florian Rabitz, 'Space Resources and the Politics of International Regime Formation' (2023) 17(1) *Intl J Commons* 243.

Seabed Authority has been criticised for prioritising the interests of commercial resource usage over the protection of the deep-sea common heritage.[96]

6 Conclusion

With the exception of the *Moon Agreement*, which is applicable only to its 17 State Parties, none of the space law treaties refer directly to space resource activities.[97] Despite this, the foundational principles laid down in the existing space law treaties are relevant for the conduct of space resource activities. Recent efforts to make progress on a legal framework for space resource activities are a step in the right direction, but more work is needed. As this chapter has demonstrated, many of the efforts to develop a framework for space resource activities have been subject to criticism about being Western-centric and unrepresentative, or inconsistent with existing international space law. The most promising effort for the development of a truly international legal framework is the UNCOPUOS Working Group on Space Resources; however, with all such international bodies, it will be some time before a legal framework is developed, debated, and agreed. In the meantime, there is a risk that space resource activities will continue unfettered by any detailed rules governing their conduct.

Lessons can be drawn from the regulation of resource activities in other areas beyond national jurisdiction, specifically Antarctica and the deep-sea bed. In particular, consideration should be given to the value of a temporary moratorium on commercial space resource activities to ensure scientific research can be conducted before any large-scale destructive resource extraction activity takes place. Additionally, as Antarctica and the deep-seabed both demonstrate, there may be a benefit in establishing a new dedicated international authority for space resource activities to grant permits for prospecting and extraction activities, and to manage benefit-sharing obligations.

In the absence of a new United Nations space treaty, and with little indication of state interest in negotiating one specific to space resource activities, it is likely that any future international framework will emerge through soft law principles. Byrd argues that the development of soft law principles would allow States to 'develop a set of evolving principles for best practices in space resource extraction that accounts for the needs of a rapidly changing industry while avoiding the lengthy treaty negotiation process'.[98] While that would result in a framework for space resource activities that is not legally binding nor legally enforceable, the establishment of international norms

---

[96] Isabel Feichtner and Surabhi Ranganathan, 'International Law and Economic Exploitation in the Global Commons: Introduction' (2019) 30(2) *EJIL* 541.

[97] Agreement Governing the Activities of States on the Moon and Other Celestial Bodies (adopted 5 December 1979, entered into force 11 July 1984) 1363 UNTS 21, art 11 ('Moon Agreement').

[98] Laura C Byrd, 'Soft Law in Space: A Legal Framework for Extraterrestrial Mining' (2022) 71(4) *Emory LJ* 801, 840.

and best practices for space resource activities may eventually crystallise into binding law, either codified as a new treaty or crystallising into customary international law. More work is needed to ensure that the interests of all States are respected and protected, that space remains the province of all humankind, and that an appropriate balance can be struck between protecting the space environment, enabling in-situ space resource utilisation, and permitting commercial extraction of space resources.

Stacey Henderson is Associate Professor in Law at Flinders Law School, College of Business, Creative Arts, Law & Social Sciences, Flinders University. Stacey researches the protective capacity of law, including international law generally, responsibility of States, and governance of new and emerging technology. Her other research interests include commercial space activities, sustained human presence in space, the responsibility to protect civilians from atrocity crimes, disarmament and arms control, international humanitarian law, international peace and security, grey zone operations in space and cyber, and information and disinformation operations.

Dr. Joel Lisk is a lawyer and academic based in Adelaide, Australia. Dr Lisk is a Lecturer in Space Law at Flinders Law School, College of Business, Creative Arts, Government & Law & Social Sciences, Flinders University. Dr Lisk undertakes research on the design and implementation of regulatory frameworks applicable to emerging operating domains with a particular focus on the use of outer space. Dr Lisk is the Principal Lawyer of a specialist legal practice that provides advice on space business matters including satellite manufacture, payload hosting, and regulatory compliance.

The Legality of the Appropriation of Space Resources: Current International Framework and Future Legal Developments

Leonardo Alberto López Marcos and Charles W. Stotler

Abstract Humankind is on the cusp of creating a true space economy, based not on terrestrial end-users of current satellite systems, but on services in space—tourism, commercial space stations, in-space manufacturing and eventually, space settlements. Putting humankind back on the Moon promises new market opportunities with an estimated value of $216 billion in 2040. To secure a permanent lunar presence and create space settlements, space resources (e.g., regolith, ice, water) will be essential. Nowadays, there is a down-mass problem, meaning there is no business case for a space resource return mission; for this reason, near-term space resource plans are designed for space: orbital, cislunar, and lunar economies. The utilization of space resources presents legal challenges under International Space Law. Various interpretations of Article II of the Outer Space Treaty, which prohibits "national appropriations" in outer space, have been put forth by States without consensus. Is exploitation lawful? Will exploitation zones be open to other countries and their commercial space companies? What are the metes and bounds of the prohibition of national appropriations, and how do they affect the activities of non-governmental actors? Is this a matter of interpretation of lex lata or the creation of lex ferenda? This article acknowledges current difficulties in the adoption of a new treaty on space resources and will focus instead on existing sources of international law that may be leveraged for the lawful extraction and utilization of space resources.

Keywords Space economy · Lunar resources · Outer space treaty · National appropriation · Space settlements · International law

L. A. L. Marcos (✉)
Department of Public International Law, Complutense University of Madrid, Madrid, Spain
e-mail: leonarlo@ucm.es

C. W. Stotler
University of Mississippi School of Law, Oxford, USA
e-mail: cstotler@olemiss.edu

S. Bhat B and A. Variath (eds.), *Legal Responses to Space Resource Utilisation*, Space Law and Policy, https://doi.org/10.1007/978-981-95-8864-0_3

1 Space Resources in Context: Balancing International Affairs and Legal Foundations

The exploitation of space resources—also known as space mining—is an emerging industry. Along with other activities, such as space tourism, space habitats, in-orbit servicing, and commercial space stations, these represent just 1% of the total space economy, while the launch and satellite industries account for 10% and 89% of the space economy, respectively.[1] Nevertheless, an increasing number of commercial projects and companies aim to extract and utilize space resources from the Moon and other celestial bodies. Space resource utilization has been identified as a cardinal aspect of a return to and permanent presence on or near the Moon.[2] Another factor driving this increase is the enormous estimated value—on the order of hundreds of trillions of dollars—of minerals locked in asteroids.[3] The economic impact of harvesting this vast wealth would be, for better or worse, tremendous for the Earth's economy. It is estimated that a single asteroid could have more value than the entire Earth's GDP.[4] Yet, the influx of such a quantity of valuable resources on Earth could destabilize the terrestrial economy.[5] For instance, the value of gold and other precious resources would fall, prices would increase, salaries would not be expected to rise, and employment would decline. Nonetheless, as the influx of such resources would

[1] Chad Anderson, *The Space Economy: Capitalize on the Greatest Business Opportunity of Our Lifetime* (John Wiley & Sons 2023) 185–202.

[2] Stephanie Schierholz and Josh Finch, 'NASA Selects Companies to Collect Lunar Resources for Artemis Demonstrations' (NASA, 3 December 2020) https://www.nasa.gov/news-release/nasa-selects-companies-to-collect-lunar-resources-for-artemis-demonstrations/. Accessed 7 October 2024.

[3] Simonetta Di Pippo, *Space Economy: The New Frontier for Development* (Bocconi University Press 2023) 78–79; Statista Research Department, 'Market value of asteroid mining in 2017 and 2025' (July 2024). https://www.statista.com/statistics/1023115/market-value-asteroid-mining/. Accessed 12 August 2024.

[4] Aaron O'Neill, 'Global Gross Domestic Product (GDP) at Current Prices from 1985 to 2029' (Statista, 2024). https://www.statista.com/statistics/268750/global-gross-domestic-product-gdp/. Accessed 12 August 2024; see also Simonetta Di Pippo (n 3) 77–80.

[5] A great, sudden influx of money into a country (in this case, the world) would flood the economy. A classic example is the discovery of oil fields under the Netherlands' sovereignty. This led to an increase in prices, a high rate of unemployment, and inflation—known as the *Dutch Disease*. See, Jonathon Moses and Bjørn Letnes, *Managing Resource Abundance and Wealth: The Norwegian Experience* (Oxford: Oxford University Press, 2017). In the case of space mining, if space resources are brought back to Earth (e.g., platinum), this could alter the already existing geopolitical order and the value of this material. Platinum is concentrated in 90% of the world's production in South Africa; therefore, the massive quantity of platinum from asteroids brought to Earth (if possible, given the current technology) would lead to a loss of value for platinum and all its related economic activities. On the other hand, if resources are not brought back to Earth but are exploited and utilized in space settlements, these settlements could even compete with the Earth's economy, potentially shifting the center of the terrestrial economy to specific permanent space settlements. See, Leonardo A López Marcos, 'Customers? In Outer Space. Creating a Real Space Economy: from the Traditional Space Economy to the Emerging Space Activities' (2015) 15 *Boletín del Observatorio Jurídico Aeroespacial* 92, 97–102.

be constant, it is expected that prices and the world's economy would eventually settle down.

The market for space resources remains unclear. This is due to several factors. Though lunar sample returns were conducted during the Apollo era, the technology for robotic harvesting of these resources is either in its infancy or commanded wholly by governments.[6] Beyond the technical difficulties of harvesting space resources is a down-mass problem: it is simply currently too expensive to bring large quantities of space resources back to Earth.[7] Alternative plans include in-space refueling and manufacturing using space resources. These plans also face technical challenges. Refueling a satellite, in and of itself—not to mention processing and use of space resources to do so—has presented considerable technical challenges.[8]

Technical challenges are not the only challenges. Though there is significant enthusiasm for exploiting resources and using them in outer space—enthusiasm from both governments and non-governmental actors alike—legal questions persist. To an extent, the questions persist because there is little consensus among States as to whether exploiting resources harvested from areas beyond national jurisdiction is an activity that requires *ex ante* regulation—an agreement developed in advance of the activities and supported by a regulatory authority—or *ex post* legal solutions—an agreement arising after the activities commence, whether that be a tacit agreement via custom or an express agreement via treaty.

As described elsewhere, Law tends to evolve through the interactions of its subjects.[9] The subjects of international law are States and international organizations, and their interactions can lead to the development of Customary International Law (CIL). In this regard, witness the slow accretion of customs for activities

---

[6] For more on lunar sample returns by both the US and USSR, see JH Allton, 'Lunar Samples: Apollo Collection Tools, Curation Handling, Surveyor III and Soviet Lunar Samples' (NASA Johnson Space Center, undated) https://www.nasa.gov/wp-content/uploads/2019/04/02_allton_corrected_apollo.pdf. Accessed 6 October 2024; see also Wynn Scott, 'NASA Shares Asteroid Bennu Sample in Exchange with JAXA' (NASA, 23 August 2024) https://www.nasa.gov/centers-and-facilities/johnson/nasa-shares-asteroid-bennu-sample-in-exchange-with-jaxa/. Accessed 6 October 2024.

[7] Matching the terrestrial platinum group of metals production of 2023 (400 metric tonnes) would require many trips. Should SpaceX's Starship come online, it is projected to be able to launch 150 metric tonnes with its fully reusable design and up to 250 metric tonnes with its expandable version. Whether the ship could return with these kinds of weights is unknown. Some authors point out that if the total production of the mining industry (metals) is considered, it would be impossible to phase out terrestrial mining. Total terrestrial mining production is about 1.7 billion metric tonnes, which would require thousands of Starship launches per day. Jarrett Dillenburger, 'The Return Problem: Can We Bring Space Resources Back to Earth?' (Medium, undated) https://jdillenburger.medium.com/the-return-problem-72377d615d6e. Accessed 13 August 2024.

[8] Filip Timotija, 'NASA shutting down $2B satellite refueling project over tech, cost challenges' *The Hill* (2 March 2024) https://thehill.com/homenews/space/4504140-nasa-shutting-down-2b-satellite-refueling-project-over-tech-cost-challenges/. Accessed 6 October 2024.

[9] This part summarizes and builds upon a previous description of the evolution of international space law. Charles Stotler, 'Towards a New Space Law: The State of the Art and the Prospects for New Regulation' in Alessandro Gili (ed), *The Sky Is Not the Limit: Geopolitics and Economics of the New Space Race* (ISPI, Ledizioni LediPublishing 2024) 155, 156–158.

on the high seas and their eventual codification in various instruments, including the 1982 UN Law of the Sea Convention.[10] Even where technology advanced rapidly, such as with the rise of international aviation in the early twentieth century, customs—some dealing with such fundamental issues as national sovereignty over airspace—preceded treaties and regulations.[11]

International Space Law (ISL) likewise began in this manner. Seminal scholars recognized several customs that preceded international conventions on space activities, including that outer space is free for exploration and use by States, that sovereignty does not extend to outer space, which is not subject to national appropriation, and that States retain jurisdiction and control over objects launched into outer space.[12] The Outer Space Treaty codified these from preceding UN General Assembly Resolutions.[13]

The Outer Space Treaty is a treaty of general principles formed in the crucible of the Cold War Space Race. As a treaty of general principles, it provides useful flexibility for the further development of international law via various means, including subsequent agreements, principles, UNGA resolutions and guidelines. The rapid development of space treaties during the 1960s and 1970s may have had both benefits and detriments.[14] The benefits include a solid legal foundation upon which States have participated in an incredible expansion of the use and exploration of outer space. Few detriments exist, but one may be the expectation of a proactive international relations environment involving binding legal instruments prior to the commencement of particular space activities. The Moon Agreement is an example: its negotiation was concluded in 1979; it entered into force in 1984; and it calls upon States to establish an

[10] United Nations Convention on the Law of the Sea (adopted 10 December 1982, entered into force 16 November 1994) 1833 UNTS 31363, as modified by the 'Agreement relating to the Implementation of Part XI of the Convention' (adopted 28 July 1994, entered into force 28 July 1996) 33 ILM 1309 (collectively 'UNCLOS').

[11] The Paris Convention of 1919 recognized the complete and exclusive sovereignty over the airspace above States' territories notwithstanding that the airplane had only been invented less than two decades prior. Convention Relating to the Regulation of Aerial Navigation (adopted 13 October 1919, entered into force 11 July 1922) 11 LNTS 173, art 1.; Michael P Scharf, 'Accelerated Formation of Customary International Law' (2014) 20(2) *ILSA Journal of International & Comparative Law* 305, 310–14.

[12] V S Vereshchetin and G M Danilenko, 'Custom as a Source of International Law of Outer Space' (1985) 13(1) *Journal of Space Law* 22, 25.

[13] Treaty on Principles Governing the Activities of States in the Exploration and Use of Outer Space, including the Moon and Other Celestial Bodies (adopted 27 January 1967, entered into force 10 October 1967) 610 UNTS 205 (*Outer Space Treaty*); Paul G Dembling and Daniel M Arons, 'The Evolution of the Outer Space Treaty' (1967) 33 *Journal of Air Law and Commerce* 419, 420–29.

[14] Subsequent agreements include: *Agreement on the Rescue of Astronauts and the Return of Objects Launched in Outer Space,* 22 April 1968, 672 UNTS 119; *Convention on International Liability for Damage Caused by Space Objects*, 29 March 1972, 961 UNTS 187 [hereinafter, "Liability Convention"]; *Convention on Registration of Objects Launched into Outer Space*, 14 January 1975, 1023 UNTS 15; and the *Agreement Governing the Activities of States on the Moon and Other Celestial Bodies*, 11 July 1984, 1363 UNTS 3 [hereinafter, "Moon Agreement"].

international regime, including appropriate procedures, prior to the commencement of resource exploitation.[15]

International conventional law tends to be durable but also inflexible and challenging to modify quickly. Moreover, Law—whether it be a judicial decision, a codified custom, or legislation—tends to occur *ex post*: it is often a reaction to circumstances after they occur. The rules of CIL for State responsibility are an example. Regulations, on the other hand, emanate from technical bodies authorized by Law to (re)visit issues and update regulatory frameworks. Mostly, this occurs at the domestic level—a subunit of government with technical expertise is mandated to produce regulations under political oversight. There are rare instances of regulation at the international level, most importantly for space activities, being the International Telecommunication Union (ITU) and its Radio Regulations.[16] Generally, regulation tends to occur *ex ante*—it is often an attempt by States to curb problems before the consequences manifest.

Outside of the context of the ITU, *ex ante* regulatory solutions for space activities largely exist at the national level. The twenty first century has played witness to a profound and dramatic increase in national space legislation and regulation. There are many related factors which contributed to this shift from proactive international law-making to national legal solutions. The actions of the U.S. loom large in this respect due to its historic prominence in space activities. The Reagan Administration's pivot away from international law coincided with a national push for commercial space activities, planting the seeds for a commercial space launch sector and the privatization of other space activities such as remote sensing. A rise in private, non-governmental space actors necessitated, pursuant to Article VI of the Outer Space Treaty, authorization and continuing supervision regimes—i.e., the implementation of ISL into domestic legal systems. That implementation, in turn led to the creation of *ex ante* regulatory regimes and the establishment of regulators. This pattern has been replicated around the world in one form or another with only a few exceptions.[17]

Though treaty-making ceased, ISL continued to grow through international principles, resolutions and guidelines.[18] These included recommendations on national

[15] Moon Agreement, supra note 14 at Art. 11(5).

[16] *Constitution and Convention of the International Telecommunications Union*, 22 December 1992, 1825 UNTS 31251 (Article 4 of the ITU Constitution binds member States to the regulations of the ITU.)

[17] Of space powers, by which we mean States possessing robust capabilities for the use and exploration of outer space, only India has not followed this pattern, choosing instead to conduct all space activities through government agencies. See, generally, Annette Froehlich and Vincent Seffinga (eds), *National Space Legislation: A Comparative and Evaluative Analysis* (Springer/ESPI 2018); SG Sreejith, 'Unmaking a National Space Legislation for India: Indigenizing Space Law through the Organic Science of the India Space Program' (2018) 83 *Journal of Air Law and Commerce* 109, 133–36.

[18] S Marchisio, 'The Evolutionary Stages of the Legal Subcommittee of the United Nations Committee on the Peaceful Uses of Outer Space (COPUOS)' (2005) 31 *Journal of Space Law* 219, 224, 231–32, 237.

legislation for space use and exploration.[19] And yet calls persist for a UN specialized agency for space.[20] The rapid development of space treaties in the 1960s and 1970s may lend to the appearance that States must act through international, multilateral forums for space activities and, therefore are somehow negligent in leaving the regulation of new space activities to national legal and regulatory regimes. Worse still is a common misconception that UN specialized agencies are international regulators. Even where States have established specialized agencies at the international level, such as the International Civil Aviation Organization or ITU, these agencies' standards and rules, though often binding, nevertheless require a national regulator for implementation and administration. Thus, an international regulator is something of a misnomer; UN specialized agencies are mostly standard-setting bodies. A notable exception may be the International Seabed Authority set up pursuant to UNCLOS to deal with deep seabed mining.[21]

At the heart of the question of *ex ante* regulation or *ex post* legal solutions for space resource exploitation is the sovereign equality of States. According to Kelsen, "... [under] general international law no State can be legally bound without or against its will..." and "...the principle of sovereign equality is the principle of autonomy of the States as subjects of international law."[22] This autonomy is not boundless. In carrying out activities in areas beyond national jurisdiction, States are bound by an obligation of due regard for the corresponding interests of other States.[23] A universal description of due regard is likely not possible, as due regard is fundamentally tied to the legal context within which States act. For example, in a dispute involving marine protection zones, the rights and obligations arising under due regard were informed by provisions of UNCLOS; for a dispute involving whaling, they were informed by the International Convention on the Regulation of Whaling.[24] Further, the determinations in those cases were based upon the actions (practices) of States in light of the obligations arising out of those varying legal contexts. Though the legal context for space resource utilization can be identified by examining the treaties, UN Principles, Resolutions and Guidelines relating to space activities, there is little

---

[19] *Recommendations on national legislation relevant to the peaceful exploration and use of outer space*, UN General Assembly, 68th Sess. (11 December 2013) UNGA Res 68/74.

[20] N Jasentuliyana, *International Space Law and the United Nations* (Kluwer Law International 1999); RS Jakhu, T Sgobba and PS Dempsey (eds), *The Need for an Integrated Regulatory Regime for Aviation and Space: ICAO for Space?* (Springer Wien 2011).

[21] UNCLOS, *supra* note 10 at Articles 151–168.

[22] Hans Kelsen, 'The Principle of Sovereign Equality of States as the Basis for International Organization' (1944) 53(2) *Yale LJ* 207, 209.

[23] See, e.g.,: Andrea J Harrington, 'Due Regard as the Prime Directive for Responsible Behavior in Space' (2023) 20(1) *Loy U Chi Intl L Rev* 57.

[24] Caroline E Foster, *Global Regulatory Standards in Environmental and Health Disputes* (OUP 2021) 90–96. (Describing international tribunals' analyses of due regard rights and obligations in *Chagos Marine Protected Area (Republic of Mauritius v United Kingdom) (Award)* (2015) 162 ILR 1 and in *Whaling in the Antarctic (Austrialia v Japan: New Zealand intervening) (Judgment)* (2014) ICJ Rep 226).

State practice for space resource extraction and no practice involving utilization or exploitation for commercial purposes.

The legal context for the exploration and use of space resources is further complicated by the mandate in Article I of the Outer Space Treaty that such exploration and use be carried out for the benefit and in the interest of all States. This provision, in addition to the rights of each State to freely explore and use space on the basis of equality and to have free access to all areas of celestial bodies, arguably invokes the general principle of law on joint stewardship over resources in areas beyond national jurisdiction. Based on sovereignty equality and the concept of joint stewardship, Criddle and Fox-Decent advocate for "mandatory multilateralism" under such circumstances, which they characterize as requiring States to "to balancing their own legal rights and authority, on the one hand, with the legal rights and authority of other individual States and the international community collectively, on the other."[25]

Thus, the question of whether space activities require *ex ante* (international) regulation or *ex post* law should be whether States can, under ISL as it currently exists, engage in space resource extraction, utilization and exploitation and afford due regard for the corresponding interests of other States in the exploration and use of outer space. Given the dearth of State practice with space resource utilization and exploitation, this chapter seeks only to identify the legal context within which the corresponding interests of States might arise. This involves an examination of the legal foundation for space resource exploration and use, which, we argue, demonstrates that States have already made choices on this question.

2 The Outer Space Treaty: Interpreting the Foundation for Space Resource Utilization and Exploitation

The starting point for any discussion of the legal context for space resource exploration and use is the Outer Space Treaty. It is unique within the *corpus iuris spatialis* as a treaty of general principles; it was designed specifically for inter-State relationships and focused on avoiding conflict in outer space.[26] It codified principles from the 1963 Declaration of Principles and other UNGA resolutions, codifying and crystallizing customs.[27] It does not expressly address or even mention space resource mining, utilization or exploitation. Nonetheless, it does contain provisions relevant to these activities.

[25] EJ Criddle and E Fox-Decent, 'Mandatory Multilateralism' (2019) 113(2) *American Journal of International Law* 272, 276.

[26] Philip De Man, 'State Practice, Domestic Legislation and the Interpretation of Fundamental Principles of International Space Law' (2017) 42 *Space Policy* 92, 93–101.

[27] *Declaration of Legal Principles Governing the Activities of States in the Exploration and Use of Outer Space*, UN General Assembly, 18th Sess. (13 December 1963) UNGA Res 1962 (XVIII) (laying the groundwork for much of the Outer Space Treaty); *Questions of general and complete disarmament,* UN General Assembly, 18th Sess. (17 October 1963) UNGA Res 1884(XVIII) (Serving as the predecessor to Article IV of the Outer Space Treaty dealing with arms control.)

Arguably, more ink has been spilled over the legality of space mining than any other space-related issue.[28] Yet, clarity on this question has evaded scholars. This is likely due to several factors. There have been very few instances of resource extraction, and utilization or exploitation for commercial purposes has never occurred. As States will be the ultimate arbiters for determining a way forward on space resource exploration and use, scholars are left to toil in something of a vacuum. The lack of clarity is also likely due to implicit biases in analyses under the Vienna Convention on the Law of Treaties (VCLT).[29] While its provisions guide interpretation, so many variables are at play in determining the ordinary meaning of a term or in identifying the most important aspects of context that might inform the meaning of a term. Scholars might import, consciously or not, their own preferences. Indeed, scholars have offered interpretations time and time again, and disagreement among States on the appropriate way forward persists.

States can offer their interpretations in a number of ways, and custom, too, may forge a path ahead. These are addressed in part 3 of this chapter. In the meantime, and notwithstanding the foregoing commentary on prior attempts at interpretation, we throw our hats in the ring in hopes of offering some novel considerations via a VCTL analysis.

[28] Philip De Man, *Exclusive Use in an Inclusive Environment – The Meaning of the Non-appropriation Principle for Space Resource Exploitation* (Springer 2016); Ricky Lee, *Law and Regulation of Commercial Mining of Minerals in Outer Space* (Springer 2012); Virgil Pop, *Who Owns the Moon?: Extraterrestrial Aspects of Land and Mineral Resources Ownership* (Springer 2009); Stephan Hobe and others (eds), *Cologne Commentary on Space Law*, vols I and III (Wolters Kluwer 2009); Maureen Williams, 'The Principle of Non-Appropriation Concerning Resources of the Moon and Celestial Bodies' in *Proceedings of the Colloquium on the Law of Outer Space* (International Institute of Space Law 1970); Ricky Lee, 'Reconciling International Space Law with the Commercial Realities of the Twenty-first Century' (2000) 4 *Singapore Journal of International & Comparative Law* 194; Ricky Lee, 'Article II of the Outer Space Treaty: Prohibition of State Sovereignty, Private Property Rights, or Both?' (2004) *Journal of International Law*; Yun Zhao, 'An International Space Authority: A Governance Model for a Space Commercialization Regime' (2004) *Journal of Space Law* 277; Henry R Hertzfeld and Frans G von der Dunk, 'Bringing Space Law into the Commercial World: Property Rights without Sovereignty' (2005–06) 6 *Chicago Journal of International Law* 81; Alexander J Simon-Butler, 'Bifurcated Sovereignty and the Territorial Conception of "the Province of All Mankind"' (2019) 43(1) *Journal of Space Law* 1; Antonino Salmeri, 'The Integration Between National and International Regulation of Space Resources Activities Under Public International Law' (2019) 43(1) *Journal of Space Law* 60; Thomas Cheney, 'There's Not Rush: Developing a Legal Framework for Space Resource Activity' (2019) 43(1) *Journal of Space Law* 106.

[29] Vienna Convention on the Law of Treaties, 23 May 1969, 1155 UNTS 331, Art. 31 [*Hereinafter*, "VCLT"].

2.1 *The Vienna Convention on the Law of the Treaties: General Rules of Interpretation*

In the context of International Law, the interpretation of treaties is centered around the 1969 Vienna Convention on the Law of Treaties.[30] The general rules of interpretation are found in Article 31 of the VCLT.[31] When interpreting the Outer Space Treaty, as with any other treaty, the VCLT provides two complementary characteristics for interpretation. The first refers to the existence of an order, meaning that, contrary to popular opinion, the task of interpretation cannot start with the *travaux préparatoires*—e.g., preparatory works, drafts. There is a hierarchy between the general rules of interpretation (Art. 31 VCLT) and preparatory works (Art. 32 VCLT). Only when Article 31 fails to provide a clear or common-sense interpretation is it necessary to refer to Article 32 VCLT. This order is important to consider since some scholars tend to first approach the drafts, preparatory works, and negotiations of some treaties to figure out the intention of the States parties.[32] The consequence of neglecting this order could result in an erroneous interpretation. The second characteristic is that the general rules of interpretation in Article 31(1) VCLT do not provide an internal hierarchy; they only provide a common sense logic for the interpretation within article 31(1) VCLT[33]: Specifically, Article 31(1) states, "A treaty shall be interpreted in good faith in accordance with the ordinary meaning to be given to the terms of the treaty in their context and in the light of its object and purpose."

Thus, the Outer Space Treaty will be interpreted by using the ordinary meaning, the object and purpose, and then the context of the treaty—this is not a sequential order.[34] This approach serves the sole purpose of looking first inside the treaty—e.g., words, preamble, annexes, other articles—and later outside it—e.g., subsequent agreements on the interpretation of the treaty, subsequent State practice in interpreting the treaty. Under this method, special attention is given to the meaning of the words, i.e., what they ordinarily mean; the object and purpose of the article or the treaty

[30] Vienna Convention on the Law of Treaties, 23 May 1969, 1155 UNTS 331, entered into force on 27 January 1969 [Vienna Convention].

[31] The interpretation provisions of the VCLT apply to treaties concluded before its coming into force in 1980 notwithstanding a provision on non-retroactivity in Article 4. This is because those rules codify pre-existing customs. See Steven Freeland, 'Peaceful purposes governing the military uses of outer space' (2016) 18(1) *European Journal of Law Reform* 35, 45; Malcolm N Shaw, *International Law* (9th edn, Cambridge University Press 2021) 814.

[32] John H Currie, Craig Forcese and Valerie Oosterveld, *International Law: Doctrine, Practice, and Theory* (Irwin Law 2008) 81–82. See also the *Competence of Assembly regarding admission to the United Nations, Advisory Opinion*, I.C. J. Reports, 1950, p. 4: "If the relevant words in their natural and ordinary meaning make sense in their context, that is an end of the matter. If, on the other hand, the words in their natural and ordinary meaning are ambiguous or lead to an unreasonable result, then, and then only, must the Court, by resort to other methods of interpretation, seek to ascertain what the parties really did mean when they used these words", p. 8.

[33] Mark Weston Janis, *International Law* (Wolters Kluwer 2012) 32.

[34] Joel A Dennerley, 'State Liability for Space Object Collisions: The Proper Interpretation of "Fault" for the Purposes of International Space Law' (2018) 29 EJIL 281.

are identified within the treaty itself. This approach seeks to understand what the Parties actually agreed upon in the text, not what they might have agreed upon (as found in preparatory works). The object and purpose are found in the preamble and annexes, as well as in an overall reading of the treaty itself. And the context that does not refer to the circumstances under which the treaty was drafted; instead, context means giving special attention to the preamble and annexes—as related to the object and purpose—and five other mechanisms outside the treaty under Article 31 VCLT, *inter alia*, agreements and instruments between State Parties to the treaty when it was concluded, subsequent agreements and practices in interpreting the treaty, and applicable rules of International Law.[35]

2.2 The Application of the Vienna Convention on the Law of the Treaties to the Outer Space Treaty.

There are two main provisions that affect the exploitation and appropriation of space resources within the Outer Space Treaty: Articles I and II. Article I expresses the principle of free exploration of outer space, while Article II contains the so-called non-appropriation principle. Article II constitutes the keystone for the lawfulness of the appropriation of space resources. It states, "Outer space, including the moon and other celestial bodies, is not subject to national appropriation by claim of sovereignty, by means of use or occupation, or by any other means."

The first rule is to examine the ordinary meaning of the terms. The term "appropriation" means claiming ownership of something.[36] The adjective "national" changes the character of or denotes/connotes a type of appropriation. Thus, this opens up what appropriation means in light of the term "National." The latter term includes

[35] Article 31(2) and Article 31(3) VCLT state that: "2. The context for the purpose of the interpretation of a treaty shall comprise, in addition to the text, including its preamble and annexes: (a) any agreement relating to the treaty which was made between all the parties in connexion with the conclusion of the treaty; (b) any instrument which was made by one or more parties in connexion with the conclusion of the treaty and accepted by the other parties as an instrument related to the treaty. 3. There shall be taken into account, together with the context: (a) any subsequent agreement between the parties regarding the interpretation of the treaty or the application of its provisions; (b) any subsequent practice in the application of the treaty which establishes the agreement of the parties regarding its interpretation; (c) any relevant rules of international law applicable in the relations between the parties." The interpreter does not examine only the treaty language itself. Under Article 31 VCLT, context includes language elsewhere in the treaty first and foremost, but it cannot be forgotten the preamble or annexes. For example, the terms "national" or "nationals" is used in several places in the Outer Space Treaty. Those are context for the purpose of understanding a 'national appropriation' under Art. II Outer Space Treaty. See for more, Malcolm N Shaw, *International Law* (9th edn, Cambridge University Press 2021) 813–18; Malcolm D Evans, *International Law* (6th edn, Oxford University Press 2024) 159–64.

[36] Oxford English Dictionary, 'appropriation' (Oxford University Press, online edn) https://www.oed.com/search/dictionary/?scope=Entries&q=appropriation. Accessed 21 August 2025.

public and private space activities, which indirectly extends the principle of non-appropriation to non-governmental entities. The term "National" appears in other articles of the Outer Space Treaty (Article VI) and in Article 11 of the Moon Agreement, which refers to both public and private activities; this closes any loophole on the use of private entities to circumvent the principle of non-appropriation under Article II of the Outer Space Treaty.[37] This phrase is qualified by the phrases that follow it, specifically, "by claim of sovereignty, by means of use or occupation, or by any other means." "Sovereignty" refers to the power that can be exercised over a State´s own part or whole territory (e.g., including people, community, etc.).[38] "Use" refers to a kind of appropriation that involves giving a purpose or use to what is the State's own.[39] "Occupation" means to control an area effectively, thus exercising sovereignty.[40] Finally, the last reference under Article II of the Outer Space Treaty is "by any other means." This is a safeguard formula, meaning that any title that could endorse the ownership of any celestial body to someone or something (it could be a governmental or non-governmental entity, or an individual), i.e., public and private titles of ownership.[41]

A joint interpretation of these terms implies a prohibition on the appropriation of any area of a celestial body by a private or public organization, including an individual. This prohibition extends to any title of ownership, such as renting the area of the celestial body or even controlling it, as this constitutes an exercise of power.[42] Still, there are some exceptions to this article that could be interesting

---

[37] Stephan Hobe, Bernhard Schmidt-Tedd, Kai-Uwe Schrogl and Gérardine Meishan Goh, *Cologne Commentary on Space Law* (Carl Heymanns Verlag 2009) 48–54.

[38] Oxford English Dictionary, 'sovereignty' (Oxford University Press, online edn) https://www.oed.com/search/dictionary/?scope=Entries&q=sovereignty. Accessed 21 August 2025.

[39] "The act of putting something to work, or employing or applying a thing, for any (esp. a beneficial or productive) purpose; the fact, state, or condition of being put to work, employed, or applied in this way; utilization or appropriation, esp. in order to achieve an end or pursue one's purpose.": Oxford English Dictionary, (Oxford University Press, online edn) https://www.oed.com/dictionary/use_n?tab=meaning_and_use#16010028. Accessed 21 August 2025.

[40] In the majority of dictionaries, the term "occupation" refers to the military action to take over a territory. For instance, the Collins Dictionary refers to it as "the control of a country by a foreign military power". Available at https://www.collinsdictionary.com/dictionary/english/occupation. Occupation is permitted. The legal question is whether exclusive occupation rises to the level of an appropriation. Occupation is something less than territorial sovereignty. It might be an exercise of sovereign powers, as in a military occupation during an armed conflict, but even that doesn't constitute the plenary power of a State that exists within its own territory.

[41] Stephan Hobe, 'Adequacy of the Current Legal and Regulatory Framework Relating to the Extraction and Appropriation of Natural Resources' in *Session 4: The Adequacy of the Current Legal and Regulatory Framework?* (2006) 204, 205–206.. From this chapter´s perspective, 'by claim of sovereignty, use or occupation or any other means' is a classic description of ways to acquire territory. 'By any other means' replaced 'discovery and conquest', which had already been outlaws by the UN Charter, so there was no need to mention them.

[42] Pursuant to Article XII and Article VIII Outer Space Treaty, control is permitted under the notion of jurisdiction over an object or facility. Additionally, exercise of power, generally, is not prohibited; Only certain exercises of certain power - national appropriation.

for further analysis.[43] Is temporary occupation allowed? Can an area be controlled seasonally? Are there potential loopholes concerning the temporary occupation of a part of a celestial body?[44]

The next method of interpretation, according to Article 31 VCLT, and the order followed here, is an examination of the object and purpose of the whole treaty. This interpretation method continues to look inside the treaty. This means considering Article II Outer Space Treaty in special reference to the spirit of the treaty, which can be observed in the Preamble, other articles (which cannot be interpreted in isolation), titles, and annexes.[45] The Preamble states:

> Recognizing the common interest of all mankind in the progress of the exploration and use of outer space for peaceful purposes,
>
> Believing that the exploration and use of outer space should be carried on for the benefit of all peoples irrespective of the degree of their economic or scientific development.

The Outer Space Treaty Preamble recognizes the importance of the use of outer space for space exploration. This inevitably includes the utilization of space resources to expand human presence in outer space. Second, the exploitation and appropriation of space resources are part of the benefits that can be derived from the exploration of outer space and can benefit States, entities, etc. The Preamble, then, demonstrates that the object and purpose of the Outer Space Treaty is to foster the use of space itself, including what it contains, i.e., space resources, connectivity, Earth Observation, to serve humankind in the *progress of exploration*.[46] When Article II Outer Space Treaty is read in light of the Preamble, the intent is clear: the non-appropriation principle refers only and exclusively to the area where space resources are found.[47]

---

[43] The main purpose of this chapter is to analyze whether the appropriation of space resources is prohibited under Article II Outer Space Treaty. Thus, there is not further analysis for the area where these resources are place.

[44] These questions must be considered in the context of Article XII Outer Space Treaty: "All stations, installations, equipment and space vehicles on the Moon and other celestial bodies shall be open to representatives of other States Parties to the Treaty on a basis of reciprocity. Such representatives shall give reasonable advance notice of a projected visit, in order that appropriate consultations may be held and that maximum precautions may be taken to assure safety and to avoid interference with normal operations in the facility to be visited." It is obvious that any installation on the Moon or any other celestial body can produce a genuine "occupation" of the area. However, if the area which is being use by other State is free to be visited and open to other States or actors it seems to respect the principle of free use and access to outer space by means of Article I Outer Space Treaty.

[45] *Maritime Delimitation in the Indian Ocean (Somalia v. Kenya), Preliminary Objections, Judgment, I.C.J. Reports 2017, 3.* When analyzing the object and purpose of the treaty (and articles), the ICJ refers to the title of the MoU, chapter, etc. In this case, in paragraph 74, 74 and 76 the ICJ holds its legal argument."

[46] The object and purpose of the Preamble has to do with what States are trying to accomplish via the treaty, in this case, foster 'progress in exploration and use' for the 'benefit and insures of all peoples'. That can be done without using something 'contained in space´- such as a satellite in orbit (the void of outer space) providing valuable services to Earthlings.

[47] By 'resources' here it also meant areas in the void of space. Permanent occupation of a GEO slot would be an appropriation, which is why the ITU developed the rules for allocation of those slots and limitations on their use (bringing into use provisions). 'Resources' tends to connote things

On the other hand, Article II Outer Space Treaty must be interpreted in conjunction with Article I of the Outer Space Treaty, which states:

> "Outer space, including the moon and other celestial bodies, shall be free for exploration and use by all States without discrimination of any kind, on a basis of equality and in accordance with international law, and there shall be free access to all areas of celestial bodies."

Some authors rightly assert that the Article II Outer Space Treaty should be interpreted as a means of legal reinforcement of Article I Outer Space Treaty without mentioning the appropriation of space resources. This means that the main purpose of Article II Outer Space Treaty is to remove any legal barrier that could obstruct the freedom of use, access, and exploration of outer space. Thus, Article II Outer Space Treaty is a necessary corollary to Article I Outer Space Treaty, in that States cannot have free access and use if another State has appropriated. If property rights over a celestial body by any title—both public and private—were allowed, this would create a legal barrier to the free use of outer space by other space actors.[48]

The first interpretation method of the context—i.e., an examination of treaty provisions, Preamble and Annexe—does not reveal any prohibition and regulation of space resources.[49] For this reason, other steps that are considered part of the context under Article 31 VCLT must be taken. In this sense, the interpretation task goes beyond the treaty. Article 31(2) includes as context:

> "(a) any agreement relating to the treaty which was made between all the parties in connection with the conclusion of the treaty; (b) any instrument which was made by one or more parties in connection with the conclusion of the treaty and accepted by the other parties as an instrument related to the treaty."

This part of Article 31 VCLT refers to two elements: any instrument—legally binding or not—and the moment that must be considered, which is when the Treaty is concluded and accepted by all State Parties.[50] Arguably, in the context of space

---

extracted - minerals, etc. Art. II Outer Space Treaty does apply too to the occupation of GEO slot or a Lagrange Point. Additionally, see Bartosz Radoslaw Malinowski, 'On the Legality of Appropriation of Space Resources' in *Proc SPIE 11176, Photonics Applications in Astronomy, Communications, Industry, and High-Energy Physics Experiments* (2019) 171, 172. Interestingly, when considering space mining benefits, these are to accrue to peoples on Earth, so the non-appropriation principle has some relevance outside the context of outer space itself.

[48] Fabio Tronchetti, 'The Moon Agreement in the 21st Century: Addressing Its Potential Role in the Era of Commercial Exploitation of the Natural Resources of the Moon and Other Celestial Bodies' (2010) 36(2) *Journal of Space Law* 489, 497–500. See, Stephan Hobe, *supra* note 30, at p. 208. See also, Bartosz Radoslaw Malinowski, *supra* note 34, at pp. 173–175. An interesting thought would be directly centered around the prohibition for States to take territory; the Outer Space Treaty provision does not apply to individuals. In addition, individuals cannot take real estate outside of the context of State territory. Therefore, Article II Outer Space Treaty applies indirectly to individuals as well. Nonetheless, extracted resources presents a more complicated syllogism.

[49] There is a slight difference between the purpose and object of the treaty itself, which involves examining the intent of the parties as expressed in the Preamble and Annexes, and the contextual interpretation, which does not focus on the purpose of the Preamble and Annexes but rather seeks to identify any regulations or provisions that reveal the meaning of the matter.

[50] Anthony Aust, *Handbook of International Law* (2nd edn, Cambridge University Press 2010) 84–87.

mining, there are no international instruments, either legally or non-legally binding, that were agreed upon by the States Parties to the treaty when the Outer Space Treaty was concluded.[51]

The next method of interpretation, along with the context, is Article 31(3). It is vital to dynamize and modernize the interpretation of the Outer Space Treaty in light of the appropriation of space resources:

> "3. There shall be taken into account, together with the context: (a) any subsequent agreement between the parties regarding the interpretation of the treaty or the application of its provisions; (b) any subsequent practice in the application of the treaty which establishes the agreement of the parties regarding its interpretation; (c) any relevant rules of international law applicable in the relations between the parties."

"Subsequent agreements" and "subsequent practice" both are important to show the dynamic interaction between the treaty itself with the evolution of International Law; the ILC has recognized the relevance of both methods of interpretation arguing that they represent an "objective evidence of the understanding of the parties as to the meaning of the treaty."[52] Article 31(3) refers to events that happen after the treaty has been concluded—i.e., agreements and subsequent practice.[53] As for agreements, this usually refers to agreements carried out by Meetings of Parties following the conclusion of the treaty; but it can also be Memorandums of Understanding (MoU), Declarations, Resolutions (non-legally binding instruments), and legally binding instruments

---

[51] By means of Art. 31(2)(a) VCLT, Arthur Goldberg, who was a former US Supreme Ct Justice who negotiated the Outer Space Treaty on behalf of the US, in his testimony to the US Senate on the ratification of the Outer Space Treaty was pressed about Art. I. The fear was that the 'benefits and interests' provision would require the US to share an American satellite with other countries. In explaining that Article I does not create binding obligations but merely sets for general principles, Goldberg responded that, "It was determined at this stage to set forth some broad principle. It was also determined to study.. the question of how communications satellites should be utilized.". *Hearing Before the Committee on Foreign Relations of the United States Senate* "Treaty on Outer Space" 90th Congress (7, 13 March and 12 April 1967) at page 14. This 'determination to study' was a reference to UNGA Res 2222 (XXI) (19 December 1966), introducing the Outer Space Treaty to the UNGA and through which the UNGA unanimously adopted the Outer Space Treaty., In Para. 4(b), the Resolution called upon States "To begin at the same time the study of questions relative to the definition of outer space and the utilization of outer space and celestial bodies, including the various implications of space communications."

[52] International Law Commission, 'Draft Conclusions on Subsequent Agreements and Subsequent Practice in Relation to the Interpretation of Treaties' (2018) https://legal.un.org/ilc/texts/instruments/english/draft_articles/1_11_2018.pdf. Accessed 21 August 2025, See conclusion 3: "Subsequent agreements and subsequent practice under article 31, paragraph 3 (a) and (b), being objective evidence of the understanding of the parties as to the meaning of the treaty, are authentic means of interpretation, in the application of the general rule of treaty interpretation reflected in article 31."

[53] *Supra* note 10.

such as Protocols and Accords.[54] Aside from the Moon Agreement,[55] in ISL there are no current international legally-binding instruments ("subsequent agreements") that refer to the regulation and clarification of the situation of space resources in COPUOS or any other such forum.[56]

If the Moon Agreement is considered a "subsequent agreement," Article 11 regulates space resources as the common heritage of humankind. As noted, the MA is a valid treaty under Article 2(1) of the VCLT. It contains specific provisions regarding space resources, such as their utilization to create a lunar economy; however, it binds a limited number of States.[57] If the Moon Agreement is considered in the interpretation of space resources as a subsequent agreement, it supports the lawfulness of the appropriation and exploitation of space resources. This is evident in Article 11.5 of the Moon Agreement, which recognizes the need to create a regulatory framework for space resources among States Parties.[58]

Nonetheless, there is a set of principles that could guide the governance of space resources by interpreting Article II Outer Space Treaty, and they fall under the formula

---

[54] The ILC refers to the "subsequent agreements" as those whose legal nature could bound or not to the States. See ILC, *supra* note 36, conclusion 10: "Agreement of the parties regarding the interpretation of a treaty 1. An agreement under article 31, paragraph 3 (a) and (b), requires a common understanding regarding the interpretation of a treaty which the parties are aware of and accept. Such an agreement may, but need not, be legally binding for it to be taken into account."

[55] The Moon Agreement was adopted unanimously by the UNGA. The Moon Agreement does not bind non-parties, but it might be State practice for the purpose of interpretation via VCLT. Additionally, this treaty is a legally binding instrument (for some states) that refers to the regulation and clarification of the situation of space resources. The nature of the legally binding of the Moon Agreement is not in question. Yet this legally binding nature is reduced to the few States Parties to it. See, Thomas D Halket, Eric A Savage, Valnora Leister and Jerry V Lephart, 'Report on the Proposed Agreement Governing the Activities of States on the Moon and Other Celestial Bodies' (1983) 23(3) *Jurimetrics Journal* 259, 265–74; Michael David and Ricky Lee, 'Twenty Years Later – The Moon Agreement and Its Legal Controversies' (1999) *Australian International Law Journal* 9, 9–32.

[56] The Outer Space Treaty did not create a Conference of Parties (COP), or Meeting of Parties (MOP) like other framework conventions under International Law. For instance, the United Nations Framework Convention on Climate Change, 9 May 1992, 1771 UNTS 107 (entered into force 21 March 1994) [UNFCCC], and the Convention on Biological Diversity, 5 June 1992, 1760 UNTS 79 (entered into force 29 December 1993) [Biodiversity Convention] (UNFCCC and Biological Diversity). These can keep alive the treaty itself by amending the treaty, creating new annexes, forum for developing new binding instruments, and so forth. See Bodansky, Daniel, Jutta Brunnée, & Lavanya Rajamani, *The Framework Convention on Climate Change* (Oxford: Oxford University Press, 2017) 143. UNCOPUOS is a committee created by United Nations that has not the same legal nature of a COP created by a Convention. Due to the current political tensions and ineffectiveness of COPUOS, the creation of space law is shifting from COPUOS to other multilateral forums (e.g., Conference of Disarmament, United Nations Security Council). See Gennady M Danilenko, 'International Law-Making for Outer Space' (2016) 37 *Space Policy* 179, 181–83.

[57] Id. 55, and see Leonardo Alberto López Marcos and Rafael Moro Aguilar, 'Some Comments on the Current State of International Space Law: An Open Window for the Future Development of Space Law' (2024) 14 *Boletín Observatorio Jurídico Aeroespacial* 34, 34–36.

[58] Article 11. 5 MA: "States Parties to this Agreement hereby undertake to establish an international regime, including appropriate procedures, to govern the exploitation of the natural resources of the moon as such exploitation is about to become feasible. This provision shall be implemented in accordance with article 18 of this Agreement."

of "subsequent practice," or even "subsequent agreements," that is, the Artemis Accords.[59] While some authors affirm the limited value of the Artemis Accords in interpreting the Outer Space Treaty in light of Article 31(3) VCLT,[60] others rightly assert the growing acceptance of the Artemis Accords in interpreting the lawfulness of space mining activities under Article II Outer Space Treaty. By assuming a correct interpretation of Article 31(3) VCLT, the Artemis Accords are still being tested by emerging acceptance, but their normative value cannot be denied in both interpreting Article II Outer Space Treaty and—outside the scope of the rules of interpretation of VCLT—creating a rule of CIL or regional law.[61]

---

[59] The ILC defines subsequent agreements as "an agreement [not necessarily legally binding] between the parties, reached after the conclusion of a treaty, regarding the interpretation of the treaty or the application of its provisions." Meanwhile, "subsequent practice" is defined as "conduct in the application of a treaty, after its conclusion, which establishes the agreement of the parties regarding the interpretation of the treaty." See ILC, supra note 36, conclusion 4. The Artemis Accords could be included in either of these two elements. However, since the Artemis Accords do not represent, as a whole, as a common understanding among States Parties, it seems that these accords fit better under "subsequent practice." This is because the Artemis Accords are also a conduct initiated by the U.S. Government that reflects a growing interpretation among States regarding Article II of the Outer Space Treaty. Even those countries that have not signed the Artemis Accords (which cannot be ratified since they are not a treaty) have positioned themselves positively, in some way, toward the appropriation of space resources for various purposes. In the case of the ILRS MoU between Russia, China, and other countries, these resources are intended to be utilized on the lunar surface, which is a form of appropriation (focused more on scientific, construction, and energy purposes rather than commercialization). Thus, regardless of whether they are signatories to the Artemis Accords, there is at least a common conduct or perception ("subsequent practice") that space resources can be appropriated under Article II of the Outer Space Treaty. See *Chinese National Space Agency supra* note 12. To put it into some perspective, as of October 2024, Artemis Accords have 45 signatories, while the ILRS signatories are 11. Joshua Faleti, 'Estonia Becomes 45th Artemis Signatory Ahead of IAC24' (SpaceWatch, 15 October 2024) https://spacewatch.global/2024/10/estonia-becomes-45th-artemis-signatory-ahead-of-iac24/#:~:text=Ibadan%2C%2015%20October%202024.,Moon%2C%20Mars%2C%20and%20beyond. Accessed 15 October 2024;

Andrew Jones, 'Senegal among new members of China's ILRS moon base project' (SpaceNews, 15 October 2024) https://spacenews.com/senegal-among-new-members-of-chinas-ilrs-moon-base-project/. Accessed 15 October 2024.

[60] Some authors conduct a superficial analysis of the VCLT without referencing ICJ cases that support their legal arguments. When interpreting an article or a treaty, it is necessary to follow a logical order rather than skipping directly to Article 32 of the VCLT. This approach first examines the treaty itself, then considers external factors, and finally revisits the treaty. There is a lack of accuracy in interpreting the VCLT, particularly when it comes to declaring without argument that the Preamble of the Outer Space Treaty (context) cannot be interpreted in light of the utilization and appropriation of space resources, while also disregarding the work of the negotiators. The same issue arises with the object and purpose of the treaty, which results in a lack of legal rigor in its interpretation. In this sense see this in Michael Byers and Aaron Boley, *Who Owns Outer Space? International Law, Astrophysics and the Sustainable Development of Space* (Cambridge University Press 2023) 139–162.

[61] Rossana Deplano, 'The Artemis Accords: Evolution or Revolution in International Space Law' (2021) 70(3) *International and Comparative Law Quarterly* 799, 805–808;

Leonardo A López Marcos, 'Análisis Jurídico sobre las Actividades de Explotación y Apropiación de los Recursos del Espacio Ultraterrestre' (2022) 9 *Boletines del Observatorio Jurídico Aeroespacial* 59, 59–63.

The interpretation presented here leads to the conclusion that national regulation of space mining activities is permissible and the appropriation of space resources is lawful. By interpreting Article II of the Outer Space Treaty, a clear distinction is made between property rights over celestial bodies (land rights) and property rights over resources on or in the land once they have been extracted. The former is expressly prohibited by Article II of the Outer Space Treaty, while the latter is allowed. Once space resources have been extracted, they can become the subject of property rights. However, it is important to note that neither a state nor a private company can claim rights over space resources prior to their extraction.[62] That would be a wrongful act under International Law.[63]

In light of this interpretation, it is significant to conclude that the Outer Space Treaty does not prohibit the appropriation and utilization of space resources and allows for national regulation of these activities, while respecting the provisions of the Outer Space Treaty.[64] In view of this situation, the optimal path under conventional law to address space resources would be to provide an international cooperation solution.[65]

---

[62] Frans G Von der Dunk, 'Property Rights Over the Moon or On the Moon? The Legality of Space Resource Exploitation on Celestial Bodies' (2023) 6(1) *Journal of Law & Innovation* 95, 95–134; Leonardo A López Marcos, supra note 44, 61–63.

[63] International Law Commission, *Responsibility of States for Internationally Wrongful Acts* (2001) https://legal.un.org/ilc/texts/instruments/english/draft_articles/9_6_2001.pdf. Accessed 22 August 2024; A Salmeri, 'The Integration Between National and International Regulation of Space Resources Activities Under Public International Law' (2019) 43(1) *J Space L* 60.

[64] Until this moment, there are four national space legislation that allow the appropriation of space resources in compliance with the Outer Space Treaty. The 2015 U.S. Commercial Space Launch Competitiveness Act (Title IV), the Luxembourgish Law of July 20th, 2017, on the exploration and use of space resources, the 2023 Space Resources Regulation Regulatory Framework on Space Activities of the United Arab Emirates, and the 2023 Japanese Act. These space laws have not been stated by other States as unlawful under International Law; this is also consistent with all lunar and space missions that have practiced exploiting and appropriating space resources (Apollo Missions, OSIRIS-Rex, Change 6, Chandrayaan). On the contrary, these space legislations provide a framework not only for scientific missions, but also for the appropriation of space resources for commercial purposes.

[65] The VCLT applies to treaties. This section has focused on the interpretation of Article II of the Outer Space Treaty. However, the role of customary international law in developing new rules that might predict future international regulation of space resources has not been addressed. Additionally, as mentioned earlier, there is currently a five-year working group at the UNCOPUOS level working towards presenting a report, resolution, or draft treaty on space resources. UNCOPUOS, 'Working Group on Legal Aspects of Space Resource Activities' https://www.unoosa.org/oosa/en/ourwork/copuos/lsc/space-resources/index.html. Accessed 15 October 2024.

3 The Prospect for the Development of International Space Law in Space Resources: A Return to Customary International Space Law

The foregoing discussion brings to the fore a spectrum of activities, some prohibited and some not. At one end of the spectrum is the establishment of territory—the full exercise of sovereignty in outer space. This is clearly prohibited. At the other end of the spectrum is the harvesting and/or extraction of resources and subsequent claims of ownership over those resources that have been removed from their place. This is the subject of this chapter. Between these two extremes are harder questions about temporary or even permanent occupation of an area of a celestial body or claims over resources prior to extraction—inchoate rights over resources still in place but identified via prospecting. These harder questions are left to another day because States are still debating property rights over resources that have been extracted.

Notwithstanding these debates, State practice demonstrates an interpretation of the Outer Space Treaty that ownership of extracted resources is permitted. The earliest examples of this practice are the Apollo lunar sample return missions. No State has protested the ownership over those resources, and NASA has on several occasions asserted its ownership title over those resources, sometimes successfully and sometimes not.[66] Moreover, China, Japan and Russia have all conducted space resource sample return missions.[67] It is unclear whether those States have asserted, through statements or otherwise, a right to own those resources, but they have retained exclusive control over those samples, which indicates ownership.

The understanding that the free access and use principle of Article I and the non-appropriation principle of Article II permit resource extraction and ownership is in keeping with a long-standing interpretation of the US as described by Brian Egan, then-Legal Advisor to the US State Department in 2016.[68] As an official statement of the US State Department, this statement is an interpretation of the Outer Space Treaty

[66] Steven Jedinak, 'Can You Legally Own Lunar Rock in the United States?' *Kershaw, Vititoe & Jedinak, PLC* (1 May 2020) https://www.monroecountylawyers.com/blog/2020/05/can-you-legally-own-lunar-rocks-in-the-united-states/ accessed 7 October 2024, (Citing: *United States v One Lucite Ball Containing Lunar Material*, 252 F.Supp.2d 1367 (S.D. Fla. 2003), determining that memorabilia containing a lunar rock and gifted to Honduras was Honduran national property alienable only pursuant to Honduran legislation; *Carlson v. United States et. al.*, Ct. #: 2:19-cv-02027, ruling in favor of plaintiff Carlson and against NASA on a determination as to whether Carlson was a good faith purchaser of value of Moon dust and thereby entitled to ownership.)

[67] Supra, note 6.

[68] Brian Egan, 'The Next Fifty Years of the Outer Space Treaty' (Remarks delivered at the annual Galloway Symposium on Critical Issues in Space Law, 7 December 2016) https://2009-2017.state.gov/s/l/releases/remarks/264963.htm. Accessed 5 October 2024 ('Egan Statement').

and it can be evidence of *opinio juris* for the purpose of custom identification.[69] Egan explained as follows:

> "The United States' position on the issue of space resource utilization dates back several decades. For example, in 1979, Secretary of State Cyrus Vance articulated what was already at that point a longstanding U.S. interpretation of Articles I and II of the Treaty ... that "this 'non-appropriation' principle applies to the natural resources of celestial bodies only when such resources are 'in place´. The prohibition on national appropriation does not, however, limit "ownership to be exercised by States or private entities over those natural resources which have been removed from their 'place' on or below the surface of the moon or other celestial bodies." Such removal, Secretary Vance further explained, is permitted by Article I of the Outer Space Treaty, which provides that "outer space, including the moon and other celestial bodies, shall be free for exploration and use by all States…"[70]

Egan continued, explaining that, "In 1980 testimony before the Senate, State Department Legal Adviser Roberts Owen reiterated that 'the United States has long taken the position that Article I of the Treaty… recognizes the right of exploitation.'"[71] This statement provides useful context for understanding the 2015 US Commercial Space Resources Exploration and Utilization Act. Egan stated that the passage of the 2015 Act "… has not altered the United States' consistent approach to the Outer Space Treaty for the past half-century."[72] Thus, the 2015 Act is more properly described as a codification of an interpretation rather than the creation of a new property right, but it can also serve as an interpretation of the Outer Space Treaty and as evidence of State practice for the formation of custom.[73] Likewise, similar legislation on space resource ownership adopted by Luxembourg, the UAE and Japan can be evidence of State practice.[74] Combining the legislation and Egan's statement, we now have evidence of both *opinion juris* and State practice—the two elements of custom—on the ownership of space resources.

---

[69] *Identification of customary international law,* UN General Assembly, 73rd Sess. (11 January 2019) UNGA Res. 73/203 (Adopting ILC Conclusion 10, stating, "Forms of evidence of acceptance as law (*opinio juris*) include, but are not limited to: public statements made on behalf of States…") [*Hereinafter*, "Conclusions of the ILC on ID of Custom"].

[70] Brian Egan, 'The Next Fifty Years of the Outer Space Treaty' (Remarks delivered at the annual Galloway Symposium on Critical Issues in Space Law, 7 December 2016) https://2009-2017.state.gov/s/l/releases/remarks/264963.htm. Accessed 5 October 2024 ('Egan Statement').

[71] Brian Egan, 'The Next Fifty Years of the Outer Space Treaty' (Remarks delivered at the annual Galloway Symposium on Critical Issues in Space Law, 7 December 2016) https://2009-2017.state.gov/s/l/releases/remarks/264963.htm. Accessed 5 October 2024 ('Egan Statement').

[72] Brian Egan, 'The Next Fifty Years of the Outer Space Treaty' (Remarks delivered at the annual Galloway Symposium on Critical Issues in Space Law, 7 December 2016) https://2009-2017.state.gov/s/l/releases/remarks/264963.htm. Accessed 5 October 2024 ('Egan Statement').

[73] Conclusions of the ILC on ID of Custom, supra note 64 (Adopting ILC Conclusion 6, stating, "State practice consists of conduct of the State, whether in the exercise of its executive, legislative, judicial or other functions.").

[74] *Respectively: Loi du 20 juillet 2017 sur l'exploration et l'utilisation des ressources de l'espace, Doc. parl. 7093 (Sess. Ord. 2016–17);* UAE Federal Law No. (12) of 2019 on the Regulation of the Space Sector; Japan Act on the Promotion of Business Activities for the Exploration and Development of Space Resources (Act No. 83 of 2021).

Perhaps more telling of the U.S.'s intentions for the progressive development of ISL via custom was Egan's reference to a 1958 paper by Professors Myers McDougal and Leon Lipson. Therein, the authors challenged the notion that an explicit agreement for space activities was necessary and described as debatable assumptions:

> "(e) That international control, operation or ownership is somehow "demanded" by the supposed intrinsic supranationality of extraterrestrial activity or else morally superior to a "national" solution.
>
> (f) That the supposed tidiness of an explicit comprehensive treaty or convention would be so far preferable to the uncertainty of the legal situation in the absence of such a treaty that it is worth paying a substantial price to achieve it."[75]

They continued, explaining that more likely will be the development of space law via, "…the slow building of expectations, the continued accretion of repeated instances of tolerated acts, the gradual development of assurance that certain things may be done under promise of reciprocity and that other things must not be done on pain of retaliation."[76] In other words, the development of CIL.

Speculation that this was the course the U.S. intends to follow emanated from the publication of the Artemis Accords.[77] Such speculation was confirmed in a recent speech by Gabriel Swiney, one of the drafters of the Artemis Accords.[78] Though Swiney's statements were made in his personal capacity and are not official statements of the U.S. government, they nonetheless confirm that the purpose of the Artemis Accords was to offer an interpretation of the Outer Space Treaty and to lay the groundwork on *opinio juris* for future custom formation. Specifically, Swiney stated:

> "So, the section in the Artemis Accords about space resources ... was explicitly written with the two prongs of customary international law in mind. We wrote it that way to create *opinio juris*. It says ... the signatories to the Accords agree that the prohibition on national appropriation does not inherently prohibit space resource utilization, basically. So, that's *opinio juris*, right there. Around the same time, some of us arranged a NASA purchase contract to buy lunar regolith, so to transfer ownership from a commercial company to NASA – super small dollar amounts – but it happened. So, there's State practice. So, those

---

[75] Myers S McDougal and Leon Lipson, 'Perspectives for a Law of Outer Space' (1958) 52 *American Journal of International Law* 407, 420.

[76] Myers S McDougal and Leon Lipson, 'Perspectives for a Law of Outer Space' (1958) 52 *American Journal of International Law* 407, 420.

[77] Walker A Smith, 'Using the Artemis Accords to Build Customary International Law: A Vision for a US-Centric Good Governance Regime in Outer Space' (2021) 86(4) *Journal of Air Law and Commerce* 661, 663–64.

[78] Gabriel Swiney, Director, Policy, Advocacy and International Division, NOAA, US Department of Commerce https://www.space.commerce.gov/about/gabriel-swiney/. Accessed 6 October 2024. (Swiney also served as Attorney Advisor in the US State Department's legal bureau and in that capacity, as head of delegation to the Legal Subcommittee of COPUOS.)

are the two prongs of customary international law. So this is a very intentional thing; it's not an accident."[79]

At the time of the NASA's initiative to purchase space resources, then-NASA Administrator Jim Bridenstein confirmed US intentions, stating, "We're going to buy some lunar soil for the purpose of demonstrating that it can be done."[80]

Even more remarkable than the confirmation that the U.S. intends to develop space law via custom is the rationale for doing so. Prior to confirming U.S. intentions behind the Artemis Accords, Swiney stated:

> "The reality is that the international community is not in the mood for treaty-making, right now. ... [T]hat sounds a little bit pessimistic but I don't think it has to be. There is ... without getting too academic... a school of thought called the New Haven School of International Law, which one of my former bosses, Harold Koh, is sort of a big advocate of and the idea there is that you make progress internationally where you can. You don't just say, well, you can't have a new treaty, so you throw up your hands and give up. No, you find areas where you can make progress, whether it's through non-binding [agreements], industry best practices, [etc.]. Those over time can become increasingly sticky, in the sense of more and more people adhering to them, more and more people following them."[81]

This rationale for using non-binding norms and custom formation as the only way forward is a bit like the tail wagging the dog. Since the Reagan Administration, the US has embraced realism and assiduously steered the international community away from initiatives for binding space law treaties. Nevertheless, the invocation of the New Haven School of International Law is cause for optimism. Myers McDougal founded the New Haven School in response to abject skepticism over not only the efficacy but also the existence of international law as law.[82] McDougal's theories on "policy-oriented jurisprudence" have been criticized as failing to transform American realist thought.[83] Nevertheless, the continued invocation of McDougal demonstrates

---

[79] Gabriel Swiney, US Department of Commerce, Statement, 4th Annual USSPACECOM Legal Conference 2024, Day 1, Panel 1, *'The Future of Normative International Law & Space Governance'* at minute 48 https://www.youtube.com/watch?v=XtxYzJHR-NQ&list=PLQLdHYu6BwM5MpzOGTQks6X9Hmr0IHSXV. Accessed 5 October 2024.

[80] Jeff Foust, 'NASA offers to buy lunar samples to set space resources precedent' *SpaceNews* (10 September 2020) https://spacenews.com/nasa-offers-to-buy-lunar-samples-to-set-space-resources-precedent/. Accessed 7 October 2024.

[81] Gabriel Swiney, US Department of Commerce, Statement, 4th Annual USSPACECOM Legal Conference 2024, Day 1, Panel 1, *'The Future of Normative International Law & Space Governance'* at minute 48 https://www.youtube.com/watch?v=XtxYzJHR-NQ&list=PLQLdHYu6BwM5MpzOGTQks6X9Hmr0IHSXV. Accessed 5 October 2024 at minute 37.

[82] Myers McDougal and W Michael Reisman, 'The Prescribing Function in World Constitutive Process: How International Law is Made' (1980) 6(2) *Yale Studies in World Public Order* 249, 249 (stating 'Domestic analogues … have so little relevance to the complexities of international politics that those who invoke them finish either by throwing up their hands and conceding that the model is inappropriate for the task (citing HLA Hart, *The Concept of Law* (1961) 208–231) or by painting themselves into the palpably absurd position that there is no international law (citing John Austin, *The Province of Jurisprudence Determined* (1954) 12, 127, 142, 201)').

[83] Martti Koskenniemi, 'The Fate of Public International Law: Between Technique and Politics' (2007) 70(1) *Modern Law Review* 1, 21.

that the seeds may have sprouted; it also offers insights into thinking within U.S. policy circles on the path forward for ISL.

In describing McDougal's approach to CIL, Michael Scharf contrasted it with that of Professor Anthony D'Amato.[84] For McDougal, the formulation is "claim and response": "Custom pioneers' (the first State to initiate a new practice) have no guarantee that their actions will actually lead to the formulation of a binding custom. Indeed, the response may be a repudiation of the claim…."[85] When McDougal addressed custom for space activities, he identified potential claims related to the enjoyment and acquisition of space resources based upon various classifications of space resources, specifically:

> "... spatial-extension resources, such as surfaces or voids, whose distinctive characteristic is their utility as media of transportation and communication; flow or renewable resources, which have "successively available quantities" becoming available at different intervals and are variously affected by human action; and stock or non-renewable resources, whose characteristic is that the "total physical quantity does not increase significantly with time" and which may be either abundant or scarce."[86]

We've not seen this kind of granularity from the U.S. or other Artemis Accords signatories vis-à-vis resource differentiation and the effects of those differentiations on the legality of appropriations under Article II. The reason for the lack of granularity may stem from the fact that under McDougal's theory, such granularity would come not from the claim but from the response—the repudiation or objection to the claim. According to Scharf:

> "... the repudiation could constitute a reaffirmation of existing law, which is strengthened by the protest. Or, the claim and repudiation could constitute a stalemate, which could decelerate the formation of new customary international law. The reaction of Third States is also relevant. Out of this process of claim and response, and third-party acquiescence or repudiation, rules emerge or are superseded."[87]

And yet, the Artemis Accords seem to follow more closely D'Amato's alternative formulation based on "articulation and act." According to Scharf:

> "In D'Amato's view, the articulation can either accompany the initial act (what McDougal called the "claim"), or it can be embodied in a treaty, draft instruments of the ILC, or resolutions of the U.N. General Assembly. Acts that follow and are consistent with the articulation will crystallize the policy into a principle that takes on life as a rule of customary international law. In other words, once there is a consensus articulation that States ought to conform to a given rule of conduct, a legal custom can emerge when some level of spontaneous compliance with the rule is manifest."[88]

---

[84] Michael P Scharf, 'Accelerated Formation of Customary International Law' (2014) 20(2) *ILSA Journal of International & Comparative Law* 305, 313.

[85] Michael P Scharf, 'Accelerated Formation of Customary International Law' (2014) 20(2) *ILSA Journal of International & Comparative Law* 313.

[86] Myers McDougal, 'Emerging Customary Law of Space' (1963–64) 58(5) *Northwestern University Law Review* 618, 637.

[87] Michael P Scharf, 'Accelerated Formation of Customary International Law' (2014) 20(2) *ILSA Journal of International & Comparative Law* 314.

[88] Michael P Scharf, 'Accelerated Formation of Customary International Law' (2014) 20(2) *ILSA Journal of International & Comparative Law* 314.

The Artemis Accords seem to be building that consensus with more and more States regularly signing, and yet important space actors, specifically Russia and China, are intentionally selecting out or prohibited from engaging in cooperative endeavors due to national restrictions.[89] Thus, the Accords are not like guidelines formulated via COPUOS or its subcommittees, as these function under a rule of consensus. Rather, "articulation" is limited and instead, a fragmented picture of custom emergence is developing through dualling initiatives such as the International Lunar Research Station (ILRS) and its associated MoU.[90] The ILRS MoU has a more concrete and technical nature; it does not create a set of principles but rather establishes a framework for the construction of a permanent scientific base on the Moon.[91] However, the ILRS MoU recognizes a legal component in Sect. 2, "Cooperation Guide." Specifically, it calls for "Joint development of the legal documents regulating relations, including the involvement in cooperation of third parties, in the framework of the creation of the ILRS."[92]

The Artemis Accords and ILRS MoU raise the specter of what the ILC has described as "particular CIL."[93] According to the ILC, "A rule of particular customary international law, whether regional, local or other, is a rule of customary international law that applies only among a limited number of States."[94] Thus, it is not binding on third States, which means it is not general in nature.[95] This naturally diminishes its power, rendering it similar to a treaty that binds only parties.[96] Particular customs are not necessarily related to geographic regions and can be related to States linked by a

---

[89] Historically, China has sought access important space collaborations, such as the International Space Station, however such collaboration is prohibited by the 'Wolf Amendment,' which prohibits even the execution of policy. See, DOD and Full Year Continuing Appropriations Act, 2011, 112 Pub L 10 (125 Stat 38) Section 1340 ("None of the funds made available by this division may be used for the National Aeronautics and Space Administration or the Office of Science and Technology Policy to develop, design, plan, promulgate, implement, or execute a bilateral policy, program, order, or contract of any kind to participate, collaborate, or coordinate bilaterally in any way with China or any Chinese-owned company unless such activities are specifically authorized by a law enacted after the date of enactment of this division.").

[90] Mike Wall, 'China recruits Pakistan and Belarus for its planned moon base' *Space.com* (24 October 2023) https://www.space.com/china-ilrs-moon-base-partners-belarus-pakistan last visited 7 October 2024. (Indicating that as of October 2023, China had seven States in its ILRS coalition).

[91] International Lunar Research Station (ILRS) Guide for Partnership, China National Space Administration (16 June 2021) https://www.cnsa.gov.cn/english/n6465652/n6465653/c6812150/content.html, Last visited 7 October 2024.

[92] International Lunar Research Station (ILRS) Guide for Partnership, China National Space Administration (16 June 2021) https://www.cnsa.gov.cn/english/n6465652/n6465653/c6812150/content.html, Last visited 7 October 2024.

[93] Draft Conclusions on Identification of Customary International Law, with Commentaries, 2018, UN Doc A/73/10, Conclusion 16.

[94] Draft Conclusions on Identification of Customary International Law, with Commentaries, 2018, UN Doc A/73/10, Conclusion 16.

[95] Draft Conclusions on Identification of Customary International Law, with Commentaries, 2018, UN Doc A/73/10, Conclusion 16, Commentary, para 3.

[96] Michael P Scharf, 'Accelerated Formation of Customary International Law' (2014) 20(2) *ILSA Journal of International & Comparative Law* 309 (Describing general customary international law

common cause, interest or activity.[97] Moreover, the State practice element consists of a "…general practice prevailing among the States concerned that is accepted by them as governing their relations *inter se*."[98] While little to no practice exists under the Artemis Accords or ILRS MoU, these dualling initiatives nevertheless threaten the fragmentation of CIL for space activities, particularly for space resource utilization on the Moon.

4 Conclusions

Debate continues among States on the appropriate way forward for the progressive development of ISL. The prospect of lunar resource utilization on the horizon has exacerbated concerns over whether States must engage in multilateralism and craft a binding agreement, norms, principles, create a regulatory authority and on and on.

The most expedient way to a viable, binding international instrument on space resource utilization might be the modification of the Moon Agreement. Though it might be a failure politically, from the perspective of law-making, it is far from a failed legal instrument. As the Moon Agreement has come into force, it is a binding agreement under Article 2(1) of the VCLT. Nothing demonstrates its legal success more than that the US formalizing in Executive Order 13,914 the mandate to the US State Department to object to any attempt to treat it as reflecting CIL. And yet, it is often described as a "failed treaty" because it has garnered only 17 State parties, none of whom are space powers. The limited membership of the Moon Agreement masks a hidden opportunity—it is much easier to modify an agreement through consensus of the mere 17 States parties than it would be to negotiate a new binding instrument through COPUOS, as attaining consensus would be nearly impossible with more than 100 member States. The idea to amend the Moon Agreement is not new; Professor von der Dunk identified provisions that might be adopted to make it a viable instrument.[99]

Regardless of the viability of amending the Moon Agreement, we can use treaty interpretation, done properly under the VCLT, to determine the current legal landscape. We can also look to the States' interpretations through various statements and acts. States are proceeding with custom as the means for the progressive development of ISL. This is evident from State practice, including national laws on resource use and exploration, and through *opinion juris* demonstrated in the Artemis Accords and ILRS MoU. This aligns with the traditional understanding of the development of

as having greater jurisprudential power than a treaty in the sense that it binds all States which have not persistently objected.)

[97] Draft Conclusions on Identification of Customary International Law, with Commentaries, 2018, UN Doc A/73/10, Conclusion 16, Commentary, para. 5.

[98] Draft Conclusions on Identification of Customary International Law, with Commentaries, 2018, UN Doc A/73/10, Conclusion 16, Commentary, para. 6.

[99] Franz G von der Dunk, 'The Moon Agreement and the Prospect of Commercial Exploitation of Lunar Resources' (2007) 32 *Annals of Air and Space Law* 91, 109–13.

law as an *ex post* reaction to situations rather than an *ex ante* attempt at regulation. Nevertheless, it bodes a fragmented international regime, based not on consensus but on particular customs developed among groups of States. Whether this situation is ideal, only time will tell.

Leonardo Alberto López Marcos holds an LL.M. in Public International Law from the University of Oslo (Norway) and a Master's Degree in Law and Legal Practice from the Valencian International University (Spain). Mr. López Marcos is a Teaching Assistant and Doctoral Research Fellow in the Public International Law section of the Department of International Law, Ecclesiastical Law and Legal Philosophy, Faculty of Law, Complutense University of Madrid (Community of Madrid) in Spain. He is an Economist and a qualified EU/Spanish Attorney at Law. Mr. López Marcos is Board Member of the Spanish Association of Air and Space Law (AEDAE) and he is the Secretary and Coordinator of the Spanish Aerospace Legal Observatory (AEDAE & UCM G-BioDin Law) and the Director of its Debate Platform. He is part of the Advisory Member Board Space and Air Law Section at the Illustrious Bar Association of Madrid – ICAM. Mr. López Marcos is a prospective member of the International Institute of Space Law (IISL). Ultimately, he is a proud former Visiting Scholar at the Center for Air and Space Law at the University of Mississippi.

Charles W. Stotler is an Assistant Professor of the Practice of Law at the University of Mississippi School of Law and a Director of its Center for Air and Space Law. He teaches courses on air, cyber and space law and serves as faculty advisor for LL.M. theses and student participation in air and space law moot court competitions. Mr. Stotler also advises a broad array of clients on law, policy and regulatory issues, both independently as well as in support of the work of Aviation Advocacy, a Swiss-based consultancy. He focuses on activities and technologies that push policy and regulatory boundaries, such as unmanned aircraft systems and novel uses of space applications. Mr. Stotler holds a Master of Laws in Air and Space Law from McGill University, a Juris Doctor from Loyola University New Orleans College of Law, a Master of Liberal Arts from St. John's College and a Bachelor of Sciences in Biology from Mount Saint Mary's University. He was called to the bar in California in 2010 and New York in 2011.

Artemis Accords Diplomacy and Resource Utilisation: Strategic Coalitions and Norms Setting Beyond Multilateralism

Adithya Variath

Abstract This chapter contextualises the Artemis Accords as a strategic instrument of contemporary space diplomacy, marking the United States' effort to depart from traditional multilateral governance toward strategic coalition-based frameworks in outer space. It argues that the Accords serve as tools of norm entrepreneurship, advancing American interpretations of space law under the guise of international cooperation and global space governance. The chapter delves into a detailed analysis of the Accords' legal structure, United Nations Committee on Peaceful Uses of Outer Space (UNCOPUOS) debates, and national space legislation of spacefaring nations. It explores how the Accords are reshaping the foundational principles of international space law to favour emerging commercial actors and entrenched space powers. This chapter examines how the Accords operate as instruments of space diplomacy, advancing the interests of signatory states while simultaneously shaping emerging norms on space resource utilisation. Framing the Artemis diplomacy alongside the International Lunar Research Station (ILRS), led by China and Russia, this chapter conducts a comparative inquiry into how these rival coalitional models reflect broader geopolitical rivalries, legal fragmentation, the de-formalisation of international law, and differentiated visions for space governance. Drawing from the concepts of diplomacy, global governance, and customary international law, the study positions space diplomacy as a distinct and underexplored force in shaping legal and political orders of space governance. The research also traces the growing participation of states and the proliferation of aligned domestic laws, assesses whether the Artemis Accords are laying the groundwork for new customary international law, not through global consensus, but through interest-based strategic diplomatic coalitions and state practice. This chapter argues that the Accords represent a shifting paradigm in space governance, where law, diplomacy, and strategic interests intersect to shape the future of international cooperation in the final frontier. Finally, this chapter analyses the strategic dimensions of coalition-building in space exploration, where political alignments, technological capabilities, and economic incentives converge to redefine access to and exploitation of extraterrestrial resources.

A. Variath (✉)
Jindal Global Law School, O.P. Jindal Global University, Sonipat, India
e-mail: adithya.variath@jgu.edu.in

S. Bhat B and A. Variath (eds.), *Legal Responses to Space Resource Utilisation*, Space Law and Policy, https://doi.org/10.1007/978-981-95-8864-0_4

Keywords Artemis accords · Space diplomacy · Coalition-based governance · Customary international law · Geopolitical rivalries · Space resource utilisation

1 Introduction

Darth Vader, a character in the famous Star Wars franchise in *Star Wars: Episode IV—A New Hope,* famously said, "The ability to destroy a planet is insignificant next to the power of the Force".[1] While Vader may have dismissed planetary destruction as trivial, the prospect of Star Wars and galactic forces is no longer confined to science fiction. Star Wars was fiction, until it wasn't. From anti-satellite tests, space-based nuclear weapons and missile systems to proposals for establishing space forces, space is steadily becoming a contested zone.[2] These developments seem like science fiction and overly futuristic, but they are the reality that humanity is living in, and the birth of many of these technological developments can be traced back to 1957.

The launch of *Sputnik-1*, the first artificial satellite by the Soviet Union, on October 4, 1957, stunned both the American experts and the global community, many of whom had believed the United States would be the first to reach such a milestone.[3] The success of Soviet technology also raised fears and concerns among the United States' power circle about falling behind in the new arms race and escalating Cold War tensions.[4] Before the launch of Sputnik, Americans often underestimated their Russian adversary. Harry Truman, who referred to Russians as "those Asiatics", once confidently declared, "Do you know when Russia will build the [atom] bomb? Never".[5] President Eisenhower tried to downplay the importance of the launch. However, he recognised the political significance of the development and immediately allocated additional funds and resources to speed up the US Space Program.[6] While the race to develop space technology was also linked with the arms race,

---

[1] George L, Star Wars: Episode IV – A New Hope (Lucasfilm 1977)

[2] Jessica W, Geneva, We Have a Problem: Space Diplomacy Goes Nuclear (Geneva, We Have a Problem: Space Diplomacy Goes Nuclear, Centre for International Governance Innovation 2025); Johnson-Freese J, Space Warfare in the 21st Century: Arming the Heavens (Space warfare in the 21st century: arming the heavens, First edition. edn, Routledge 2016); Oelrich I, van Hooft P and Biddle S, 'Anti-satellite warfare, proliferated satellites, and the future of space-based military surveillance' (2024) 47 Journal of Strategic Studies 916; Silvester B and Jah M, 'A call for orbital accountability: DA-ASAT tests as harmful interference and contamination' (2025) Acta Astronautica.

[3] Dickson P, Sputnik: the shock of the century (Sputnik: the shock of the century, Walker Publishing Company, Inc. 2001); Herrala ME, Suutari P and Mikkonen S, 'Pianist Sviatoslav Richter: The Soviet Union Launches a 'Cultural Sputnik' to the United States in 1960', Music, Art and Diplomacy (1 edn, Routledge 2016)

[4] Rigden JS, 'Eisenhower, scientists, and Sputnik' (2007) 60 Physics today 47

[5] Mieczkowski Y, Eisenhower's Sputnik Moment: The Race for Space and World Prestige (Cornell University Press 2013)

[6] Launius RD and McCurdy HE, Presidential leadership in the space program, 1994)

as entry to outer space also meant further proliferation of intercontinental ballistic missiles (ICBMs) and further militarisation of the new domain.[7]

Even before the United States and Soviet Union started investing in space technology, Nazi Germany had been close to developing ICBMs near the end of the Second World War.[8] However, less is talked about the German contribution to space technology in the pages of history due to concerns about slaves being used in research labs and the recruitment of Nazi Germany's scientists by the United States *Operation Paper Clip* and the USSR *Operation Osoaviakhim* into their space programmes.[9] While the Soviet launch of Sputnik-1 was not just the beginning of a string of technological innovations, it also opened a fresh geopolitical debate and diplomatic grappling over the absence of clarity of the legal regulations to govern space.[10] In times of the anxieties of the Cold War, the evolution of space norms emerged not solely from abstract regulatory demand but from complex and often contested diplomatic negotiations.[11] Against this backdrop, the idea of diplomacy in outer space evolved not merely as a facilitator of international consensus but as a strategic arena, giving rise to what is now termed "space diplomacy".

In this context, space diplomacy can be defined as the integration of knowledge, innovation, and regulation between nations for a better application of and co-existence in space.[12] Space diplomacy and the strategic use of space science and technology to advance foreign policy goals and bolster national capabilities have emerged as a new tool in modern international relations.[13] As outer space becomes a theatre of strategic rivalry and scientific cooperation, the governance frameworks that once held sway, chiefly those born of the Cold War multilateral consensus, are being tested.

---

[7] Barnes-Brown A, The space race: Cold War's cosmic battle for the stars (All about history book of the space race, First editon. edn, Future PLC 2019); Huntress WT and Marov MY, 'Rockets', Soviet Robots in the Solar System (Springer Praxis Books, Springer New York 2011); Jasani B, 'Contribution of space technology to the arms race', Outer Space – A New Dimension of the Arms Race, vol 35 (1 edn, Routledge 1982); Phelan D, Cold war space sleuths: The untold secrets of the Soviet space program (Cold war space sleuths: the untold secrets of the Soviet space program, 1st 2013. edn, Springer 2013)

[8] Biddle W, Dark side of the moon: Wernher von Braun, the Third Reich, and the space race (Dark side of the moon: Wernher von Braun, the Third Reich, and the space race, W. W. Norton 2009)

[9] Jacobsen A, Operation Paperclip: the secret intelligence program to bring Nazi scientists to America (Secret intelligence program to bring Nazi scientists to America, Back Bay Books/Little, Brown and Company 2015); Neufeld MJ and Junker De, 'Overcast, Paperclip, Osoaviakhim - Looting and the Transfer of German Military Technology', The United States and Germany in the Era of the Cold War, 1945–1990 (2004); Wingate S, 'From Death Camps to the Space Race: An Interview with Patrick Hicks' (2020) 32 WLA: war, literature & the arts 1

[10] Pépin E, Legal problems created by the Sputnik (Legal problems created by the Sputnik, Institute of International Air Law, McGill University 1957)

[11] Lachs M, 'Introduction' in Lachs M (ed), The Law of Outer Space (Brill | Nijhoff 2010)

[12] J. Borgen C, McGuinness ME and Stewart DP, 'Space law and diplomacy', Research Handbook on Law and Diplomacy (Research Handbooks in Law and Politics series, Edward Elgar Publishing 2022)

[13] White F, 'Space Diplomacy and the 'Overview Effect'' (2023) 18 The Hague journal of diplomacy 464

Against this backdrop, NASA's Artemis Accords (*hereinafter* 'Accords'), tied to its lunar exploration programmes, represent a new shift from traditional treaty-based multilateralism to a more strategic coalition model. This evolving diplomatic posture raises fundamental questions about the resilience of space law's normative foundations, the rise of unilateral agendas masquerading as global values, and responses from dissenters and most notably, the changing nature of space law-making.

1.1 Forging Legal Consensus and Evolution of Space Norms

With the beginning of the space race in the 1950s, the need to prevent or at least regulate the evil effects of space exploration was felt immediately.[14] The race to develop space technology was also linked with the arms race, as entry to outer space also meant further proliferation of ICBMs and further militarisation of the new domain.[15] While both the United States and USSR knew that co-existence in space was inevitable, so was conflict and disagreement. Primarily, there were two schools of thought on how to 'peacefully use' outer space. Russia initially posited that the use of the term 'peaceful purpose' necessarily meant 'non-military', while the US has always maintained that it meant 'non-aggressive'.[16] While eventually, Russia, too, came to subscribe to the US-led interpretation, owing to the development of its own indigenous space-fighting capacity.[17] The Cold War tensions and ambition to militarise the new strategic domain also made of early space governance debates at the United Nations level chaotic and difficult.

The *first* phase of the evolution of international space law began in 1958 when the United Nations General Assembly adopted its first resolution on outer space.[18] This marked the starting point for a series of resolutions that laid the groundwork for regulating space activities. Among these, the most influential was UN General Assembly Resolution 1962, adopted in 1963, which established the fundamental principles of space law and served as a precursor to the 1967 Outer Space Treaty (*hereinafter* 'OST').[19] As these resolutions were non-binding, this first stage in the

[14] Bhat B S, An Introduction to International Space Law (Oxford University Press 2024)

[15] Jasani B, 'Contribution of space technology to the arms race', Outer Space – A New Dimension of the Arms Race, vol 35 (1 edn, Routledge 1982)

[16] Bhat B S and Gupta R, 'Sui Generis Interpretation of key elements of Space Demilitarization for ensuring Peaceful Uses' (2023) 10 Indonesian Journal of international and Comparative Law 113

[17] Bridge RL, 'International Law and Military Activities in Outer Space' (1980) 13 Akron Law Review

[18] General Assembly, Question of the Peaceful Use of Outer Space (United Nations 1958)

[19] Jessup PC and Taubenfeld HJ, 'The United Nations Ad Hoc Committee on the Peaceful Uses of Outer Space' (1959) 53 American Journal of International Law 877;, 'Report of the Committee on the Peaceful Uses of Outer Space to the 17th Session of the United Nations General Assembly (Excerpt)' (1962) 1 International Legal Materials 373;, 'Report of the Committee on the Peaceful Uses of Outer Space to the 18th Session of the United Nations General Assembly' (1963) 2 International Legal Materials 1058

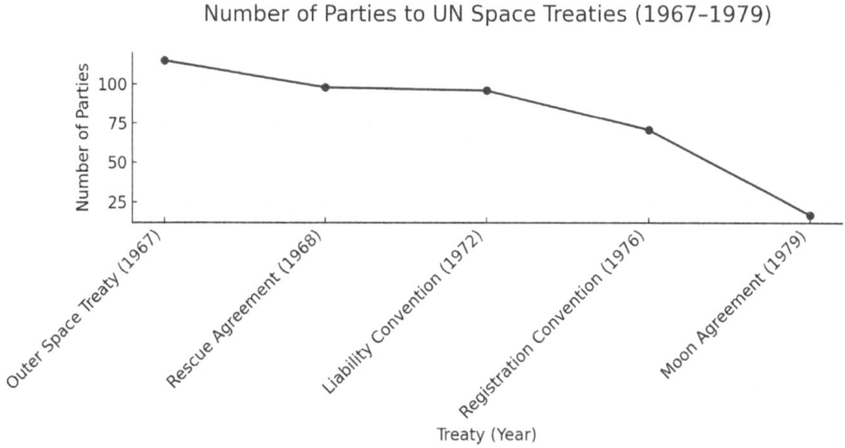

Fig. 1 Five United Nations Space Treaties and the falling number of Parties

development of space law is often referred to as the "soft law" phase.[20] However, with the growing scale and complexity of space activities, it became evident that enforceable legal norms were necessary.

This led to the emergence of the *second* phase, marked by the adoption of binding international treaties, beginning with the Outer Space Treaty of 1967. As of 2025, it has 116 parties.[21] In the following years, four additional treaties were negotiated under the auspices of the United Nations Committee on the Peaceful Uses of Outer Space (UNCOPUOS), which functions through consensus-based decision-making. It was also a phase when the space race ended with the United States' moon landing. Regrettably, the momentum of treaty-making slowed and eventually halted by 1979, with the last treaty having only 17 parties due to the inability of states to reach further consensus (See Fig. 1).[22]

It can also be deduced from the above figure that the hard law-making in space governance only lasted for a decade from 1967 to 1979. This stagnation, coupled with intensifying geopolitical rivalries, has since raised growing concerns about the inadequacy of the existing legal framework.[23] Even though the treaty-making process stopped in 1979, the developments around space technological developments, from satellite communications, lunar mining, space resource utilisation, remote sensing, and space exploration, are changing the way human civilisation is looking at space.

---

[20] Nakamura K, Norm Formation in Space Law (Brill | Nijhoff 2024)

[21] UNOOSA, 'Status of International Agreements relating to Activities in Outer Space' (United Nations Office for Outer Space Affairs, 2025) <https://www.unoosa.org/oosa/de/ourwork/spacelaw/treaties/status/index.html> accessed 20 August 2025

[22] UNOOSA, 'Status of International Agreements relating to Activities in Outer Space' (United Nations Office for Outer Space Affairs, 2025) <https://www.unoosa.org/oosa/de/ourwork/spacelaw/treaties/status/index.html> accessed 20 August 2025

[23] Jasentuliyana N, 'Treaty Law and Outer Space: Can the United Nations Play an Effective Role?' (1986) 11 Annals of Air and Space Law 219

The global space economy is projected to reach a $1.8 trillion market size by 2035.[24] The impact of the space economy is far beyond the numbers, as space infrastructure is a key enabler for many other industries. However, the most pivotal change in the space economy was the entry of private actors and commercial ventures into space.[25] While the entry of the new space actors has enabled radical technological transformations, there is growing concern about how they must be managed responsibly to ensure space operations remain sustainable and equitable. The inadequacy of old space treaties paved the way for the *third* phase of space governance.[26] Due to a lack of consensus and global interest in making legally binding treaties, the third phase mirrored the first phase as the United Nations started going back to soft laws and non-binding resolutions. However, their adherence, implementation, and respect remain fragmented and inadequate.[27] This phase further exposed how the States have shown disinterest in having legally binding norms at the global level while making substantial domestic laws on strengthening and promoting space technologies. The case of international institutional inefficiency at the United Nations also left a vacuum in global leadership in space governance. In October 2022, Biden-Harris US National Security Strategy emphasised the need for the United States to take the lead in reforming outer space governance and laying the groundwork for future space norms.[28]

Drawing back from the Cold War trend, space also becomes a domain to exert dominance. For instance, the political rivalry between the United States and China also influences which of them will shape the next normative order of space governance. This has led to what this chapter contextualises as the *fourth* phase of space law-making, one wherein the United Nations is not the primary source from wherein the law flows, but the space powers have taken it upon themselves to make the new space norms. The trend begins with the Artemis Accords, a document setting out guidelines on space exploration drafted by the United States Department of State and National Aeronautics and Space Administration (*hereinafter* 'NASA') in 2020.[29] The document was signed on 13 October 2020 by eight space exploration nations. Interestingly, the new political commitment now signed by 54 states was drafted by 'NASA', which is not a law-making/drafting body by nature. Since it was not drafted by the United Nations, the making of the document has not followed the established norms of treaty-making. It was unilaterally drafted by the United States and opened

---

[24] McKinsey & Company, 'Space: The $1.8 trillion opportunity for global economic growth' (McKinsey & Company, 2024) <https://www.mckinsey.com/industries/aerospace-and-defense/our-insights/space-the-1-point-8-trillion-dollar-opportunity-for-global-economic-growth> accessed

[25] Masson-Zwaan T, 'New States in Space' (2019) 113 AJIL unbound 98

[26] Zhao Y, 'Soft Laws Relating to Space Activities' in Ukey D, Bhat B S and Variath A (eds), International Space Law in the New Space Era (Oxford University Press 2024)

[27] Bhat B S, An Introduction to International Space Law (Oxford University Press 2024)

[28] The White House, National Security Strategy 2022)

[29] NASA, Artemis Plan: NASA's Lunar Exploration Program overview (Artemis Plan, National Aeronautics and Space Administration 2020)

for signature by other States.[30] The document promises to fill several legal deficiencies of the Outer Space Treaty 1967, which will be elaborated on in the later parts. However, the growing interest in States to sign the Accords is a case of how the United States capitalises on its space diplomacy prowess to build global support for its interpretation of space law. This chapter blends diplomacy, law, and strategic interests to study how the Accords illuminate broader transformations in how rules for outer space are created, contested, and legitimised. Situating the Accords within the broader trajectory of international law and diplomacy, this chapter examines their dual role as instruments of space diplomacy and as mechanisms for constructing new norms in the emerging domain of space resource governance.

1.2 Artemis Accords in Context

The development and the politics of the Artemis Accords are a compelling case study to explore modern space diplomacy and the growing predominance of American unilateralism and interest in the multilateral governance of outer space. Multilateralism and unilateralism are closely linked yet distinct concepts in international relations. Ruggie defines multilateralism through three key characteristics: indivisibility, generalised organising principles, and diffuse reciprocity.[31] In contrast, unilateralism refers to instances where a dominant state disregards multilateral norms in favour of a self-serving foreign policy.[32] The extent to which a state can act unilaterally is largely influenced by its power; thus, unilateralism tends to be the preferred approach of major powers, especially hegemonic states.[33]

This chapter employs an analytical framework to examine the Artemis Accords as a diplomatic instrument through which the United States constructs a unilateral vision and presents it as multilateral norms, thereby influencing global space governance outside traditional multilateral forums such as UNCOPUOS. The signatories to the Accords are like-minded partners of the United States, which asserts normative leadership in a domain increasingly defined by strategic competition. The United States argues that it is to elaborate on the existing space treaties, but the Accords also erode some fundamentals of space norms.[34] This duality, of promoting the principles

---

[30] Riordan N, Machoň M and Csajková L, 'Space Diplomacy and the Artemis Accords' (2023) 18 The Hague journal of diplomacy 380

[31] Ruggie JG, 'Multilateralism: The Anatomy of an Institution' (1992) 46 International Organization 561; Ruggie JG, Multilateralism matters: the theory and praxis of an institutional form (Multilateralism matters: the theory and praxis of an institutional form, Columbia University Press 1993)

[32] Wedgwood R, Patrick S and Forman S, 'Unilateral Action in a Multilateral World', Multilateralism and US Foreign Policy (Lynne Rienner Publishers 2002)

[33] Wallace, 'Multilateralism and U.S. foreign policy: ambivalent engagement', Multilateralism and US foreign policy: ambivalent engagement / (Lynne Rienner 2002)

[34] Din Au, The Artemis Accords: The End of Multilateralism in the Management of Outer Space? (Routledge, Taylor & Francis Group 2022)

of the Outer Space Treaty on one hand and breaching it subtly on the other, positions the Artemis Accords as a tool of American space diplomacy.

History tells tales of the United States using international law as a tool for United States diplomacy to project a global narrative that aligns with American strategic and legal interpretations.[35] In areas ranging from the use of force, the law of the sea, to cyberspace and trade, the United States has selectively invoked international legal principles often grounded in US domestic interests, values, or exceptionalism.[36] For instance, using mechanisms such as unilateral declarations of customary international law, the United States has asserted its positions as normative benchmarks for global practice.[37] Soft law instruments ranging from bilateral agreements to voluntary codes of conduct further reinforce the American hegemonic legal narrative by encouraging alignment without binding commitments. This shows how the United States has effectively blurred the lines between unilateralism and multilateralism, often recasting its geopolitical agenda as the embodiment of the international legal order.[38] The signing of the Accords by 54 States has also expanded the diplomatic reach of the United States.[39] The Artemis Accords mark a decisive shift in space diplomacy, exemplifying how powerful states now engage in strategic norm entrepreneurship. It is a reflection of the political ability of powerful actors to act independently when multilateralism proves slow or politically encumbered.

This chapter treats international legal texts as discursive artefacts, reflecting broader geopolitical imaginaries and strategic narratives. The Artemis Accords are read not simply as legal documents, but as tools of normative projection and diplomatic persuasion. In doing so, the research blends doctrinal legal analysis with discourse analysis, examining both what the law says and how it is invoked, contested, and operationalised by actors. To avoid confirmation bias and normative determinism, the study incorporates contrasting perspectives, especially the criticisms levelled by

---

[35] Crook J, 'Contemporary Practice of the United States relating to International Law' (2013) 107 The American Journal of International Law 431

[36] Cuddy B, International Law and US Foreign Relations (Oxford University Press 2019)

[37] Kolasa J, 'Unilateral Acts of a State in the Process of Forming Customary International Law' (2018) 8 Wroclaw review of law, administration & economics 59

[38] Forman S and Patrick S, Multilateralism and U.S. Foreign Policy: Ambivalent Engagement (Center on International Cooperation Studies in Multilateralism, Lynne Rienner Publishers 2023); Ikenberry GJ, 'Is American Multilateralism in Decline?' (2003) 1 Perspectives on Politics 533; Rathbun BC, 'The 'Magnificent Fraud': Trust, International Cooperation, and the Hidden Domestic Politics of American Multilateralism after World War II' (2011) 55 International studies quarterly 1; Rathbun BC, Trust in International Cooperation: International security institutions, domestic politics, and American multilateralism (1st edn, Cambridge University Press 2012); Schuller FC and Grant TD, 'Executive Diplomacy: Multilateralism, Unilateralism and Managing American Power' (2003) 79 International affairs (London) 37; Van Oudenaren J, 'Transatlantic Bipolarity and the End of Multilateralism' (2005) 120 Political Science Quarterly 1

[39] This study confines its analysis to the list of signatories of the Artemis Accords as of 14 May 2025, which at that point comprised 54 member states. All analytical elements, charts, tables, and percentages are derived from this temporal snapshot. The scope does not extend to subsequent accessions, acknowledging that the number of signatories is expected to grow as space diplomacy progresses. For convenience, this set of signatories is referred to as the status 'as of April 2025' throughout the research.

non-signatory states. Counterfactual reasoning is used to explore alternate pathways of space governance and critique the Artemis diplomacy. These methodological strategies reflect the commitment to reflexivity and pluralism that is central to qualitative research in international law and diplomacy.

2 Legitimacy and the Politics of Artemis Accords: A New Era of Space Diplomacy?

International law and diplomacy have long had overlapping roles in international relations, with the former providing normative frameworks and the latter offering the practical mechanisms through which these frameworks are established and enforced.[40] In its conventional understanding, international law is the set of rules, formally agreed by treaty or understood as customary, by which nation-states interact with each other in an international society.[41] However, international law has been used by actors beyond its original purpose. International law is also a tool of diplomacy. It can be used to create narratives, build consensus, and even put forward ideas to a global community. Across the history of the United States foreign relations, international law has provided both an animating vision, and ideology, for various American projects of world order.[42] It has been a practical tool for the advancement of the United States' power and interests.[43] International law offers a foundation and framework of expectations for politico-diplomatic conduct between nations.

Historically, the global commons have been subject to processes of privatisation that are often marked by inequality and injustice.[44] International law in general and space law in specific are heavily influenced by the powers and interests of non-sovereign actors, especially with the rise of powerful private actors.[45] As Teeple notes,

[40] Bolewski W, Diplomacy and international law in globalized relations (Diplomacy and international law in globalized relations, Springer 2007); Chatterjee C, International law and diplomacy (International law and diplomacy, Paperback edition, with revisions. edn, Routledge 2007); Esiemokhai EO, Commentaries on international law and diplomacy. Volume 1 (Commentaries on international law and diplomacy Volume 1, AuthorHouse Publishing 2013); Jacovides A, 'The UN Decade On International Law: Progress And Promises', International Law and Diplomacy (2011)

[41] Brownlie I, Principles of public international law (Principles of public international law, Clarendon P. 1966); Helmersen ST, 'The ICJ Statute Article 38(1)', The Application of Teachings by the International Court of Justice (Cambridge University Press 2021); Linderfalk U, 'Legal positivism and legal idealism', Understanding Jus Cogens in International Law and International Legal Discourse (Elgar International Law series, Edward Elgar Publishing 2020)

[42] Chimni BS, 'The Past, Present and Future of International Law: A Critical Third World Approach' (2007) 8 Melbourne journal of international law 499; Falk RA, Legal order in a violent world (Legal order in a violent world, Princeton University Press 2019)

[43] Cuddy B, International Law and US Foreign Relations (Oxford University Press 2019)

[44] Chimni BS, 'Marxism and International Law: A Contemporary Analysis' (1999) 34 Economic and Political Weekly 337

[45] Byrnes A, Hayashi M and Michaelsen C, International law in the new age of globalization (International law in the new age of globalization, Nijhoff 2013); Droubi S and D'Aspremont

international law concerning the global commons remains largely influenced by the imperatives of corporate property rights.[46] The evolving framework of international law and institutions increasingly reflects the imperatives of globalisation, which may be understood as the shift in the locus of capital accumulation from the nation-state to the global sphere.[47] This transition, Teeple argues, has historical precedents. Since the early 1980s, powerful capitalist states guided by hegemonic, transnationalised segments of their national bourgeoisies and supported by similarly aligned factions within developing countries have driven a transformation of international economic law.[48] As Robinson observes, these legal changes have aimed at achieving two key goals: "(i) the global expansion and entrenchment of capitalist relations through the elimination of local barriers, and (ii) the erosion of redistributive international legal norms that were premised on state-led market interventions".[49] This dates back to the East India Company and how the British Empire lobbied to legitimise the interest of its private corporations through the law of the Nations.[50] History often echoes and repeats in unusual ways. The parallel between the East India Company and contemporary space enterprises lies in their shared focus on resource extraction. Where the Empire once legitimised and advanced the claims of the East India Company, today the United States plays a similar role, legitimising and formalising the interests of its commercial space actors.

The pressure to respond to events and to find solutions to the unique problems that confront a superpower inevitably leads to a view of international law as a tool.[51] With regard to the making of international space law, the challenges were more nuanced.[52] To begin with, the very idea of outer space as a global commons is also contested.

J, International organisations, non-state actors, and the formation of customary international law (International organisations, non-state actors, and the formation of customary international law, Manchester University Press 2020)

[46] Teeple G, 'Globalisation as the Triumph of Capitalism: Private Property, Economic Justice and the New World Order' in Schrecker T (ed), Surviving globalism: the social and environmental challenges (Macmillan 1997)

[47] Teeple G, 'Globalisation as the Triumph of Capitalism: Private Property, Economic Justice and the New World Order' in Schrecker T (ed), Surviving globalism: the social and environmental challenges (Macmillan 1997)

[48] Teeple G, 'Globalisation as the Triumph of Capitalism: Private Property, Economic Justice and the New World Order' in Schrecker T (ed), Surviving globalism: the social and environmental challenges (Macmillan 1997)

[49] Robinson W, 'Globalisation: nine theses on our epoch' (1996) 38 Race and Class 13

[50] Miles K, 'The Origins of International Investment Law: Empire, Environment and the Safeguarding of Capital', The Origins of International Investment Law (2013); Porras M, 'Constructing International Law in the East Indian Seas: Property, Soverignty, Commerce and War in Hugo Grotius De lure Praedae - The Law of Prize and Booty, or "On How to Distinguish Merchants from Pirates"' (2006) 31 Brooklyn Journal of International Law 741; Rasulov A, 'A Marxism for International Law: A New Agenda' (2018) 29 European Journal of International Law 631.

[51] Crook JR, 'Practicing International Law for the United States' (1996) 6 Florida State University Journal of Transnational Law & Policy

[52] Jenks AL, Collaboration in Space and the Search for Peace on Earth (Collaboration in Space and the Search for Peace on Earth, 1 edn, Anthem Press 2021)

While the Brundtland Commission supports this view, the Outer Space Treaty does not explicitly endorse it.[53] Critics argue that such a designation could hinder private investment or fail to ensure sustainability.[54] Reflecting this divide, a 2020 US executive order rejected the global commons label for outer space.[55] However, the first customary rule of space law is that it is non-excludable, which means that a State cannot exclude others from using it.[56] However, in its true sense, the concentrated technological powers have made space one of the most exclusive domains.[57] During the making of space law in the 1960s, the political actors rarely approached space as either truly "global" or genuinely "common". Unlike earlier multilateral efforts that governed global commons, such as the 1959 Antarctic Treaty and the 1958 UN Conference on the Law of the Sea, which were primarily driven by colonial extractive interests under the pretext of scientific and environmental cooperation, the formulation of space law emerged in a distinctly different geopolitical setting.[58] With the rise of newly independent sovereign states, the drafting process became more inclusive, allowing broader participation. However, the major powers continued to shape the core content and structural framework of the regime.[59]

Today, 80 countries have established space programmes, nearly double the number from the year 2000.[60] Remarkably, over half of these spacefaring nations belong to the Global South.[61] As the strategic and economic significance of outer space becomes increasingly evident, nations are accelerating investments in critical space technologies and engaging in a renewed race to explore and utilise the cosmos. Geopolitically, space, like other venues of diplomacy, is also a domain that exhibits geopolitical and socio-economic strategic contestations. For this chapter, space diplomacy refers to the process of dialogue undertaken by various actors within established or evolving contexts, structures, institutions, or forums, leading to outcomes that manifest as either cooperation or conflict. Over the years, more and more States that were not dominant when the space law was drafted way back in the 1960s have gained power

---

[53] Pic P, Evoy P and Morin J-F, 'Outer Space as a Global Commons: An Empirical Study of Space Arrangements' (2023) 17 International journal of the commons 288

[54] Boonen C and others, 'Governing as commons or as global public goods: two tales of power' (2019) 13 International journal of the commons 553

[55] Executive Office of the President, Executive Order 13914 of April 6, 2020: Encouraging International Support for the Recovery and Use of Space Resources (2020)

[56] Tepper E, 'Structuring the Discourse on the Exploitation of Space Resources: Between Economic and Legal Commons' (2019) 49 Space Policy 101290; van Eijk C, 'The exclusive making of space law' (2025) Leiden journal of international law 1

[57] ICJ, North Sea Continental Shelf (Germany/Netherlands), Judgment of 20 February 1969 (Judge Lachs, Dissenting Opinion), 1969)

[58] Howkins A, 'Defending polar empire: opposition to India's proposal to raise the 'Antarctic Question' at the United Nations in 1956' (2008) 44 Polar Record 35; Molenaar EJ, 'Participation in the Antarctic Treaty' (2021) 11 Polar Journal 360

[59] Tzouvala N, Capitalism as Civilisation: A history of international law (Capitalism as civilisation: a history of international law, Cambridge University Press 2020)

[60] World Population Review, Countries with Space Programs 2025, 2025)

[61] World Population Review, Countries with Space Programs 2025, 2025)

and diplomatic strength today. For instance, India was largely lurking in the shadows of the USSR Space Program, has now become a major player in space.[62] For example, India's anti-satellite (ASAT) test in March 2019 made it the fourth country (after the United States, Russia, and China) to have a demonstrated ASAT capability.[63] The rise of new actors is also changing the geopolitical landscape of space diplomacy.[64]

2.1 Theory and Practice of Space Diplomacy

Diplomacy is conventionally understood to be a set of institutions and processes by which states represent themselves and their interests to one another in a State system or society.[65] The historical record and contemporary trends both suggest that the answers to the questions "what is diplomacy?" and "who may properly be said to conduct it?" vary considerably over time and across space.[66] One of the fundamental shifts has been the changing nature of stakeholders in the new era of diplomacy. Scholars write that we live at a time when the participation of "non-state" actors in international relations is on the increase.[67] An important implication of this view is that something, which we may usefully call diplomacy, can be seen to exist, in principle at least, separately from a sovereign state system.

In an era of intensifying space activity, diplomacy in outer space plays a critical role in fostering cooperation, preventing conflict, and promoting the peaceful use of this shared domain. The scope of space diplomacy spans several key areas, most recently, to resource utilisation and mining. Daniel Deudney, for instance, warns that unchecked space expansionism carries profound and potentially catastrophic risks.[68] However, he suggests that a moderated approach could unite societies around an 'Earth-centred pro-space agenda', one that prioritises nuclear security and environmental protection over military conquest and space colonisation.[69] One of the clearest expressions of diplomacy in outer space is what scholars refer to as science-space

---

[62] Jakhu RS, Nyampong YOM and Pelton JN, 'National Space Laws and the Exploitation of Natural Resources from Space', Space Mining and Its Regulation (Springer 2016)

[63] Rajagopalan RP, 'ASAT weapons: A real threat to the future of Space' https://www.orfonline.org/expert-speak/asat-weapons; Variath A, 'Time For An Indian National Space Security Policy?' (2024) 4 Lex ad Coelum;

[64] Dobos B, 'Outer Space as a Socioeconomic Field', Geopolitics of the Outer Space (Contributions to Political Science, Springer International Publishing AG 2018)

[65] Jönsson C and Langhorne R, Diplomacy (Diplomacy, Sage Publications 2004); Nicolson H, Diplomacy (Diplomacy, 3d edn, London 1969)

[66] Sharp P, Diplomacy (Oxford University Press 2017)

[67] Kissinger H, Diplomacy (Diplomacy, 1st Touchstone edn, Touchstone 1995); Donald M, Diplomacy (Diplomacy, 1988)

[68] Deudney D, Dark Skies: Space Expansionism, Planetary Geopolitics, and the Ends of Humanity (1 edn, Oxford University Press 2020)

[69] Deudney D, Dark Skies: Space Expansionism, Planetary Geopolitics, and the Ends of Humanity (1 edn, Oxford University Press 2020)

diplomacy, a category of diplomacy centred on international cooperation or competition in scientific and technological endeavours, or economic space diplomacy.[70] For example, India's space programme has emerged as a potent instrument of space diplomacy through its remarkable ability to achieve ambitious goals with modest budgets.[71]

There is also scholarship around the theme of space exploration in itself, becoming a tool for soft power.[72] Joseph S. Nye introduced the concept of *soft power*, the ability to shape the preferences of others through attraction rather than coercion.[73] Soft power is rooted in cultural values and is reflected in the conduct of foreign policy.[74] Its effectiveness depends on how these values are communicated and perceived by foreign audiences, and whether they are seen as appealing or legitimate.[75] The recent surge in human activity in outer space has sparked renewed interest in exploring the conditions that foster cooperation among the diverse actors operating in this increasingly complex domain. Historically, space programmes were developed with political, military, intelligence, and research objectives in mind. It was during the Cold War era that space became a new arena of geopolitical competition, prompting agreements to prevent the placement of weapons of mass destruction in orbit and to promote the peaceful uses of outer space.[76] In this context, the power of a select few states to have the capability to explore outer space is also a soft power that attracts States that do not possess it.[77]

Today, the stakes are even higher as national security and economic infrastructure increasingly depend on space-based systems, which are vulnerable to cyber threats with potentially global consequences. Space influences a wide array of domains, including economic, scientific, cultural, political, climate-related, and industrial interests, and involves a diverse set of actors and processes.[78] The global space economy is currently valued at nearly USD 450 billion and is expected to surpass

---

[70] Cross MaKD and Pekkanen SM, 'Introduction. Space Diplomacy: The Final Frontier of Theory and Practice' (2023) 18 The Hague Journal of Diplomacy 193

[71] Rajagopalan RP, 'India's Space Program: Increasing Proactivism' (2024) 19 Asia policy 40

[72] Hoerber T and others, Routledge Handbook of Space Policy (Routledge Handbook of Space Policy, 1st edn, Taylor & Francis Group 2024); Oehlers A, Partnerships and Soft Power in Space, 2020)

[73] Nye JJS, Soft Power: The Means To Success In World Politics (Soft Power: The Means To Success In World Politics, 1 edn, PublicAffairs 2009)

[74] Melissen J, The New Public Diplomacy: Soft Power in International Relations (The New Public Diplomacy: Soft Power in International Relations, 1st 2005. edn, Palgrave Macmillan UK 2005); Ohnesorge HW, Soft power: The forces of attraction in international relations (Soft power: the forces of attraction in international relations, Springer 2020)

[75] Hayden C, The rhetoric of soft power: Public diplomacy in global contexts (The rhetoric of soft power: public diplomacy in global contexts, Lexington Books 2012)

[76] Levine AJ, 'The "Missile Gap" and Space Race', After Sputnik (1 edn, Routledge 2018)

[77] Marboe I, Soft Law in Outer Space: The Function of Non-binding Norms in International Space Law (Soft Law in Outer Space: The Function of Non-binding Norms in International Space Law, Böhlau Verlag 2012)

[78] Cross MaKD and Pekkanen SM, 'Introduction. Space Diplomacy: The Final Frontier of Theory and Practice' (2023) 18 The Hague Journal of Diplomacy 193

USD 1 trillion by the 2040s (or even earlier). In 2024, global government expenditure for space programmes hit a record of approximately USD 135 billion.[79] The United States Government spent around USD 79.7 billion on its space programmes in that year, making it the country with the highest space expenditure in the world.[80] The United States was followed by China, with government expenditure on space programmes of over USD 19 billion.[81] States have used space power as a form of 'smart power', the ability to combine hard and soft power into a successful strategy.[82] Space diplomacy emerges as a critical lens for understanding broader developments in outer space. Through the Accords, the United States leverages its advanced space technology as a tool of soft power to build strategic coalitions in outer space governance.

2.2 Artemis Accords: Decline of Formal Multilateralism and the Turn to Strategic Use of Coalitions

More than fifty years have passed since Gene Cernan's historic walk on the Moon during the Apollo 17 mission in 1972.[83] In 2025, the world stands on the cusp of a new era of lunar exploration with NASA's Artemis Programme. It is a series of five ambitious lunar missions to return humans to the Moon.[84] The name *Artemis* comes from the mythological Greek goddess of the Moon and twin sister to Apollo, drawing a connection to the Apollo missions that first landed humans on the Moon.[85] The crewed spacecraft is named Orion, after one of the most recognisable constellations in the sky. In Greek mythology, Orion is known as Artemis' hunting companion.[86] While these are giant leaps in space, the United States is also using its diplomatic powers to build consensus and support for its mission.

---

[79] Statista Research Department, Global governmental spending on space programs of leading countries 2024: Government expenditure on space programs in 2022 and 2024, by major country (in billion U.S. dollars) (Statista Ltd 2025)

[80] Statista Research Department, Global governmental spending on space programs of leading countries 2024: Government expenditure on space programs in 2022 and 2024, by major country (in billion U.S. dollars) (Statista Ltd 2025)

[81] Statista Research Department, NASA's approved budget by sector 2024: NASA's approved budget from FY 2017 to FY 2024, by sector (in million U.S. dollars) (Statista Ltd 2024)

[82] Luzin P, Soft Power and Cooperation for Space Security (Oxford University Press 2024)

[83] Reneau A, 'Background and Early History of Space Exploration', Moon First and Mars Second (SpringerBriefs in Space Development, Springer International Publishing AG 2021)

[84] Ehrenfried MDv, The Artemis Lunar Program: Returning People to the Moon (The Artemis Lunar Program: Returning People to the Moon, 1st 2020. edn, Springer Nature 2020); Smith M and others, The Artemis Program: An Overview of NASA's Activities to Return Humans to the Moon (IEEE 2020)

[85] Dunbar B, What is Artemis? (NASA 2019)

[86] Dunbar B, What is Artemis? (NASA 2019)

On October 13, 2020, the Administrator of NASA and representatives of seven other space agencies, namely Australia, Canada, Italy, Japan, Luxembourg, the United Arab Emirates, and the United Kingdom, signed the Artemis Accords.[87] On 8 April 2025, Bangladesh became the 54th nation to sign the accords.[88] Section 1 of the Artemis Accords calls it a 'political commitment', to "increase the safety of operations, reduce uncertainty, and promote the sustainable and beneficial use of space for all humankind". Several scholars have raised concerns about the legal implications of the Accords.[89] Hobe, for example, cautions that the Artemis Accords represent a subtle attempt by the United States to legitimise a departure from the principles of the OST.[90] Similarly, von der Dunk characterises the Accords as a political instrument aimed at cultivating international support for the US interpretation of the Treaty, especially concerning the appropriation and utilisation of lunar resources.[91]

The 'Method of Operation' of the Accords, subsequently published by NASA in 2024, reinforces that the Accords are a practical set of 'non-binding' principles, guidelines, and best practices.[92] The non-binding nature of the Accords makes it a 'soft law' as per international law. It says that the Accords are grounded in the OST and other existing instruments to ensure safe, transparent, and sustainable civil space exploration. Interestingly, the Method of Operation published by NASA does not mention the 1979 'Moon Agreement' drafted by the United Nations.[93] It is ironic that the Artemis Accords, focusing mainly on lunar activities, decided to completely shadow the specific 1979 Moon Agreement drafted by the UNCOPUOS. It is primarily because the United States has not signed the Moon Agreement as it viewed its provisions, particularly those declaring lunar resources the "common heritage of mankind" and calling for an international regime to govern their exploitation, as incompatible with its national interests and commercial ambitions.[94]

International space law has traditionally evolved through the UNCOPUOS, with foundational norms established via multilateral treaties and UN-endorsed soft law

---

[87] NASA, NASA, International Partners Advance Cooperation with First Signings of Artemis Accords (NASA 2020)

[88] Dooren JM, 'NASA Welcomes Bangladesh as Newest Artemis Accords Signatory' (NASA, 2025) <https://www.nasa.gov/news-release/nasa-welcomes-bangladesh-as-newest-artemis-accords-signatory/#:~:text=NASA%20Welcomes%20Bangladesh%20as%20Newest%20Artemis%20Accords%20Signatory,-Jennifer%20M.&text=On%20April%208%2C%202025%2C%20Bangladesh,and%20sustainable%20exploration%20of%20space.> accessed

[89] Deplano R, 'The Artemis Accords: Evolution Or Revolution In International Space Law?' (2021) 70 International and Comparative Law Quarterly 799

[90] Stirn A, 'Do NASA's Lunar Exploration Rules Violate Space Law?' <https://www.scientificamerican.com/article/do-nasas-lunar-exploration-rules-violate-space-law/>

[91] Stirn A, 'Do NASA's Lunar Exploration Rules Violate Space Law?' <https://www.scientificamerican.com/article/do-nasas-lunar-exploration-rules-violate-space-law/>

[92] NASA, Method of Operation (NASA 2024)

[93] NASA, Method of Operation (NASA 2024)

[94] Aloia V, 'Regulation of Commercial Mining of Space Resources at National and International Level: An Analysis of the 1979 Moon Agreement and the National Law Approach' (2019) 62 International Institute of Space Law 459

instruments.[95] The emergence of the United States-led Artemis Accords, negotiated outside this established multilateral framework, marks a notable departure from this tradition.[96] Although the Accords acknowledge the value of multilateralism and explicitly reference the significance of UNCOPUOS, they were neither discussed within the Committee nor accompanied by any publicly available *travaux préparatoires*. One of the earliest details about the Accords emerged through anonymous US government sources cited in a Reuters exclusive published on May 6, 2020, which characterised the initiative as a "pact for moon mining".[97] This revelation prompted a swift and critical response from the head of the Russian space agency, who, in a (now-deleted) tweet, likened the Accords to a modern-day lunar invasion, drawing parallels with Operation Iraqi Freedom and labelling the effort a new "Coalition of the Willing".[98] Russia has also expressed strong reservations about the Artemis Accords, citing their perceived bias in favour of US interests, particularly in relation to the exploitation of the Moon's natural resources.[99] China has signalled its disapproval of the unilateral approach taken by the United States in shaping norms for space governance. Accords also capitalise on the structural problems of the Outer Space Treaty and the international legal order.

2.3 Resource Utilisation: Deconstructing Outer Space Treaty and Constructing Accords

The 'Preamble' of the Accords underscores the deep political commitment of the United States to promote "greater coordination and cooperation between and among established and emerging actors in space".[100] The Artemis Accords have provisions ranging from peaceful purposes to deconfliction. While scholars argue that the Accords derail the fundamentals of space law, the Accords have reaffirmed their commitment to some of the United Nations space treaties.[101] Section 1 of the

---

[95] Freeland S and Pecujilic AN, 'Do You Like Your Regulation—Hard or Soft: The Antarctic Treaty and the Outer Space Treaty Compared' (2018) 30 National Law School of India Review

[96] Von der Dunk FG, 'The Artemis Accords as a Tool of Cooperation' (2021) 64 International Institute of Space Law 145

[97] Roulette J, 'Star Trek, not Star Wars:' NASA releases basic principles for moon exploration pact (Reuters 2020)

[98] David L and Billings L, NASA Proposes New Rules for Moon-Focused Space Race (Scientific American 2020)

[99] Stirn A, 'Do NASA's Lunar Exploration Rules Violate Space Law?' <https://www.scientificamerican.com/article/do-nasas-lunar-exploration-rules-violate-space-law/>

[100] NASA, The Artemis Accords principles for cooperation in the civil exploration and use of the moon, mars, comets and asteroids for peaceful purposes (2020)

[101] Deplano R, 'The Artemis Accords: Evolution Or Revolution In International Space Law?' (2021) 70 International and Comparative Law Quarterly 799

Accords states that it "provides for operational implementation of important obligations contained in the Outer Space Treaty and other instruments".[102] However, the problem lies beyond the text. The reality is that the Accords has put out a new way to interpret some of the critical provisions of the Outer Space Treaty. It has pushed for the United States' interpretation of how it looks at the Outer Space Treaty. Most notably, the provision on space resource utilisation and promoting commercial mining of space resources.

In 2006, NASA geologist Harrison Schmitt, the last person to walk on the Moon in 1972, published his influential book *Return to the Moon*, which outlines a vision for harnessing the Moon's potential to meet Earth's future energy demands.[103] In this work, he presents a persuasive scientific argument for renewed lunar exploration, highlighting the Moon's vast untapped resources, particularly the highly valuable isotope Helium-3 (^{3}He), which could be worth billions.[104] There are also reports proving the presence of water, rare earth minerals and metals that could be mined for the benefit of humanity.[105] As per the 1979 Moon Agreement, these resources are the common heritage of mankind and should not be appropriated by only a select few states. Since the Moon Agreement has only 17 parties, the other law to regulate space resources is the OST, which has been subject to manipulation and misinterpretation by powerful states.

The Outer Space Treaty in Article II states that "Outer space, including the moon and other celestial bodies, is not subject to national appropriation by claim of sovereignty, by means of use or occupation, or by any other means". In the interpretation of Article II, the United States have often deviated from its core intent behind it. If Article II of the OST is dissected, it mentions three important things: *first*, outer space and other celestial bodies are not subject to 'national appropriation', *second*, there cannot be 'any claim' of sovereignty, and *third*, no claims by means of use or occupation or 'by any other means'. However, States have interpreted it to mean that Article II is limited to 'national' appropriation and not 'commercial' or 'individual' appropriation.

Despite this categorical prohibition under Article II, the United States has consistently advanced an alternative interpretation, arguing that while national appropriation is barred, commercial or individual appropriation by private actors is not.[106] This

---

[102] NASA, The Artemis Accords principles for cooperation in the civil exploration and use of the moon, mars, comets and asteroids for peaceful purposes (2020)

[103] Schmitt H, Return to the Moon: Exploration, Enterprise, and Energy in the Human Settlement of Space (Springer New York 2007)

[104] Jakhu RS, Nyampong YOM and Pelton JN, 'National Space Laws and the Exploitation of Natural Resources from Space', Space Mining and Its Regulation (Springer Praxis Books, Springer International Publishing AG 2016); Variath A, 'Lunar Mining and the Erosion of the Common Heritage Principle: What's the Future of the Moon Agreement?' (Cambridge International Law Journal Blog 2024) <https://cilj.co.uk/2024/10/20/lunar-mining-and-the-erosion-of-the-common-heritage-principle-whats-the-future-of-the-moon-agreement/> accessed 22 January 2025

[105] NASA, Lunar Helium-3 and Fusion Power (1988)

[106] Tronchetti F, 'The Non-Appropriation Principle as a Structural Norm of International Law: A New Way of Interpreting Article II of the Outer Space Treaty' (2008) 33 Air & space law 277

interpretation found legislative form in the US Commercial Space Launch Competitiveness Act of 2015, which explicitly affirmed the right of US citizens to commercially explore, recover, possess, and sell space resources.[107] By framing such rights as compatible with international obligations, the US effectively initiated a domestic divergence from the foundational principles of the OST.[108]

The Artemis Accords extended this reinterpretation through diplomacy. The United States has used the Accords diplomacy to normalise the US narrative as globally acceptable. The Artemis Accords represent a significant geopolitical and legal development in the governance of space resources, particularly in how they seek to reshape the interpretation of the OST to favour commercial interests. Section 10 of the Accords explicitly states that "the extraction of space resources does not inherently constitute national appropriation under Article II of the OST", thus embedding the US legal position into an international framework. This has allowed the US to globalise its stance on resource appropriation by encouraging other countries to adopt similar positions in their domestic laws. Countries such as Luxembourg, the UAE, Japan, and India have followed suit, each passing national legislation and policies that affirm private ownership and commercial utilisation of space resources.[109] Luxembourg's *2017 Law on the Exploration and Use of Space Resources*, for instance, bluntly declares that "space resources are capable of being owned". While Japan's *2021 Space Resources Act*, UAE's *2019 Federal Law No. (12): The Regulation of the Space Sector* and India's *2023 Space Policy* mirrors the Artemis language on the commercial entitlement of space resources. This framework facilitates the crystallisation of state practice, potentially leading to the emergence of customary state practices backed by domestic laws favouring a liberal-commercial understanding of space governance. This cascading wave of national regulations aligned with the Artemis framework reveals a diplomatic strategy: to normalise and legitimise the US interpretation of the OST through widespread legal convergence and coalition-building. Consequently, the Artemis Accords function as both a legal and diplomatic tool for redefining norms in space governance.[110] They do not merely facilitate cooperation in lunar exploration; they institutionalise a new vision of resource entitlement in space, one centred on commercial viability and market access. As more countries adopt this model, the traditional multilateral consensus embedded in the OST is being quietly but significantly contested. As Accords influence the making of domestic

---

[107] U.S. Commercial Space Launch Competitiveness Act of 2015: report of the Committee on Commerce, Science and Transportation on S. 1297 (US Commercial Space Launch Competitiveness Act of 2015, U.S. Government Publishing Office 2015)

[108] Pereira M, 'Commercial Space Mining: National Legislation vs. International Space Law' (2020) 63 International Institute of Space Law 47

[109] Salmeri A, 'The Multi-Level System of Space Mining: Regulatory Aspects and Enforcement Options' (2022) 65 International Institute of Space Law 533; DePagter MM, 'Who Dares, Wins:" How Property Rights in Space Could be Dictated by the Countries Willing to Make the First Move' (2022) Chicago Journal of International Law.

[110] Giri C, 'The global planetary exploration roadmap: Cooperation and contest', India in the Second Space Age of Interplanetary Connectivity, vol 1 (1 edn, Routledge 2022)

laws in sync with the US vision, the chances of domestic laws further entrenching the American interpretation of the OST can also become a global normative practice.

The Artemis Accords are emblematic of this new era of norm entrepreneurship in space diplomacy, where powerful states use diplomatic instruments to advance legal interpretations favourable to their strategic and economic interests.[111] This marks a pivotal shift in space governance, where the legitimacy of international norms is no longer solely derived from universal consensus but increasingly through the proliferation of domestic laws and strategic coalitions. The result is a fragmented but influential reconfiguration of space law, one that tests the resilience of traditional multilateral frameworks and raises critical questions about equity and the future of the commons in the final frontier.

3 The Accords Diplomacy: Multilateralism or Strategic Coalitions for Resource Control?

The increasing number and diversity of space actors over the past decade have raised concerns about the efficiency and efficacy of existing normative structures of outer space governance. Key recommendations from UNISPACE III, the World Summit on Sustainable Development, and the UN General Assembly Resolution on transparency and confidence-building measures in space formed the foundation of US foreign policy during President Trump's first administration.[112] The private space industry was positioned to play a significant role in shaping the future of the United States' space policy. It was during Trump's first term that they employed both public and private diplomatic strategies to forge an international coalition, led by the US, that involved private sector partners in a human return to the Moon, with plans for subsequent missions to Mars and beyond.[113]

In 2020, Artemis was thought to be the United States' partners with space aspirations, forming a small bloc similar to QUAD in the era wherein space has become a strategic domain. It was also a signal to the private actors to build a form of global consensus to validate their commercial interests.[114] With eight powers formally agreeing to the political commitment, the Artemis Accords kicked off as a minilateral grouping of the space powers. Miles Kahler coined the term 'minilateralism' back in 1992, and the term was subsequently popularised by the journalist Moises

[111] Von der Dunk FG, 'The Artemis Accords as a Tool of Cooperation' (2021) 64 International Institute of Space Law 145

[112] Riordan N, Machoň M and Csajková L, 'Space Diplomacy and the Artemis Accords' (2023) 18 The Hague journal of diplomacy 380

[113] Logsdon JM, 'Artemis must learn from Apollo' (2019) 57 Aerospace America 36

[114] De Zwart M and Lisk J, The Artemis Accords and Subsequent Developments (Oxford University Press 2024)

Naim.[115] It meant the dramatic proliferation of informal arrangements and their growing importance in contemporary global governance.[116]

Minilateralism has become the popular *modus operandi* of twenty-first-century international relations among states.[117] Naim argues, "minilateralism is quite fluid as it indicates a grouping of a small number of like-minded states pursuing mutual goals and is considered a nimble and more targeted approach than multilateralism", where "the smallest possible number of countries needed to have the largest possible impact on solving a particular problem".[118] In contrast to the traditional multilateral approach, minilateralism offers agility, ad hoc flexibility, informal engagement, and a space for diplomatic innovation and experimentalism.[119] There has been a myriad of minilateral arrangements across the world, such as the QUAD, the BRICS bloc.[120] Following this trend, minilateralism in space was waiting in line, considering the geostrategic and economic impact of space.

In the minilateral system, two notable developments are emerging. First, there is a marked shift away from broad global cooperation toward the formation of strategic alliances, where states seek, as Moisés Naím has aptly noted, the smallest group required to accomplish a specific goal.[121] Second, there is a striking move away from traditional treaties, which have long been the cornerstone of international relations, toward informal, non-binding agreements.[122] These "soft law" instruments represent a relatively new and increasingly popular approach to diplomacy and global governance. The idea of minilateralism pushed and advocated by the US has not been globally accepted, as many perceive it as a unilateral move to globalise its interests. In 2023, China's State Council Information Office released a white paper, 'A Global Community of Shared Future: China's Proposals and Actions'. The paper argued that

---

[115] Kahler M, 'Multilateralism with small and large numbers' (1992) 46 Int Org 681

[116] Patrick S, 'The New "New Multilateralism": Minilateral Cooperation, but at What Cost?' (2015) 1 Global Summitry 115

[117] Falkner R, 'Towards minilateralism' (2015) 5 Nature climate change 805; Richey M and Guseinova O, 'Disputed geometries of great power politics: US-China perspectives on minilateralism' (2024) 78 Australian journal of international affairs 828; Richey M and Ohn D, "It's fine in practice, but how about in theory?' State-of-the-art minilateralism between expectations and reality' (2024) 78 Australian journal of international affairs 782; Wuthnow J, 'U.S. 'Minilateralism' in Asia and China's Responses: A New Security Dilemma?' (2019) 28 The Journal of contemporary China 133.

[118] Naim M, 'Minilateralism: the magic number to get real international action' (2009) 173 Foreign Policy 136

[119] Damtie YA, 'From Multilateralism to Minilateralism: Regional Cooperation Trends in the Horn of Africa ' (Wilson Center, 2024) <https://www.wilsoncenter.org/blog-post/multilateralism-minilateralism-regional-cooperation-trends-horn-africa> accessed

[120] Paik W and Park JJ, 'The Quad's Search for Non-Military Roles and China's Strategic Response: Minilateralism, Infrastructure Investment, and Regional Balancing' (2021) 30 The Journal of contemporary China 36

[121] Naim M, 'Minilateralism: the magic number to get real international action' (2009) 173 Foreign Policy 136

[122] Naim M, 'Minilateralism: the magic number to get real international action' (2009) 173 Foreign Policy 136

terms like 'camps' and 'cliques' often refer to emerging US-led minilateral alliances in the Indo-Pacific.[123]

The growing interest among States in signing the Artemis Accords reflects not a process of multilateralism or even minilateralism, but rather a strategic coalition. While minilateralism is typically characterised by a small, exclusive group of like-minded states focused on pragmatic cooperation, the Artemis Accords demonstrate openness, broad participation, and a clear normative agenda. With 54 signatories from every continent, the Accords are inclusive and explicitly aim to shape international space norms within the framework of existing treaties like the OST. This model mirrors earlier efforts where like-minded States aligned themselves behind a dominant power, in this case, the United States, based on shared strategic or ideological interests. A historical parallel is the "coalition of the willing" formed during the 2003 US-led invasion of Iraq, when 30 nations publicly supported Washington's decision despite contested evidence and the bypassing of multilateral UN processes.[124] Similarly, the Artemis Accords are not the product of an inclusive global consensus but a new form of Washington-led governance in outer space, advancing a particular vision for the exploration and use of the Moon and beyond.

Although initially led by the United States, the initiative fosters collaborative engagement and seeks to influence global discourse beyond its signatories. To illustrate the growing consensus of this initiative, Table 1 presents a chronological and regional breakdown of the 54 signatory states from 2020 to 2025. This geographical diversity underscores the widespread endorsement of cooperative norms in outer space governance, reflecting both geopolitical interests and emerging spacefaring capabilities across the Global North and South.

As of April 2025, the Artemis Accords have attracted 54 signatory countries, demonstrating a significant global shift toward endorsing US-led principles of space exploration and governance. Europe leads with 27 signatories, representing exactly 50% of the total, followed by Asia with 10 countries (18.5%), South America with 7 (13%), North America with 5 (9.3%), Africa with 3 (5.6%), and Oceania with 2 (3.7%). This geographic breadth reflects both strategic outreach by the United States and a growing consensus, especially among Western and emerging spacefaring nations, on key legal norms such as the peaceful use of space, interoperability, transparency, and notably, the legitimacy of extracting and utilising space resources. In this way, the Artemis Accords are not just a political agreement; they are a dynamic instrument capable of laying the groundwork for new space customs in an era marked by rapid commercialisation and renewed exploration.

---

[123] Wu X and and Kim D, 'China's perception of minilateralism and Chinese-style multilateralism' (2024) 78 Australian Journal of International Affairs 869

[124] United States Congress Document, Economic and military support for the U.S. efforts in Iraq: the coalition of the willing, then and now: hearing before the Subcommittee on International Organizations, Human Rights, and Oversight of the Committee on Foreign Affairs, House of Representatives, One Hundred Tenth Congress, first session, May 9, 2007 (Economic and Military Support for the US Efforts in Iraq, 2007)

Table 1 Artemis accords signatories by year and region

Year	Continent	Countries
2020	Europe	Italy, Luxembourg, United Kingdom
2020	North America	Canada, United States
2020	Asia	Japan, United Arab Emirates
2020	Oceania	Australia
2021	Europe	Poland
2021	North America	Mexico
2021	Asia	South Korea
2021	Oceania	New Zealand
2021	South America	Brazil
2022	Europe	Romania, France
2022	Asia	Israel, Bahrain, Singapore, Saudi Arabia
2022	South America	Colombia
2022	Africa	Nigeria, Rwanda
2023	Europe	Czech Republic, Spain, Germany, Iceland, Netherlands, Bulgaria
2023	Asia	India
2023	South America	Ecuador, Argentina
2023	Africa	Angola
2024	Europe	Belgium, Greece, Switzerland, Sweden, Slovenia, Lithuania, Slovakia, Armenia, Estonia, Cyprus, Denmark, Austria, Liechtenstein
2024	North America	Dominican Republic, Panama
2024	South America	Uruguay, Peru, Chile
2024	Asia	Thailand
2025 (As of April 2025)	Europe	Finland
2025 (As of April 2025)	Asia	Bangladesh

4 Artemis Diplomacy and Normative Entrepreneurship

The normative value of the Accords can originate only from thinking about it from the perspective of its customary value. Between 1980 and 2020, no new multilateral space treaties were adopted under the United Nations framework, indicating a stagnation in normative development and collective agreement in outer space governance.[125] This long-standing impasse was disrupted by the Accords, which, despite being a non-binding political instrument, reignited global engagement and interest. Accords has attracted a growing number of signatories from diverse geopolitical regions, signalling a renewed international interest in cooperative rulemaking for space activities. The Accords have thus revitalised space governance discussions and emerged as a pragmatic response to fill the void left by the failure of the Moon Agreement.[126] Beyond reviving global dialogue on space governance, the Artemis Accords hold the potential to shape the next generation of international customs in outer space.

4.1 The Possible Rise of New Customary Norms for Space Governance

The Artemis Diplomacy reflects the United States' strategic pivot from multilateral consensus-building to targeted, interest-based partnerships that can actively shape emerging legal standards through coalition-building and agenda-setting. As more countries adopt and operationalise the principles outlined in the Accords, such as transparency, interoperability, peaceful use, and the preservation of heritage sites, these practices may gradually evolve into widely accepted norms, even influencing customary international law.[127] Over time, consistent state practice underpinned by a sense of legal obligation (or *opinio juris*) may emerge, particularly if non-signatory states also begin to follow similar practices.[128]

[125] Clerc P, 'Towards a new legal ecosystem for the exploitation of space' in Lesley Jane Smith, Ingo Baumann and Wintermuth S-G (eds), Routledge Handbook of Commercial Space Law, vol 1 (1 edn, Routledge 2024)

[126] Matusiak M, 'Artemis Accords – A new era of space law or an initiative of the world's superpower?' (2023) Studia Iuridica 168

[127] NASA, The Artemis Accords principles for cooperation in the civil exploration and use of the moon, mars, comets and asteroids for peaceful purposes (2020)

[128] Black-Branch JL, 'Customary International Law, Opinio Juris and State Practice Regarding the Treaty on the Prohibition of Nuclear Weapons 2017: Toward Universality?', The Treaty on the Prohibition of Nuclear Weapons (Cambridge University Press 2021); Kammerhofer J, 'Uncertainty in the formal sources of international law: customary international law and some of its problems' (2004) 15 European Journal of International Law 523; Singh P, 'A Custom's Nine Lives: Decolonial Continuities of Opinio Juris in Customary International Law' (2024) 23 Chinese journal of international law 689.

Explicitly crafted outside the formal treaty framework, the Artemis Accords are ineligible for registration under Article 102 of the United Nations Charter.[129] It is also a diplomatic step to avoid scrutiny and the test of international law. Their non-binding nature is central to their legal character and political appeal, making them flexible and strategically adaptable. However, this very flexibility raises questions about their legitimacy as instruments capable of contributing to customary international law.[130] Accords certainly are capable of influencing State practice. Scholars propose that the Artemis Accords could generate state practice and *opinio juris*, thereby catalysing the formation of a new body of customary international space law rooted in a US-led governance model.[131] However, these propositions are not without criticism. The formation of customary international law necessitates not only widespread and consistent state practice but also a belief in legal obligation.[132] As affirmed in the *North Sea Continental Shelf Cases*, a practice followed merely for convenience or political gain does not meet the threshold for custom; states must act out of a felt obligation under international law.[133] In this regard, the Artemis Accords fall short. Another argument is that the signatory states enter the framework with a clear understanding that it is a political, not legal, commitment.[134] This undermines the very foundation required for customary law formation.

Artemis diplomacy does more than merely invite participation, as it sets expectations. It can be argued that by positioning adherence to the Accords as a prerequisite for collaboration with NASA and access to the Artemis Program's technological ecosystem, the United States is incentivising the diffusion of its normative vision. This conditional engagement creates a ripple effect wherein even states not party to the Accords may find themselves conforming to its principles or political alignment. This normative entrenchment strategy reflects the broader phenomenon whereby diplomacy shapes international law.[135] As new technologies disrupt established regimes, diplomatic initiatives, especially those backed by major powers and

[129] Agar J and others, 'The Role of the United Nations in Promoting Transparency in the International Treaty Framework: a view through the registration and publication of treaties under article 102 of the charter of the united nations', The Oxford Handbook of United Nations Treaties (Oxford Handbooks, Oxford University Press 2019)

[130] Johnston KA, 'The Nature and Context of Rules and the Identification of Customary International Law' (2021) 32 European journal of international law 1167

[131] Deplano R, 'The Artemis Accords: Evolution Or Revolution In International Space Law?' (2021) 70 International and Comparative Law Quarterly 799; Smith WA, 'Using the Artemis Accords to Build Customary International Law: A Vision for a U.S. - Centric Good Governance Regime in Outer Space ' (2021) 86 Journal of Air Law and Commerce.

[132] Besson S and others, 'The Sources of International Law: An Introduction', The Oxford Handbook of the Sources of International Law (Oxford Handbooks, Oxford University Press 2017)

[133] ICJ, North Sea Continental Shelf (Germany/Netherlands), Judgment of 20 February 1969 (Judge Lachs, Dissenting Opinion), 1969)

[134] NASA, Artemis Accords Method of Operation (NASA 2024)

[135] Chow WM and Levin DH, 'The Diplomacy of Whataboutism and US Foreign Policy Attitudes' (2024) 78 Int Org 103; Cohen HG and Meyer T, International law as behavior (International law as behavior, Cambridge University Press 2021); Falk RA, The status of law in international society (The status of law in international society, 1st edn, Princeton University Press 1970)

significant incentives, play a central role in recalibrating legal standards.[136] The Accords, therefore, is not just a political instrument, but are instruments of legal norm entrepreneurship. Through strategic alignment, consistent promotion, and operationalisation of shared principles, the Accords may serve as the blueprint for an emerging body of customary international space law.

While it is true that diplomacy can shape norms, not all diplomatic initiatives possess equal normative authority or legitimacy. The Accords, despite their alignment with certain OST principles, reflect a selective and instrumentalist reading of international space law. Their attempt to fill regulatory gaps must be critically examined in light of their unilateral origins and the absence of a broad-based consensus. Artemis Accords symbolise a shift in space diplomacy, from treaty-based multilateralism to strategic soft law coalitions, but whether this shift contributes constructively to the evolution of international law remains a matter of contestation.

4.2 The Story of Persistent Objectors: China and Russia's Counter-Diplomacy

The USSR and the United States have had a complex history of space exploration. During the period of Cold War tensions, space was a domain that represented conflicts and competition. The space war between the space powers led to an unhealthy relationship at the beginning.[137] Over the years, both of them joined hands on several space exploration initiatives.[138] The phase of competition was replaced by a phase of cooperation.[139] The initial phase of United States–Russia space collaboration began with the establishment of the Gore-Chernomyrdin Commission, initiated by President Bill Clinton and Russian Prime Minister Viktor Chernomyrdin during the *Vancouver Summit* in April 1993.[140] Later that year, in December 1993, the United

---

[136] Aguirre-Bastos C, Ittekkot V and Baweja JK, 'Internationalization of Science and Diplomacy, Concepts and Practices: Lessons for Developing Countries', Science, Technology and Innovation Diplomacy in Developing Countries (Research for Development, Springer 2023)

[137] Von Welck SF, 'Outer space and cosmopolitics' (1986) 2 Space policy 200; von Welck SF, 'Dominance in space — a new means of exercising global power?' (1988) 4 Space policy 319

[138] NASA, 'Union of Soviet Socialist Republics-United States: agreement on cooperation in exploration and use of outer space (announced January 21, 1971): summary of results of discussions on space cooperation between the Academy of sciences of the USSR and the US national aeronautics and space administration, Moscow, January 18-21, 1971' (1971) 10 International legal materials 617

[139] Crane RD, 'Soviet Attitude Toward International Space Law' (1962) 56 Am j int law 685; Krasnyak O, 'The Apollo-Soyuz Test Project: Construction of an Ideal Type of Science Diplomacy' (2018) 13 The Hague journal of diplomacy 410

[140] Logsdon JM and Williamson RA, 'U.S.-Russian Cooperation in Space: A Good Bet' (1995) 11 Issues in Science and Technology 39; Sredin V, 'Gore-Chernomyrdin commission: summing up' (1998) 44 International affairs (Moscow) 71; White House, Gore-Chernomyrdin Commission IV: Space Committee Report (1994)

States formally invited Russia to join as a full partner in the development of the International Space Station, an offer that Russia accepted.[141] This marked the beginning of a transformative chapter in space relations, as the former Cold War adversaries transitioned into partners, embarking on a series of cooperative missions and laying the groundwork for international collaboration in space.

Despite long-standing disagreements between the United States and Russia on various issues of international law, cooperation in outer space remained relatively stable for decades. However, tensions began to surface in both realms as military ambitions and commercial interests started to influence space policy and governance. The emergence of China as a formidable space power further disrupted this balance.[142] The rapid expansion of Chinese space corporations increasingly clashed with US commercial interests, especially amid broader trade disputes between the two nations.[143] Moreover, China's assertive diplomacy in other domains of the global commons, such as its controversial sovereignty claims in the South China Sea, sparked apprehension about how major powers might approach governance in outer space.[144]

With regard to Accords, Chinese commentators have described the Artemis Accords as an "enclosure movement", reflective of a continued "Cold War mentality" in US foreign policy.[145] This perception is shaped, in part, by the Wolf Amendment passed by the U.S. Congress in 2011, which prohibits NASA from engaging in bilateral cooperation with China or its affiliated entities unless explicitly authorised by Congress or the FBI.[146] This legislative constraint has effectively precluded routine collaboration between NASA and China's space agency. Against the backdrop of growing Sino-Russian strategic ties, Moscow has increasingly viewed Beijing as a more reliable and ideologically aligned partner in outer space governance.[147] Although Russia initially expressed interest in participating in the

---

[141] Logsdon JM and Millar JR, 'US–Russian cooperation in human spaceflight: assessing the impacts' (2001) 17 Space Policy 171;, 'Russia Is Invited to Join Space Station Program' (1993) The Washington Post (1974-) A7

[142] US Subcommittee on Space, Are we losing the space race to China?: hearing before the Subcommittee on Space, Committee on Science, Space, and Technology, House of Representatives, One Hundred Fourteenth Congress, second session, September 27, 2016 (U.S. Government Publishing Office 2017)

[143] Pollpeter K, Building for the future: China's progress in space technology during the tenth 5-year plan and the U.S. response (Strategic Studies Institute, U S Army War College 2008)

[144] Veerle N, Transatlantic Approach to China in the Global Commons: Convergence and Divergence in Seas and Outer Space (Transatlantic Approach to China in the Global Commons: Convergence and Divergence in Seas and Outer Space, Royal United Services Institute RUSI 2022)

[145] Ji E, Cerny MB and Piliero RJ, 'What Does China Think About NASA's Artemis Accords?' (2020) Diplomat (Rozelle, NSW)

[146] Kohler H, 'The eagle and the hare: U.S.-Chinese relations, the 'Wolf Amendment', and the future of international cooperation in space' (2015) 103 The Georgetown Law Journal 1135

[147] Allison G, 'China and Russia: A Strategic Alliance in the Making' (The National Interest, 2018) <https://nationalinterest.org/feature/china-and-russia-strategic-alliance-making-38727> accessed; Robert DB, No Limits?: The China-Russia Relationship and U.S. Foreign Policy (No Limits?: The China-Russia Relationship and US Foreign Policy, Council on Foreign Relations 2024); Vidal

Artemis programme, it later distanced itself, dismissing the initiative as excessively "US-centric".[148]

Russia's space agency chief, Dmitry Rogozin, strongly criticised the United States for excluding Russia from the early stages of the Artemis negotiations.[149] Drawing an analogy with American military interventions, Rogozin remarked on Twitter that "the principle of invasion is the same, whether it be the Moon or Iraq. The creation of a 'coalition of the willing' is initiated. Only Iraq or Afghanistan will come out of this".[150] Ironically, Russia's use of example to criticise the Accords is notable given that it has been brought before the International Court of Justice for its actions in Ukraine.

China and Russia's opposition to the Artemis Accords stems from some principal concerns. Both countries criticise the Accords for lacking a genuinely multilateral negotiation process. The content of the Accords was formulated primarily between the United States and the initial seven invited partners, with all subsequent signatories required to accept the pre-established language without the opportunity for substantive negotiation. This unilateral approach has drawn sharp criticism from Beijing and Moscow, who argue that space governance should be shaped through inclusive, multilateral platforms such as the UNCOPUOS.[151] The rising military capacity of the United States in the space sector was also a concern for Chinese military power. The United States has established its Space Force and has established the strategic importance of ASAT tests.[152] The ASAT tests by China and Russia also worried the United States' strategic interests.[153] The Artemis Accords became the final nail in the coffin.

While the fear of the Artemis Accords reshaping the interpretation of the OST and the growing number of interested parties raised concerns about whether the Artemis diplomacy can shape a new customary law. The United States has done similar diplomatic practices by promoting American interests as a global norm, thereby helping the unilateral interest gain the status and legitimacy by making it a customary practice,

F and Privalov R, 'Russia in Outer Space: A Shrinking Space Power in the Era of Global Change' (2024) 69 Space Policy 101579

[148] Davenport C, 'Seven nations join the U.S. in signing the Artemis Accords, creating a legal framework for behavior in space' (2020) The Washington post (Washington, DC 1974 Online)

[149] Davenport C, 'Seven nations join the U.S. in signing the Artemis Accords, creating a legal framework for behavior in space' (2020) The Washington post (Washington, DC 1974 Online)

[150] Roulette J, Exclusive: Trump administration drafting 'Artemis Accords' pact for moon mining - sources (Reuters 2020)

[151] Ji E, Cerny MB and Piliero RJ, 'What Does China Think About NASA's Artemis Accords?' (2020) Diplomat (Rozelle, NSW)

[152] Pope R, 'Space weapons and the increasing militarisation of outer space: Whether the legal framework is fit-for-purpose' (2021) 27 Auckland University law review 263; Variath A, 'Space economy and the geopolitics of South Asia's space security trilemma' (2024) 113 Round table (London) 202

[153] Marrow M, 'White House Announces Asat Test Ban' (2022) 25 Inside the Pentagon's Inside Missile Defense

for example, the Wassenaar Arrangement or the Freedom of Navigation Program.[154] Unlike in domestic legal systems, where customs typically evolve through consistent judicial recognition, international law allows for the possibility of a norm crystallising rapidly or spontaneously, for instance, *instant* customary rules.[155] A key historical example of instant custom in international law is the Truman Proclamation of 1945, through which the United States unilaterally declared sovereign rights over the natural resources of the continental shelf adjacent to its coasts.[156] The Truman Proclamation was not backed by a treaty or multilateral agreement but was a unilateral act justified on the basis of technological advancements.[157] What followed was a wave of similar proclamations by other coastal states, including Argentina, Mexico, Venezuela, and Saudi Arabia, many of which cited the US example. Over time, these national practices crystallised into customary international law and were later codified in Articles 76 and 77 of UNCLOS, which recognise the rights of coastal states over the continental shelf.[158]

Similarly, the Artemis Accords diplomacy promotes the fear of it becoming a new custom. As per international law, the way to avoid this from becoming a norm is to persistently object to it. The doctrine of the 'persistent objector' in international law, while widely discussed today, has roots that trace back well before its formal articulation in modern legal literature. Ian Brownlie, in his seminal 1966 work 'Principles of Public International Law', introduced the term 'persistent objector' and treated it as an established feature of international law.[159] Nevertheless, the concept itself predates Brownlie's formulation. The International Law Association has noted that Cornelius van Bynkershoek's 1721 writings exemplify an early recognition of this principle, indicating its acceptance as far back as the eighteenth century.[160]

---

[154] Crawford J, Chance, order, change: the course of international law: general course on public international law (Hague Academy of International Law 2014); Miller RA and Bratspies RM, Progress in international law (Progress in international law, Martinus Nijhoff Publishers 2008)

[155] Scharf MP, Customary International Law in Times of Fundamental Change: Recognizing Grotian Moments (Cambridge University Press 2013); Sender O and Wood SM, 'Between 'Time Immemorial' and 'Instant Custom': The Time Element in Customary International Law' (2021) 42 Grotiana (1980) 229

[156] Hollick AL, US Foreign Policy and the Law of the Sea (Princeton University Press 2017); Scharf MP, Customary International Law in Times of Fundamental Change: Recognizing Grotian Moments (Cambridge University Press 2013)

[157] Watt DC, 'First steps in the enclosure of the oceans: The origins of Truman's proclamation on the resources of the continental shelf, 28 September 1945' (1979) 3 Marine Policy 211; Lucky A, 'The Issues Concerning the Continental Shelf: Reflections' (2015) 17 International community law review 95

[158] Nguyen LN, 'Coastal States' Rights and Obligations under the Law of the Sea' in Nguyen LN (ed), The Development of the Law of the Sea by UNCLOS Dispute Settlement Bodies (Cambridge University Press 2023)

[159] Green JA, 'PART I: THE ORIGIN AND LEGAL SOURCE OF THE PERSISTENT OBJECTOR RULE', The Persistent Objector Rule in International Law (Oxford University Press, Incorporated 2016); Brownlie I, Principles of public international law (Principles of public international law, Clarendon Press, 1966)

[160] Green JA, The persistent objector rule in international law (The persistent objector rule in international law, 1st edn, Oxford University Press 2016)

The persistent objector doctrine holds that a state which persistently and consistently opposes the emergence of a customary international norm during its formative stage is not bound by that norm once it crystallises, provided the objection is continuously maintained.[161] As B.S. Chimni notes, the rule was developed, in part, to protect the interests of Western capitalist states during the Cold War era.[162] Russia and China have consistently and persistently primarily objected to the Accords, making it a case of counter-diplomacy. It is also an effort to stop the Artemis coalition from pushing its agenda as a new custom in space governance.

In March 2021, China and Russia formally announced the International Lunar Research Station (ILRS), an ambitious joint venture to establish a scientific base near the Moon's south pole, directly challenging the United States-led Artemis Accords both in terms of vision and geopolitical alignment.[163] It is a planned lunar base currently being led by Roscosmos and the China National Space Administration.[164] Much like Artemis, ILRS targets the Moon's highland polar regions, which offer continuous sunlight and potential access to water ice within permanently shadowed craters, vital resources for sustained lunar habitation and operations.[165]

The ILRS project's scale and timeline closely mirror Artemis' goals, but with a distinctly different governance model. Where Artemis champions transparency, interoperability, and open participation under United States leadership, ILRS presents a counter-framework led by China and Russia that emphasises strategic autonomy and multipolarity in space governance.[166] The project not only consolidates growing Sino-Russian cooperation but also attracts a coalition of participants, as of 2025, of 17 countries (See Table 2), thus building an alternative normative and technological ecosystem in lunar exploration.[167]

Despite its framing as a cooperative alternative to Western-led lunar programmes, the ILRS continues to show limited global traction. As of 2025, the ILRS has drawn participation from 17 countries, with pronounced regional imbalances.[168] Asia leads with seven members: China, Thailand, Kazakhstan, Pakistan, Iran, Bahrain, and Turkey, followed by Europe with four: Russia, Azerbaijan, Belarus, and Serbia.

---

[161] Brownlie I, Principles of public international law (Principles of public international law, Clarendon Press, 1966)

[162] Chimni BS, 'Customary International Law: A Third World Perspective' (2018) 112 Am j int law 1

[163] International Lunar Research Station (ILRS) Guide for Partnership (China National Space Administration and the State Space Corporationv "Roscosmos" 2021)

[164] Sanmartí M, 'The scramble for the Moon' (2021) 46 New Zealand International Review 11

[165] Xia Y and others, 'Lunar base infrastructure construction: Challenges and future directions' (2025) 176 Automation in Construction 106251

[166] Wu X, The International Lunar Research Station: China's New Era of Space Cooperation and Its New Role in the Space Legal Order (Elsevier 2023)

[167] PRC, 'International Lunar Research Station attracts more partners: CNSA' (State Council: The People's Republic of China 2025) <https://english.www.gov.cn/news/202504/24/content_WS680976d7c6d0868f4e8f2045.html> accessed

[168] Tronchetti F, 'China's Plan for an International Lunar Research Station: A Path Towards Multilateralism or the Beginning of the End?' (2023) 66 International Institute of Space Law 227

Table 2 Regional representation of ILRS members

Region	Countries	Count
Asia	China, Thailand, Kazakhstan, Pakistan, Iran, Bahrain, Turkey	7
Europe	Russia, Azerbaijan, Belarus, Serbia	4
Africa	South Africa, Egypt, Senegal	3
South America	Venezuela, Peru	2
Central America	Nicaragua	1

Africa contributes three members: South Africa, Egypt, and Senegal. South America and Central America are represented by only one country each: Peru and Nicaragua, respectively.

Notably, the ILRS lacks participation from major spacefaring regions such as North America, Western Europe, Oceania, and large parts of Latin America, highlighting its constrained international appeal and the geopolitical undercurrents shaping its membership. Moreover, the timing of membership reveals that a significant number of countries only joined in 2023 or 2024, suggesting a slow and recent expansion rather than consistent international uptake since its launch in 2021 (unlike the Accords).[169] This regional imbalance and gradual growth pattern highlight the ILRS's current challenge in building a diverse and globally inclusive lunar partnership. In essence, ILRS challenges the Artemis Accords by offering a rival vision of lunar exploration, one that is state-centric, led by non-Western powers, and potentially resistant to the legal and political principles embedded in the Artemis framework.[170] The competition is not merely about technology or territory, but about who gets to write the next chapter of space law and governance.

As China and Russia consolidate their lunar ambitions through the ILRS and seek to rally international support, the Artemis bloc is not only competing with technological enterprise but also with a rival normative order. It is important to underscore that there are currently no overlapping members between the Artemis Accords and the ILRS. The ILRS largely comprises nations from the Global South and Eurasia with strategic or emerging space interests. In contrast, the Artemis Accords have signatories from across all inhabited continents, including major spacefaring powers and many middle- and high-income countries. This significant disparity in participation not only underscores the limited global traction of the ILRS framework but also illustrates the broader diplomatic success of the Artemis initiative. The expanding consensus around the Artemis Accords suggests a strong alignment with American interpretations of space governance.

---

169 PRC, 'International Lunar Research Station attracts more partners: CNSA' (State Council: The People's Republic of China 2025) <https://english.www.gov.cn/news/202504/24/content_WS680976d7c6d0868f4e8f2045.html> accessed

170 Xu F and Ou J, 'Promoting international cooperation on the International Lunar Research Station: Inspiration from the ITER' (2023) 203 Acta Astronautica 341

5 Conclusion

The rise of the United States to a position of preeminence in international society has had a profound impact on international law-making and the American vision of world order.[171] The growing influence of the United States also enabled it to consolidate its position in space governance. The strength of American space diplomacy lies in its capacity to pull together actors that have space aspirations. However, the Artemis Accords cannot be called as truly multilateral in nature, as it mark a significant shift in space governance, illustrating a departure from traditional multilateralism toward a more flexible, coalition-based model. Rather than emerging from broad-based, United Nations negotiation platforms, the Accords represent a strategic alignment of states that voluntarily adhere to United States-led principles, highlighting the increasing relevance of interest-driven alliances in shaping the rules of engagement in outer space. This strategic coalition approach reflects the pragmatism of modern diplomacy, where like-minded states converge around shared policy objectives without the procedural constraints of formal multilateralism. However, Accords is also a case of the de-formalism of international law where states, particularly powerful ones, prefer informal, flexible, and interest-based arrangements that carry normative weight by fostering legal convergence and influencing the evolution of customary international law. This diplomatic strategy represents a recalibration of space governance.

At the heart of this framework lies the diplomatic architecture of the Accords, which enables the diffusion of American-centric norms related to transparency, sustainability, and the utilisation of space resources.[172] In doing so, the Accords serve as tools of public diplomacy and norm entrepreneurship, redefining the mechanisms through which international legal standards in space may evolve, and potentially influencing the formation of new customary international law. When the United States invites a state to join the Artemis initiative, the invitation often carries the weight of geopolitical leverage, creating a subtle but significant pressure for smaller or strategically dependent states to align. It renders the decision less voluntary and more a function of asymmetrical power dynamics. Judicial verdicts reached by the International Court of Justice in *Nicaragua v. United States of America and Democratic Republic of Congo v. Uganda* established the precedent of "indirect intervention", where a country can violate international law, but using direct or indirect

[171] Boyle FA, The Future of International Law and American Foreign Policy (Brill | Nijhoff 2023); Sachs JD, A New Foreign Policy: Beyond American Exceptionalism (A New Foreign Policy: Beyond American Exceptionalism, 1st edn, Columbia University Press 2018)

[172] Bartóki-Gönczy B and Nagy B, 'The Artemis Accords' (2023) 62 International Legal Materials 888

methods.[173] Under such a definition could public diplomacy, aptly categorised as an overt yet indirect means of influence is to be explored.[174]

The parallel development of initiatives like the ILRS by China underscores the competitive yet coexisting visions of global space governance. Together, these efforts highlight a fragmented but dynamic international legal landscape, where the future of space diplomacy is shaped less by universal consensus and more by competing blocs of influence, each promoting distinct normative orders. This evolving structure will have long-term implications for how authority, legitimacy, and legal obligation are constructed in the governance of the final frontier. As instruments of norm-setting, the Accords diplomacy illustrates how space exploration is no longer just a scientific endeavour but a diplomatic arena where legal, political, and ethical standards are negotiated.[175] The Artemis Accords reveal how coalition-driven pragmatism and universal norms intersect, underscoring that space resource utilisation is a profoundly political issue that will define the future of space governance. This evolving dynamic signals the future of space governance, one where traditional theories of international cooperation, multilateralism, minilateralism, and strategic coalitions will be both tested, reshaped, and rewritten. These changes are likely to define the contours of global space politics, marking a shift toward new models of norm entrepreneurship and institutional innovation in the final frontier.

Adithya Variath is an Assistant Professor and Associate Dean at the Jindal Global Law School, O.P. Jindal Global University. Adithya was an Elizabeth Moir Scholar at the Oxford India Centre for Sustainable Development at the Oxford Department of International Development, University of Oxford. Before moving to Oxford, he served as an Assistant Professor of Law and founding Coordinator of the Centre for Research in Air and Space Laws at Maharashtra National Law University Mumbai. He is the recipient of the 2025 Sachdeva Award for Space Law. Adithya's research focuses on space and aviation laws, emerging technologies, and global institutional design.

[173] ICJ, North Sea Continental Shelf (Germany/Netherlands), Judgment of 20 February 1969 (Judge Lachs, Dissenting Opinion), 1969)

[174] Arceneaux PC, 'International Law Provides New Context for Public Diplomacy Scholarship' (USC Center on Public Diplomacy, 2020) <https://uscpublicdiplomacy.org/blog/international-law-provides-new-context-public-diplomacy-scholarship> accessed

[175] AJIL, 'Signatories of the U.S.-Led Artemis Accords Meet in Person for the First Time' (2023) 117 American Journal of International Law 133

Space Mining for the Benefit of All? Advancing International Cooperation through Standards and Interoperability in the Artemis Accords and International Lunar Research Station

Art Cotterell

Abstract Space mining on the lunar surface, particularly in situ resource utilisation (ISRU), raises pressing questions: will access, or lack thereof, to enabling technologies alleviate or entrench global inequities and inequalities? The foundational principle of international space law and governance—that outer space is for the benefit of all nations, irrespective of their degree of economic or scientific development—will remain unrealised, and indeed undermined, unless opportunities are pursued to expand access to enabling infrastructure and systems. The technological gap, particularly between the nations and industries of the Global North and Global South, will also likely widen. Yet, ISRU cannot be discussed in a geopolitical vacuum. The spectre and rhetoric of a second 'space race' persists, reflecting earthly tensions. The United States, through the *Artemis Accords*, and China and Russia, via the International Lunar Research Station, are leading separate initiatives and membership blocs to establish a more permanent human presence on the lunar south pole, which is speculated to contain scarce, life-sustaining resources for such missions. In parallel, the United Nations Committee on the Peaceful Uses of Outer Space is advancing a draft set of recommended principles for space resource activities. Commitments to standards and interoperability are emerging across all three fronts, to varying degrees. This chapter argues these commonalities can be leveraged to operationalise benefit-sharing for all nations and—drawing on lessons from the Cold War—foster international cooperation, despite heightened competition.

Keywords Space mining · In situ resource utilisation (ISRU) · Global inequities · Artemis Accords · International Lunar Research Station · International cooperation

---

A. Cotterell (✉)
Faculty of Law and Justice, University of New South Wales, Sydney, NSW, Australia
e-mail: a.cotterell@unsw.edu.au

S. Bhat B and A. Variath (eds.), *Legal Responses to Space Resource Utilisation*, Space Law and Policy, https://doi.org/10.1007/978-981-95-8864-0_5

1 Introduction

Since the Cold War, humanity's 'earthly ways' have been launched off-Earth, one rocket at a time.[1] The first 'space race' to the Moon—between the United States (US) and the former Union of Soviet Socialist Republics (USSR)—was a quest for strategic, economic, ideological and technological dominance on Earth. Great power politics persist. Today, the rivalry between the US and the People's Republic of China (China), is frequently framed as the second 'space race', with both aiming to land crewed missions on the lunar south pole by decade's end.[2]

Neither superpower is going alone. The US leads the *Artemis Accords*, joined by a growing coalition of signatory nations. China, in a partnership with the Russian Federation (Russia), heads the International Lunar Research Station (ILRS) and is backed by a smaller number of partner nations.[3] As Ben-Itzhak observes, two 'space blocs' are emerging, reflecting global alliances and fissures;[4] a phrase to describe these lunar initiatives that this chapter will continue to use. Yet, reservations exist over whether coexistence will be peaceful or lead to conflict.

That is because this time the goal is not a temporary visit, but a sustained presence by establishing bases to support scientific research, onward exploration and lay the foundations for an in-space economy. The first-to-arrive will gain an advantage in shaping its governance and securing—even monopolising—prime sites where scarce and valuable resources may exist. Both programmes will rely on in situ resource utilisation (ISRU) by extracting and utilising lunar resources to reduce dependency on Earth and overcome the cost and logistics of transporting supplies.[5] Mining regolith

---

[1] Myres S McDougal and Leon Lipson, 'Perspectives for a Law of Outer Space' (1958) 52 *American Journal of International Law* 407, 407.

[2] Abbey A Donaldson, 'NASA Shares Orion Heat Shield Findings, Updates Artemis Moon Missions' (NASA, 5 December 2024) <https://www.nasa.gov/news-release/nasa-shares-orion-heat-shield-findings-updates-artemis-moon-missions/>; The State Council of the People's Republic of China, 'China to Realize Manned Lunar Landing by 2030' (29 May 2023) <https://english.www.gov.cn/news/202305/29/content_WS64748c46c6d03ffcca6ed781.html>.

[3] National Aeronautics and Space Administration (NASA), 'NASA, International Partners Advance Cooperation with First Signings of Artemis Accords' (13 October 2020) <https://www.nasa.gov/news-release/nasa-international-partners-advance-cooperation-with-first-signings-of-artemis-accords/>; China National Space Administration (CNSA) and Roscosmos, 'Joint Statement Between CNSA and Roscosmos Regarding Cooperation for the Construction of the International Lunar Research Station' (29 April 2021) <https://www.cnsa.gov.cn/english/n6465652/n6465653/c6811967/content.html>.

[4] Svetla Ben-Itzhak, 'Space Blocs: The Future of International Cooperation in Space is Splitting Along Lines of Power on Earth' (*The Conversation*, 22 April 2022) <https://theconversation.com/space-blocs-the-future-of-international-cooperation-in-space-is-splitting-along-lines-of-power-on-earth-180221>.

[5] National Aeronautics and Space Administration (NASA), 'Overview: In-Situ Resource Utilization' <https://www.nasa.gov/overview-in-situ-resource-utilization/>.

could provide material to 3D print landing pads and habitats.[6] Water ice in permanently shadowed regions (PSRs) could be split into oxygen and hydrogen to sustain human and plant life and produce rocket fuel for return trips to Earth and missions to Mars.[7]

With time, these bases could support commercial operators looking to mine resources and return them to Earth, with the potential for profits worth billions.[8] Whether mining asteroids for precious metals or the Moon for helium-3—a non-radioactive isotope rare on Earth but speculated to be more abundant on the lunar surface, and proposed as a potential fuel for future nuclear fusion reactors, that could help support net zero emission targets[9]—such ventures carry consequences far beyond geopolitics. This chapter, however, will mainly focus on the activities of nations with respect to ISRU on the lunar surface—referred to here as 'space mining'.

Space mining also risks launching another 'earthly way' off-Earth—that of rising global inequities and inequalities. This activity raises threshold intersecting legal, ethical and philosophical questions: should Earth's nearest celestial neighbour be viewed through an extractivist lens that promotes 'space colonisation' and ultimately, capitalism in the cosmos?[10] Will intergenerational equity be safeguarded?[11] Should, like Antarctica, a moratorium be imposed on mining the Moon?[12] These deliberations deserve far more attention than they receive. Since mining lunar resources is likely to occur in the future, this chapter pragmatically—and optimistically—asks: how can international cooperation foster opportunities for all nations to benefit from this activity?

---

[6] Athanasios Goulas, Jon GP Binner, Russell A Harris, Ross J Friel, 'Assessing Extraterrestrial Regolith Material Simulants for In-Situ Resource Utilisation Based 3D Printing' (2017) 6 *Applied Materials Today* 54.

[7] Aaron Paz, *Technology Assessment for Producing Propellant from Lunar Water* (2023) <https://ntrs.nasa.gov/api/citations/20230010039/downloads/Technology%20Assessment%20for%20Producing%20Propellant%20from%20Lunar%20Water.pdf>.

[8] PricewaterhouseCoopers, *Lunar Market Assessment* (2021) <https://www.pwc.com.au/industry/space-industry/lunar-market-assessment-2021.pdf>.

[9] Interlune, 'Interlune' (2025) <https://www.interlune.space/>; AstroForge, 'Astroforge' (2025) <https://www.astroforge.com/>.

[10] Audra Mitchell, Sarah Wright, Sandie Suchet-Pearson, Kate Lloyd, Laklak Burarrwanga, Ritjilili Ganambarr, Merrkiyawuy Ganambarr-Stubbs, Banbapuy Ganambarr, Djawundil Maymuru and Rrawun Maymuru, 'Dukarr Lakarama: Listening to Guwak, Talking Back to Space Colonization' (2020) 81 *Political Geography* 102218; Peter Dickens, *Capital and the Cosmos: War, Society and the Quest for Profit* (Palgrave, 2023).

[11] Space Generation Advisory Council, *Submission to the Working Group on the Legal Aspects of Space Resource Activities* (2025) <https://www.unoosa.org/documents/pdf/copuos/lsc/space-resources/DraftPrinciplesContributions/SGAC_Submission_to_the_LSC_Working_Group_on_the_Legal_Aspects_of_Space_Resource_Activities.pdf>.

[12] AC Grayling, *Who Owns the Moon? In Defence of Humanity's Common Interests in Space* (Oneworld Publications, 2024).

After all, the oft-called 'magna carta' of international space law[13]—the *Outer Space Treaty* (OST) of 1967, negotiated through the United Nations (UN) Committee on the Peaceful Uses of Outer Space (COPUOS)—speaks to us from an ever-present past. Article I declares:

> The exploration and use of outer space, including the Moon and other celestial bodies, shall be carried out for the benefit and in the interests of all countries, irrespective of their degree of economic or scientific development, and shall be the province of all [hu]mankind.[14]

Operationalising this provision—referred to here as the 'benefit-sharing principle'—often focuses on the resources and the profits derived from them.[15] Comparatively minimal attention has been given to expanding access to enabling technologies for space mining. Beyond the 'hype', lack of, or limited, access to such technologies risks exacerbating existing global inequities and inequalities, which will likely accelerate with the entrance of private actors from dominant space-faring powers.[16] A manifest risk exists that the existing space technological gap between nations and their industries in the Global North and Global South will widen further.[17]

This chapter proposes that open standards and interoperability can expand access to ISRU-related technologies and infrastructure for all nations and their industries, thereby helping to operationalise the benefit-sharing principle. Achieving this will require international cooperation—and could emerge from seemingly unlikely fronts, as commitments to standardisation and interoperability are threaded across the *Artemis Accords*, the ILRS *and* the recent draft set of recommended principles emerging from UN COPUOS. As this author has previously emphasised, such cooperation must persist—and can coexist—even amid geopolitical tensions between the US and China and their respective space blocs.[18] Finding points of consensus in this

---

[13] Ram S Jakhu, 'Evolution of the Outer Space Treaty' in Ajey Lele (ed), *Fifty Years of the Outer Space Treaty: Tracing the Journey* (Pentagon Press, 2017) 13–19, 13.

[14] *Treaty on Principles Governing the Activities of States in the Exploration and Use of Outer Space, Including the Moon and Other Celestial Bodies* (opened for signature 27 January 1967, entered into force 10 October 1967) 610 UNTS 205 (OST) art I.

[15] See, for example, Ben McKeown, Andrew G Dempster, Serkan Saydam and Jeff Coulton, 'The Space Resources Fund: A Solution to the Space Resources Benefit Sharing Dilemma?' (2025) 71 *Space Policy* 101663; Morgan Sterling Saletta and Kevin Orrman-Rossiter, 'Can Space Mining Benefit All of Humanity? The Resource Fund and Citizen's Dividend Model of Alaska, the "Last Frontier"' (2018) 43 *Space Policy* 1.

[16] KR Moore, J Segura-Salazar, L Bridges, P Diallo, K Doyle, C Johnson, P Foster, N Pollard, N Whyte and O Wright, 'The Out-Of-This-World Hype Cycle: Progression Towards Sustainable Terrestrial Resource Production' (2022) 186 *Resources, Conservation and Recycling* 106519; JA Dallas, S Raval, JP Alvarez Gaitan, S Saydam and AG Dempster, 'Mining Beyond Earth for Sustainable Development: Will Humanity Benefit from Resource Extraction in Outer Space?' (2020) 167 *Acta Astronautica* 181.

[17] Chaitanya Giri and Kyle Hiebert, 'Closing the Global North–South Gap in the Second Space Age' (*Centre for International Governance Innovation*, August 2024) <https://www.cigionline.org/static/documents/no.298_update.pdf>.

[18] Art Cotterell, 'Space Isn't All About the "Race"—Rival Superpowers Must Work Together for a Better Future' (*The Conversation*, 10 October 2024) <https://theconversation.com/space-isnt-all-about-the-race-rival-superpowers-must-work-together-for-a-better-future-240543>.

technical area may advance benefit-sharing for all nations—and, in turn, generate positive spill-over cooperation and even help lower tensions on Earth.

As to the structure: part 2 considers the origins and evolution of the benefit-sharing principle in the OST and subsequent non-binding developments. It spotlights the active, principled negotiating of Global South nations in securing this provision. Part 3 introduces the concepts of standards and interoperability, showing how, since the first 'space race', technical commitments have fostered dialogue. It explains why this heritage is relevant to mining lunar resources in today's geopolitically febrile environment and how it can help advance benefit-sharing. Part 4 examines the *Artemis Accords*, the ILRS and multilateral processes within the UN COPUOS Working Group on Legal Aspects of Space Resource Activities, noting the emergence of standardisation and interoperability as priorities across them. Part 5 concludes by exploring how to translate such commitments into practice, including opportunities for space bloc-to-bloc cooperation, the potential role of overlapping memberships and the significant barrier to any direct US–China cooperation.

2 Benefit-Sharing and Addressing the Space Technological Gap

The concern with asymmetrical access to space technologies is not new, having gained prominence during the peak of treaty-making for international space law. While it is generally noted that the OST and the treaties that followed through UN COPUOS were negotiated amidst the Cold War,[19] this also occurred against the backdrop of a growing movement towards formal decolonisation.[20] Having experienced the ongoing harms of colonial exploitation and extractivism, which overall left the Global South lacking the material resources for space exploration, newly independent nations did not want history repeating off-Earth.[21] Aganaba, explains that nations from the Global South were active agents in UN COPUOS negotiations and 'secured legal protections and necessary safeguards for their interests', reflecting a tradition of Third World Approaches to International Law (TWAIL).[22]

The benefit-sharing principle in Article I speaks to this advocacy. It reflects an intention that outer space not only benefits 'first arrivals', but all nations, both then

---

[19] Albert K Lai, *The Cold War, the Space Race and the Law of Outer Space* (Routledge 2021).

[20] Peter Sutch and Peri Roberts, 'Outer Space and Neo-Colonial Injustice: Distributive Justice and the Continuous Scramble for Dominium' (2019) 46(11) *International Journal of Social Economics* 1291.

[21] Cristian van Eijk, 'Unstealing the Sky: Third World Equity in the Orbital Commons' (2022) 47(1) *Air and Space Law* 25.

[22] Timiebi Aganaba, in Sathesh Raj, 'TWAIL & CAIL: Decolonising Global Space Governance to Empower Global South in New Space' (Space Generation Advisory Council, 23 December 2021) <https://spacegeneration.org/twail-cail-decolonising-global-space-governance-to-empower-global-south-in-new-space>.

and into the future.[23] Indeed, it was Brazil's 'strong and well-articulated' proposal that advocated for this provision within the 'operative' part of the OST, along with the inclusion that outer space was 'for the benefit and in the interests of all countries, irrespective of their degree of economic or scientific development'.[24] As van Eijk notes, the view was held among such nations that: 'If space engaged global interests… all states should regulate and benefit from it'.[25] Space governance, to this day, owes much to this principled stance emerging from the Global South.

Reflecting on these attempts to right historic global wrongs through multilateralism, international space law at that time was 'startling in ambition and progressive in content'.[26] The *Moon Agreement*, the last binding instrument concluded through UN COPUOS in 1979, is the most obvious example. It contains provisions declaring that the Moon and its natural resources are the 'common heritage of humanity' and that, at such a future time when exploitation of natural resources on the lunar surface is feasible, an international regime is to be established.[27] One of the purposes of that unrealised regime was to provide for:

> An equitable sharing by all States Parties in the benefits derived from those resources, whereby the interests and needs of the developing countries, as well as the efforts of those countries which have contributed either directly or indirectly to the exploration of the moon, shall be given special consideration.[28]

While space-faring major powers like the US were at the table when the *Moon Agreement* was negotiated, they never joined. Thus, reflecting a shift 'from solidarity with the Third World to... neoliberalism'.[29] Such provisions were seen as antithetical to the eventual establishment of a property rights regime for such resources; an approach seen as necessary to incentivise future commercial activity.[30]

Now, the *Moon Agreement* is one of the most under-ratified treaties within the UN, with a declining membership base, leaving it 'hanging by a thread'.[31] Even though the lack of widespread support for the *Moon Agreement* indicates that a commitment to equitable benefit-sharing will likely be unrealised on a widespread scale, the benefit-sharing principle enshrined in the OST remains, and must be materially pursued.

---

[23] van Eijk (n 21) 34.

[24] Ram Jakhu, 'Legal Issues Relating to the Global Public Interest in Outer Space' (2006) 32(2) *Journal of Space Law* 31, 37–38.

[25] van Eijk (n 21) 35.

[26] Sutch and Roberts (n 20) 1291.

[27] *Agreement Governing the Activities of States on the Moon and Other Celestial Bodies* (opened for signature 18 December 1979, entered into force 11 July 1984) 1363 UNTS 3 (Moon Agreement) art 11(1), 11(5).

[28] ibid. art 11(7)(d).

[29] Frans G von der Dunk, 'Property Rights Over the Moon or On the Moon? The Legality of Space Resource Exploitation on Celestial Bodies' (2023) 6 *Journal of Law & Innovation* 95, 107.

[30] Matthew Johnson, *Property Rights in Outer Space: Mining, Techno-Utopian Imaginaries, and the Privatisation of the Off-World Frontier* (Routledge 2024).

[31] Stefan-Michael Wedenig and Jack Wright Nelson, 'The Moon Agreement: Hanging by a Thread?' (*McGill Institute of Air & Space Law*, 26 January 2023) <https://www.mcgill.ca/iasl/article/moon-agreement-hanging-thread>.

Gaining access to space technology permeated subsequent negotiations that culminated in the adoption of the *Declaration on International Cooperation in the Exploration and Use of Outer Space for the Benefit and in the Interest of All States, Taking into Account the Needs of Developing Countries* (the *Benefits Declaration*) in 1996.[32] This represented a renewed push by Global South nations to gain access to space technology through international cooperation, including via technology transfer, by operationalising benefit-sharing.[33]

Whether that principle imposed a mandatory or voluntary obligation on states had long been contested. This non-binding instrument ultimately landed on the side of the latter. While it sought to advance international cooperation by '[f]acilitating the exchange of expertise and technology among States on a mutually acceptable basis',[34] it critically stated that States were 'free to determine all aspects of their participation in international cooperation'.[35] Reflecting on the *Benefits Declaration*, von der Dunk states it:

> clarified that the obligation of international cooperation effectively amounts to a *bona fide* 'obligation-of-effort', and in the conspicuous absence of any reference to material-benefit-sharing, any benefits to be shared with others are comprehensively at the discretion of the States concerned.[36]

Consequently, the benefit-sharing principle is treated as a moral rather than legal obligation—and is often overlooked.[37] To address this, Deplano proposes adopting a 'novel legal construct to international space law': 'an original proportionality test which can be used to determine the equitable amount of benefit-sharing which is due to the international community'.[38] Yet, such efforts to make space governance more 'inclusive' remain largely scholarly. Benefit-sharing remains discretionary. As the former director of the UN Office for Outer Space Affairs (UNOOSA), Nandasiri Jasentuliyana, remarked: 'access to the benefits of space technology… is ultimately a question of the nature of international co-operation among States'.[39]

---

[32] *Declaration on International Cooperation in the Exploration and Use of Outer Space for the Benefit and in the Interest of All States, Taking into Account the Needs of Developing Countries* (adopted 13 December 1996) UNGA Res 51/122 (*Benefits Declaration*).

[33] Nandasiri Jasentuliyana, 'Article I of the Outer Space Treaty Revisited' (1989) 17(2) *Journal of Space Law* 129.

[34] *Benefits Declaration* (n 32) art 5(c).

[35] ibid. art 2.

[36] von der Dunk (n 29) 102.

[37] Rossana Deplano, 'Inclusive Space Law: The Concept of Benefit Sharing in the Outer Space Treaty' (2023) 72(3) *International and Comparative Law Quarterly* 671, 672.

[38] ibid. 700.

[39] Jasentuliyana (n 33) 138.

3 Operationalising Benefit-Sharing Via Standards and Interoperability

Given that the benefit-sharing principle remains at the discretion of nations, it is crucial to identify which approaches to technology can incentivise and sustain mutually beneficial cooperation among nations. To this end, Aganaba suggests an 'opportunity to fill in the meaning' of both international cooperation and 'space benefit' exists,[40] asking:

> How can the existing framework enable all countries to foster capabilities in a way that is of mutual significance to all? It is proposed that this leads us to analyzing closely the hierarchy of possible benefits and how we can imagine a cycle of positive feedback to build increasing cooperation between those on the margins of space activity and those gaining the greatest benefit from space.[41]

One opportunity to realise this 'cycle of positive feedback' lies in the need for standards and interoperability, which have a long-standing role in fostering international cooperation—and can be leveraged to advance the principle of benefit-sharing.

3.1 Overview of Standards and Interoperability

'The truth is that a day without standards is inconceivable'—as a past World Standards Day[42] message reminds us—is as relevant to ambitions to return to the Moon as it is to life on Earth. Understanding their potential to increase access to space mining technology requires first examining the concepts of standards and interoperability.

The term 'standards' might first bring to mind behavioural norms. However, in this chapter, the term 'standards' refers to technical or interoperable standards, which are detailed documents describing an agreed-upon technical solution by subject matter experts, often through standard-setting organisations (SSOs) or standard-development organisations (SDOs).[43] Standards have become increasingly necessary as technologies have become increasingly complex and interdependent, and consumers expect plug-and-play options.[44]

---

[40] Timiebi Aganaba-Jeanty, 'Common Benefit from a Perspective of "Non-Traditional Partners": A Proposed Agenda to Address the Status Quo in Global Space Governance' (2015) 117 *Acta Astronautica* 172, 173.

[41] ibid. 182–183.

[42] Bernard H Falk, Liew Mun Leong and Pekka Tarjanne, *World Standards Day Message: Standards in Daily Life* (International Organization for Standardization, 14 October 1998) 1 <https://www.iso.org/files/live/sites/isoorg/files/archive/pdf/en/wsd_message_1998.pdf>.

[43] Intellectual Property Office, *Technical Standards and Standard Development Organisations* (22 July 2024) <https://www.gov.uk/guidance/technical-standards-and-standard-development-organisations>.

[44] Paul Zubrinich and others, Proprietary vs Open Standards (4iP Council November 2018) 3 <https://www.4ipcouncil.com/application/files/3615/4357/3178/4iP_Council_-_Proprietary-vs-Open-Standards_-_Nov18.pdf>.

In everyday life, standards are often taken for granted, going unnoticed until a friction occurs. For example, cell phones communicate seamlessly among differing brands and across global borders due to the existence of standards.[45] Similarly, Wi-Fi standards 'connect us invisibly to each other', enabling devices like laptops to access networks across different providers, even during international travel.[46] These taken-for-granted conveniences highlight that the 'direct consequence' of standardisation is to 'achieve a common ground on the specification of technology for ensuring interoperability and ease of innovation on sophisticated technologies'.[47] An oft-cited definition of interoperability is 'the ability of two or more systems or components to exchange information and to use the information that has been exchanged'.[48]

Standards are developed by international, national and regional standard-setting bodies, typically by voluntary, consensus-based processes.[49] The International Organization for Standardization (ISO), as the name suggests, is an international SSO/SDO, and national counterparts exist too. Government departments, agencies and regulators also develop standards. Businesses and industry consortia likewise develop standards independently to address internal or market needs.

Adding to the complexity, standards can be broadly classified as open or proprietary. Open standards, developed through consensus-driven processes—generally by SSOs/SDOs but sometimes by industry groups—are designed to foster interoperability and widespread adoption. In contrast, proprietary standards are 'developed and controlled by one company or a small group', often limiting accessibility and interoperability to maintain market dominance and drive profits.[50] For instance, Apple's Lightning cable is a well-known proprietary standard, controlled by the tech giant. This means interoperability with non-Apple devices and charging compatability was limited until Apple started using the USB-C industry standard.[51]

This chapter is orientated towards technical, open standards developed through internationally cooperative processes, to maximise participation and benefit-sharing. This is because international standards are guided by a consensus-driven approach; responsive to market need and global experts from industry, government, academia

---

[45] Intellectual Property Office (n 43).

[46] International Organization for Standardization (ISO), *Benefits of ISO Standards* <https://www.iso.org/benefits-of-standards.html>.

[47] Soumya Prakash Patra and KD Raju, 'Standardization and Standard Essential Patents for Public Good: Application in Automotive Industry' (2021) 27 *SASI* 53, 54.

[48] Institute of Electrical and Electronics Engineers (IEEE), *IEEE Standard Computer Dictionary: A Compilation of IEEE Standard Computer Glossaries* (IEEE 1991) IEEE Std 610, 114; see, for example, Federal Communications Commission, Interoperability (Web Page) <https://www.fcc.gov/general/interoperability>.

[49] International Organization for Standardization (ISO), *National, Regional and International Standards and How They Relate to Regulatory Regimes* (Web Page) <https://www.iso.org/sites/ConsumersStandards/1_standards.html>.

[50] Paul Zubrinich and others (n 44) 3.

[51] Tyrone Berger, 'iPhone Switching to USB-C is a Win for Consumers and the Environment – But to What Extent?' (*The Conversation*, 15 September 2023) <https://theconversation.com/iphone-switching-to-usb-c-is-a-win-for-consumers-and-the-environment-but-to-what-extent-213549>.

and civil society; and ostensibly seek to address the concerns of nations in the Global South.[52] Adoption of international standards by nations is voluntary. Once adopted at a national level, standards may remain voluntary or become mandatory for affected industries.

The benefits of adopting international standards are numerous, including facilitating integration into the global marketplace, reducing trade barriers, fostering economic growth, increasing competition and consumer choice, improving safety outcomes, encouraging domestic innovation through the adoption of international best practices, reducing the costs of innovation by enabling access to existing technologies rather than high-cost bespoke options and providing a 'level playing field' for multiple actors.[53]

The development of technical standards can intersect with the intellectual property (IP) system. Patents may be deemed by SSOs/SDOs as standard-essential patents (SEP). As Brinsmead clarifies, IP is 'considered 'essential' to the practice of a standard when it is not technically possible to implement the standard without infringing the intellectual property in question'.[54] SEP holders generally commit to licensing under Fair, Reasonable and Non-Discriminatory (FRAND) terms, in an attempt to balance accessibility and compensation.[55]

3.2 History of Standards and Interoperability in Fostering International Cooperation

Since the first 'space race', international cooperation has persisted 'against all odds'.[56] Standardisation and interoperability have formed part of that space heritage. During the Cold War, interoperability was required to achieve the historic and symbolic Apollo–Soyuz Test Project—the first international human spaceflight partnership—which saw the in-orbit rendezvous and docking of spacecraft from the US and USSR.[57] This joint mission, designed to test compatibility for potential rescue operations in space, was the culmination of over three years of collaboration between scientists and technicians from both nations to enable spacecraft interoperability.[58]

[52] ISO, *Developing Standards* <https://www.iso.org/developing-standards.html>; ISO, *ISO Action Plan for Developing Countries: 2021–2025* (2020) <https://www.iso.org/files/live/sites/isoorg/files/store/en/PUB100374.pdf>.

[53] Intellectual Property Office (n 43); Prakash Patra and Raju (n 47) 54.

[54] Simon Brinsmead, *Essential Interoperability Standards: Interfacing Intellectual Property and Competition in International Economic Law* (Cambridge University Press 2021) 81.

[55] World Intellectual Property Organization (WIPO), *Standard Essential Patents* <https://www.wipo.int/en/web/patents/topics/sep>.

[56] Mai'a K Davis Cross, *International Cooperation Against All Odds: The Ultrasocial World* (Oxford University Press, 2024) 140.

[57] ibid. 161.

[58] ibid.; Edward Clinton Ezell and Linda Neuman Ezell, *The Partnership: A History of the Apollo–Soyuz Test Project* (National Aeronautics and Space Administration 1978); 'Telephone Conversation

Stewart and Dittmer refer to this as 'more-than-human diplomacy', where focusing on technical cooperation and the compatibility of materials and systems can bring nations in conflict together and serve as a precursor to more ideologically charged political cooperation.[59] At the time, UN COPUOS heralded the Apollo–Soyuz preparations as inspiring future cooperation.[60] And, it did just that: unlocking further cooperative endeavours necessitating interoperability, of which the ISS stands as a contemporary successor.[61]

The ISS has remained an orbiting, tangible testament to sustained cooperation and coexistence in orbit since 2000, underpinned by an *Intergovernmental Agreement* (IGA) among the partner space agencies of the US, Russia, Europe, Japan and Canada.[62] While generally rising above tensions below, this commitment was tested at the start of the Ukraine war, with Russia threatening to withdraw.[63] This cooperative endeavour—enabling over 3000 life-enhancing experiments to be conducted in microgravity[64]—would not have been possible without ongoing technical cooperation, especially between the US and Russia.[65] Its modular design and the sharing of systems among its partners and third parties from across government and industry has required unprecedented levels of standardisation and interoperability.[66]

Instances where partner technologies were incompatible—and thus not interoperable—led to the development of new standards by the partner space agencies, such as the International Docking System Standard (IDSS). This standard 'establishes a standard docking interface to enable collaborative endeavours between the international

of President Gerald R Ford with Apollo–Soyuz Test Project Crews Following Rendezvous and Docking of the Spacecraft' (Gerald R Ford Presidential Library & Museum, 17 July 1975) <https://www.fordlibrarymuseum.gov/library/speeches/750412.htm>.

[59] William Stewart and Jason Dittmer, 'More-Than-Human Space Diplomacy: Assembling Internationalism in Orbit' (2023) 18(2–3) *The Hague Journal of Diplomacy* 219, 220–221, 233–234.

[60] UN Doc A/C.1/PV.2052 (14 October 1975) 27.

[61] Davis Cross (n 56) 163; Stewart and Dittmer (n 59) 223.

[62] Davis Cross (n 56) 162–163; *Agreement among the Government of Canada, Governments of Member States of the European Space Agency, the Government of Japan, the Government of the Russian Federation, and the Government of the United States of America concerning Cooperation on the Civil International Space Station*, opened for signature 29 January 1998, TIAS 12927 (entered into force 27 March 2001) (ISS IGA).

[63] Wendy Whitman Cobb, 'Russia's Withdrawal from The International Space Station Could Mean the Early Demise of the Orbital Lab—And Sever Another Russian Link with the West' (*The Conversation*, 27 July 2022) <https://theconversation.com/russias-withdrawal-from-the-international-space-station-could-mean-the-early-demise-of-the-orbital-lab-and-sever-another-russian-link-with-the-west-187754>.

[64] NASA, *Five Space Station Research Results Contributing to Deep Space Exploration* (21 January 2022) <https://www.nasa.gov/missions/station/five-space-station-research-results-contributing-to-deep-space-exploration/>.

[65] Stewart and Dittmer (n 59) 236–238.

[66] Antonino Salmeri, '#SpaceWatchGL Opinion: One Size to Fit Them All: Interoperability, the Artemis Accords and the Future of Space Exploration' (*SpaceWatch.Global*, 8 November 2020) <https://spacewatch.global/2020/11/spacewatchgl-opinion-one-size-to-fit-them-all-interoperability-the-artemis-accords-and-the-future-of-space-exploration/>.

space fairing [sic] community while also supporting possible crew rescue operations'.[67] Building upon this, the ISS partners released, after consultation, the International Deep Space Interoperability Standards (IDSIS) in 2019.[68] These standards have broad applicability beyond the ISS, covering the 'early phase of exploration planning' for avionics, communications, docking, environmental control and life support systems, power and rendezvous.[69] The aim is to enable stakeholders from industry, government, academia and international organisations to 'independently develop compatible systems for deep space exploration that would be compatible aboard any spacecraft, irrelevant of the spacecraft developer'.[70]

This recognition of the need for agreed standards and interoperable technologies remains a cornerstone of international space diplomacy and cooperation, even within a geopolitically charged environment. The *Global Exploration Roadmap*, released in August 2024 by the International Space Exploration Coordination Group (ISECG), which comprises major and emerging space-faring nations, embodies this sentiment. It sets out a 'coordinated vision' for long-term space exploration and seeks to 'encourage and embrace' global participation.[71] Priority areas identified for future international cooperation include interoperable medical devices, complementary positioning, navigation and timing (PNT) standards and power standards.[72]

3.3 Applying this to the Mining of Lunar Resources

Interoperability will be 'inherent' in the 'systems of systems' required for space mining.[73] Challenges, particularly for ISRU, include the need for agreed standards to integrate systems for resource extraction, processing, utilisation and more.[74] Such 'seamless interoperability' must occur 'across a broad spectrum of technological platforms, each with its unique specifications and operational parameters'.[75] DeMinico et al. identify multiple benefits:

---

[67] International Space Station Partner Agencies, 'International Docking System Standard' (April 2022) <https://www.internationaldockingstandard.com/>.

[68] International Deep Space Standards Partners, *Frequently Asked Questions: International Space Station Interoperability Standards* <https://internationaldeepspacestandards.com/faq/>.

[69] ibid.

[70] ibid.

[71] International Space Exploration Coordination Group (ISECG), *Global Exploration Roadmap* (August 2024) 1 <https://www.globalspaceexploration.org/wp-content/isecg/GER2024.pdf>.

[72] ibid. 5, 54–55.

[73] Charles Hibbitts and Athonu Chatterjee, 'The In-Situ Production and Use of Metals on the Moon' (ASCEND 2022 Conference, Las Vegas and online, 24–26 October 2022) AIAA 2022–4376 <https://doi.org/10.2514/6.2022-4376>.

[74] United Nations Office for Outer Space Affairs, *Luxembourg—Input to the Working Group on Legal Aspects of Space Resource Activities* (A/AC.105/C.2/2024/CRP.29, 17 April 2024).

[75] ibid.

> ISRU in particular presents a well-bounded framework for ensuring interoperability and accessibility for a wide range of partners, the best value for customers, and the greatest opportunity for rapid incremental technology development. The lunar ISRU campaign directly leverages power, materials handling and processing, excavation, and multiple types of robotics.[76]

Now is the opportune time to identify 'appropriate areas for modularity and standards' for ISRU, to enable testing in space analog situations on Earth before returning to the lunar surface.[77] Standard-setting by consensus is time-intensive and needs to commence sooner, not later.

Beyond technical necessity, reaching agreement on standards and interoperable technologies for space mining activities can help realise the benefits of space for all nations. A risk exists that whoever reaches the lunar surface first could consolidate dominance, by occupying prime sites on the lunar south pole that are advantageous for ISRU. For example, craters in the PSRs may contain limited deposits of water ice that are necessary for human survival and sustenance.[78]

If monopolised—whether through mining operations in the PSRs or solar generators on the Peaks of Eternal Light—Batcha et al. argue 'correctives' are needed, as this could be counter to the benefit-sharing principle.[79] Such 'correctives' could include the sharing of infrastructure among nations to harness these resources, necessitating such lunar bases be designed for interoperable technologies.[80] Pino, an Open Lunar Foundation fellow, notes that a 'shared and standardized lunar power infrastructure' could reduce the cost to develop and transport costly custom power systems, and standards could 'scale and secure sustainable resources utilization'.[81]

Scholars associated with the Space Generation Advisory Council (SGAC)—a network for young people and professionals—have identified the tripartite relationship among standardisation, interoperability and benefit-sharing. Thus, this is a matter of intergenerational equity, of safeguarding the benefits of space for present

---

76 Mathew DeMinico, James P Mastandrea and Wesley T Fuhrman, *Modular Open Systems Approach (MOSA) for a Robust Commercial Lunar Ecosystem* (Nuclear and Emerging Technologies for Space (NETS-2022), Cleveland, 8–12 May 2022) 3 <https://ntrs.nasa.gov/api/citations/20220004762/downloads/MOSA%20Paper%20NETS.pdf>.

77 ibid.

78 Amelia Batcha, Chinmayee Govinda Raj, Shayna Hume, Ashley Kowalski, Atila Meszaros, Annaliese Meyer, Paolo Pino, Matej Poliaček, Antonino Salmeri and Jahnavi Shah, *All for One and One for All: Recommendations for Sustainable International Lunar Base Utilization and Exploration Approaches* (71st International Astronautical Congress—The CyberSpace Edition, 2020) 6–7 <https://orbilu.uni.lu/bitstream/10993/44645/1/M.%20Poliacek%2c%20A.%20Salmeri%20%26%20al%2c%20All%20for%20one%20and%20one%20for%20all%20-%20Recommendations%20for%20Sustainable%20International%20Lunar%20Base%20Utilisation%20and%20Exploration%20Approaches.pdf>.

79 ibid.

80 ibid.

81 Paulo Pino, *An Open Dashboard for Lunar Power Candidate Standards* (Open Lunar Foundation, 2024) 3, 6 <https://static1.squarespace.com/static/659ddc41121da9469c35e2b1/t/67073034b35ce5223be326ca/1728524341483/1724784651-lunar-power-standards-whitepaper-compressed_1.pdf>.

and future generations. Writing on reclaiming lunar resources, Samanga issues 'a clarion call for international cooperation is made, prompting states to facilitate greater interoperability'.[82] Similarly, Salmeri opines: 'interoperability in itself is in the interest and for the benefit of all Countries, given that in principle every actor could benefit from its implementation', since it 'fosters contributions from international players at all levels of financial and technological development'.[83]

Standardisation supports the development of interoperable commercial off-the-shelf (COTS) technologies. Kearney predicts: 'the standards of the future are deeply linked to off-the-shelf commercial capabilities as well as interoperability capabilities'.[84] This highlights the need for global public and private participation in standard-setting processes. Maximising the use of COTS technologies in space is likewise beneficial, as it can 'reduce the cost of manufacture, use and replacement', according to de Zwart.[85]

Such downward pressures on entry costs could help narrow the space technological gap. For emerging space-faring nations and their industries, the ability to acquire COTS technologies, rather than having to develop costly, time-intensive bespoke solutions, could help lower technological and financial barriers to entry. For instance, several South American nations have utilised COTS technologies for Earth observation satellites.[86]

Moves towards standardisation and interoperability can benefit all nations by addressing first-order priorities, such as preserving human life. Mining the Moon must not prioritise profits over people. Historically, the need for agreed standards and interoperable space technologies has been 'predominately' driven by humanitarian needs—notably the rescue of personnel, as trialled by the Apollo–Soyuz Test Project.[87] In the harsh and inhospitable environment of space, Li argues: 'interoperability can have a powerful effect in ensuring successful cooperation and mutual assistance in outer space'.[88] Life support systems, docking mechanisms and in-space refuelling infrastructure are identified as priority areas for standardisation to preserve

---

[82] Ruvimbo Samanga, 'Reclaiming Lunar Resources: Paving the Way for an International Property Rights Regime for Outer Space' in James SJ Schwartz, Linda Billings and Erika Nesvold (eds), *Reclaiming Space: Progressive and Multicultural Visions of Space Exploration* (Oxford University Press, 2023) 251, 254–255.

[83] Salmeri (n 66).

[84] Mike Kearney, 'The Broad Landscape of Civil Space Standards' (2017) *Defense Standardization Program Journal* (January/March) 15 <https://www.dsp.dla.mil/Portals/26/Documents/Publications/Journal/170101.pdf?ver=2017-03-30-124832-847>.

[85] Melissa de Zwart, 'To the Moon and Beyond: The Artemis Accords and the Evolution of Space Law' in Melissa de Zwart and Stacey Henderson (eds), *Commercial and Military Uses of Outer Space* (Springer, 2021) 65, 72.

[86] João Dallamuta, Leonel Fernando Perondi and Mônica Elizabeth Rocha de Oliveira, 'Space Missions in South America: Profile and Evolutionary Perspective of their Development' (2023) 206 *Acta Astronautica* 9.

[87] Rossana Deplano, 'The Artemis Accords: Evolution or Revolution in International Space Law?' (2021) 70(3) *International and Comparative Law Quarterly* 799, 803.

[88] Alex S Li, 'Opening Outer Space: Safety and Stability through Open Standards and Open Source' (2022) 126(3) *Penn State Law Review* 667, 680, 701.

safety and stability off-Earth.[89] Such systems 'should be universally interoperable with each other'.[90]

Reducing the potential for tensions to erupt on the lunar surface is another first-order priority. Space mining is predicted to give rise to future conflicts on the lunar surface over prime sites and resources, especially in the context of the 'space race' between the US and China. The Hague International Space Resources Governance Working Group (HISRG Working Group), in its 2019 *Building Blocks for the Development of an International Framework on Space Resource Activities* (*Hague Building Blocks*), identifies the need for technical standards for safety zones around space resource activities.[91] The *Hague Building Blocks* were developed to inform future space law and governance within international fora and by nations. The purpose of safety zones, an emerging concept for space law and governance that gained traction in the *Artemis Accords*, is to prevent harmful interference by notifying and coordinating activities in close proximity on the lunar surface by nations, such as space mining, thereby reaffirming the principle of due regard in the OST.[92]

Reflecting upon this, Salmeri compellingly claims that the development of shared standards and common infrastructure would in fact inherently pay due regard to the corresponding interests of all States' and 'significantly minimize the risk of harmful interference by addressing the issue at the root, within the design of space systems.[93] As a result, this could help supplement the shortcomings of relying solely on international space law and governance to maintain safety on the lunar surface, as adherence to rules and norms can break down, whether due to geopolitical pressures and/or shifting national policies and politics.[94]

4 Growing Commitment to Standards and Interoperability

While current plans for lunar exploration are often associated with differing geopolitical alignments, commonalities can be observed in standards and interoperability. This next part examines how these commitments are expressed in the *Artemis Accords*, the ILRS and the UN COPUOS Working Group on Legal Aspects of Space Resources.

---

[89] ibid.

[90] Space Generation Advisory Council, *Effective and Adaptive Governance for a Lunar Ecosystem* (10 May 2021) 36 <https://www.unoosa.org/documents/pdf/copuos/lsc/space-resources/LSC2023/PermObsResponses/SGAC_-_EAGLE_Report_merged.pdf>.

[91] The Hague International Space Resources Governance Working Group, *Building Blocks for the Development of an International Framework on Space Resource Activities* (Universiteit Leiden, November 2019) <https://www.universiteitleiden.nl/en/law/institute-of-public-law/institute-of-air-space-law/the-hague-space-resources-governance-working-group>.

[92] Deplano (n 87); Alexander Q Gilbert, 'Implementing Safety Zones for Lunar Activities under the Artemis Accords' (2023) 10(1) *Journal of Space Safety Engineering* 103.

[93] Salmeri (n 66).

[94] Li (n 88) 680–681.

4.1 NASA Artemis Accords

'We're going to the Moon to stay' became a commonly invoked refrain after NASA launched the Artemis programme in 2020.[95] Unlike the time-limited Apollo missions, Artemis envisions a sustained human presence, to be realised through a phased technical roadmap of missions, with NASA now targeting early 2028 for a crewed landing on the lunar surface.[96] As with previous NASA endeavours like the ISS, the Artemis programme is behind schedule and subject to financial pressures.[97] The proposed first lunar space station, the Gateway, was halted as part of budget considerations.[98]

To realise this ambition, NASA has sought extensive international cooperation. In October 2020, NASA launched the *Artemis Accords*, with eight founding signatory nations and inviting others to join.[99] As of March 2026, the membership has expanded to sixty-one signatories (including the US), representing established and emerging space-faring nations from across the Global North and Global South with varying space capabilities.[100] Of significance, China and Russia are not signatories. The *Wolf Amendment*, which restricts federally funded US agencies such as NASA from directly cooperating with China on space activities without congressional approval, has acted as a de facto prohibition on cooperation between Beijing and Washington on space since 2011—and would prevent China from joining even if it wished to do so.[101]

The *Artemis Accords* are non-binding (soft law), meaning they do not constitute a binding treaty (hard law). Section 1 frames them as a 'political commitment' to 'establish a common vision via a practical set of principles, guidelines, and best practices to enhance the governance of the civil exploration and use of outer space',[102]

---

95 See, for example, NASA, 'We Are Going' (14 May 2019) <https://science.nasa.gov/resource/we-are-going/>; Government Matters, 'NASA Deputy Administrator: "We're going back to the moon to stay this time"'(YouTube, 20 September 2021) <https://www.youtube.com/watch?v=SL6IY2dUHX0>.

96 NASA, 'Artemis' (Web Page, 2025) <https://www.nasa.gov/humans-in-space/artemis/#missions>.

97 US Government Accountability Office, *Artemis Programs: NASA Should Document and Communicate Plans to Address Gateway's Mass Risk* (GAO-24-106,878, US Government Accountability Office, July 2024) <https://www.gao.gov/assets/gao-24-106878.pdf>; Jeff Foust, 'NASA Halts Work on Gateway to Develop Lunar Base' (*Space News*, 24 March 2026) <https://spacenews.com/nasa-halts-work-on-gateway-to-develop-a-lunar-base/>.

98 Jeff Foust, 'NASA Budget Proposal Draws Strong Criticism' (*Space News*, 15 May 2025) <https://spacenews.com/nasa-budget-proposal-draws-strong-criticism/>.

99 NASA (n 3).

100 NASA, *Signatories by Date* <https://www.nasa.gov/wp-content/uploads/2024/10/signatories-02.pdf>.

101 Cotterell (n 18); Hannah Kohler, 'The Eagle and the Hare: US–Chinese Relations, the Wolf Amendment, and the Future of International Cooperation in Space' (2014) 103 *Georgetown Law Journal* 1135.

102 *The Artemis Accords: Principles for Cooperation in the Civil Exploration and Use of the Moon, Mars, Comets, and Asteroids for Peaceful Purposes* (signed by multiple countries, 13

to be supplemented by implementing instruments like Memoranda of Understanding (MOUs) for cooperative activities.[103]

It affirms compliance with the OST and many of the principles 'provide for operational implementation of important obligations contained in the [OST] and other instruments'.[104] The *Moon Agreement* is conspicuously omitted—unsurprising given long-standing US opposition.

The principles cover: peaceful purposes, transparency, emergency assistance, registration of space objects, release of scientific data, preserving outer space heritage, space resources, deconfliction of activities, orbital debris—and relevantly, interoperability. Section 5 stresses:

> The Signatories recognize that the development of interoperable and common exploration infrastructure and standards, including but not limited to fuel storage and delivery systems, landing structures, communications systems, and power systems, will enhance space-based exploration, scientific discovery, and commercial utilization. The Signatories commit to use reasonable efforts to utilize current interoperability standards for space-based infrastructure, to establish such standards when current standards do not exist or are inadequate, and to follow such standards.[105]

Given Artemis' reliance on contributions from national partners and industry for key activities, standardisation and interoperability are 'both necessary and common sense'.[106] While section 5 does not explicitly namecheck ISRU, such standards and interoperable infrastructure and systems will be indispensable to its success. ISRU, however, is foreshadowed with a permissive provision that such activities do not inherently constitute national appropriation.[107]

This also links to section 4 on transparency,[108] particularly since NASA has previously encouraged signatories to 'utilize open international standards'.[109] Open standards are widely regarded as 'best practice' for a robust space ecosystem,[110] fostering 'safety and stability'.[111]

October 2020) <https://www.nasa.gov/wp-content/uploads/2022/11/Artemis-Accords-signed-13Oct2020.pdf> (*Artemis Accords*), s 1.

[103] ibid. s 2.

[104] ibid.

[105] *Artemis Accords*, s 5.

[106] Christopher Johnson, 'The Space Law Context of the Artemis Accords: Part 2' *SpaceWatch.Global* (28 May 2020) <https://spacewatch.global/2020/05/spacewatchgl-feature-the-space-law-context-of-the-artemis-accords-part-2/>.

[107] ibid. 10.

[108] Memme Onwudiwe and Kwame Newton, 'Africa and the Artemis Accords: A Review of Space Regulations and Strategy for African Capacity Building in the New Space Economy' (2021) 9(1) *New Space* 38, 44.

[109] Mike Gold, 'Artemis Accords—Enabling International Partnerships for Lunar Exploration' (FPC Briefing, US Department of State, 15 May 2020) <https://2017-2021.state.gov/briefings-foreign-press-centers/artemis-accords-enabling-international-partnerships-for-lunar-exploration/>.

[110] Fred Slane, 'AIAA and Space Standards' (January/March 2017) *Defense Standardization Program Journal* 20, 21.

[111] Li (n 88) 671.

Interoperability itself is described as a 'new concept' in space law and governance, relevant to the 'safety, efficiency and cost' of exploration.[112] It 'reflects established practice in the field of international cooperation in outer space' since the Cold War.[113] This was critical to the ISS' operation, albeit with an 'inherent limit to the degree of interoperability that [could] be achieved'.[114] NASA's decision to embed this principle from the outset reflects lessons learnt.

Economic imperatives further drive this move. As Kearney explains, they have 'increasingly forced international civil space agencies into cooperative spaceflight ventures'.[115] This has spurred the growth of standardisation and interoperability in such multi-stakeholder partnerships. It reflects the 'commercial realities' of the Artemis programme, where greater reliance on COTS technologies is expected to help reduce costs.[116]

However, translating a high-level principle into workable practice is complex. In May 2024, NASA convened a workshop of signatories to discuss implementing principles like interoperability, building on agreed parameters for launch dates, landing sites and activities.[117] NASA had previously planned for the now-paused Gateway to employ the IDSIS.[118] Yet, as NASA Kennedy Space Center engineering personnel, De Jesus Gomez Jr. et al., have observed:

> Although the IDSIS plays a key role in deep space, there have not been any defined standards for surface systems that will require a similar need for the Artemis campaign.[119]

Surface systems—for landers, habitats, transport and ISRU—will also need to communicate, share data and distribute power to sustain operations.[120] The same engineers identify benefits:

> Standardization… commonality, reusability, and interoperability, enables technology infusion, reduces operations and logistics costs, and promotes a multi-use philosophy. Standardization of hardware and software interfaces, components, processes, procedures, tools, and repair kits creates an effective, cost-efficient, and sustainable architecture throughout its lifecycle. It also provides common frameworks for infusing new capabilities as they mature and implementing automation and autonomy across the architecture. Standardization results in an

---

[112] de Zwart (n 85) 72.

[113] Deplano (n 87) 803.

[114] Salmeri (n 66).

[115] Kearney (n 84) 9.

[116] de Zwart (n 85) 72.

[117] Jennifer M Dooren, 'NASA Engages in Artemis Accords Workshop to Advance Exploration' (24 May 2024) <https://www.nasa.gov/news-release/nasa-engages-in-artemis-accords-workshop-to-advance-exploration/>.

[118] NASA, *Gateway: The Cislunar Springboard for International and Sustainable Human Deep Space Exploration* 4 <https://www.lpi.usra.edu/lunar/artemis/resources/WhitePaper_2023_GatewayCislunarSpringboard.pdf>.

[119] Jaime De Jesus Gomez Jr., Don Pittman, Gabor Tamasy, Chad Caron, Michael Dupuis and Mark Lewis, 'Surface Systems and Interface Standardization' (51st International Conference on Environmental Systems, Minnesota, 2022) 1, 3.

[120] ibid. 2.

> open surface architecture that is affordable, robust, and flexible while promoting partnerships and cost-sharing.[121]

LunaNet speaks to such potential. Developed by NASA and ESA, LunaNet will be a 'network of cooperating networks (network of networks, akin to the terrestrial Internet) upon which providers can deliver communications, [PNT], and other services for users in transit to, around, and on the Moon'.[122] With Artemis targeting sites at the lunar south pole—often in shadow or topographically complex terrain without a continuous direct line-of-sight to Earth—this necessitates the deployment of a communication relay system.[123] LunaNet will provide information service alerts, including on space weather and solar storms, and incorporate a search and rescue capability for astronauts. The LunaNet Interoperability Specification, released in January 2025, is a 'starting point' for 'mutually agreed-upon standards' to support an 'open, evolving, cooperative lunar communications and navigation architecture'.[124]

In short, the *Artemis Accords* mark a 'new path forward for space lawmaking', signalling an ongoing trend from treaty-making towards soft law instruments, supplemented by national legislation and contract law.[125] Deplano observes, they represent both an 'evolution' and a 'revolution' in international space law: affirming compliance with existing principles, interpreting them (as with national appropriation) and introducing new ones, like interoperability.[126]

Consequently, criticism has followed. Initially, 'observers' from China referred to the *Artemis Accords* as a 'preposterous attempt' to unilaterally establish norms for lunar resource utilisation.[127] Officials from Russia similarly criticised their perceived

---

[121] ibid. 2.

[122] NASA, *LunaNet Interoperability Specification Version 5* (29 January 2025) 2 <https://www.nasa.gov/wp-content/uploads/2025/02/lunanet-interoperability-specification-v5-baseline.pdf>.

[123] Richard Reinhart, 'Lunar Communications and Navigation: Lunar Communications, Position, Navigation and Timing (PNT) Architecture' (*Glenn Space Technology Symposium*, 17 July 2024) 9 <https://ntrs.nasa.gov/api/citations/20240008336/downloads/GRC%20Symposium-Lunar%20Infrastructure-Reinhart-fin2.pdf>.

[124] Katherine Schauer and Danny Baird, 'LunaNet: Empowering Artemis with Communications and Navigation Interoperability' (7 October 2021) <https://www.nasa.gov/humans-in-space/lunanet-empowering-artemis-with-communications-and-navigation-interoperability/>.

[125] Brian R Israel, 'Space Resources in the Evolutionary Course of Law Making' (2019) 113(1) *American Journal of International Law Unbound* 114; Rachel Neef, 'Artemis Accords: A New Path Forward for Space Lawmaking?' (2021) *42 Adelaide Law Review* 569.

[126] Deplano (n 87); de Zwart (n 85).

[127] Elliot Ji, Michael B Cerny and Raphael J Piliero, 'What Does China Think About NASA's Artemis Accords?' (*The Diplomat*, 8 September 2020) <https://thediplomat.com/2020/09/what-does-china-think-about-nasas-artemis-accords>.

US centrism, stating that '[a]ttempts to expropriate outer space… hardly set the countries (on course for) fruitful cooperation'.[128] Scholarly commentary has expressed concern about bypassing UN COPUOS and undermining multilateralism.[129]

Notably, China has since tempered criticism made by 'observers' in its first formal comments on commercial activities on the Moon.[130] In a 2024 statement to the UN COPUOS Working Group on Legal Aspects of Space Resource Activities, it asserted that scientific investigation and research involving space resources should be 'eligible for benefit sharing'[131] and emphasised the importance of mitigating any potential negative impact that commercial space resource activities might have on scientific endeavours.[132]

4.2 International Lunar Research Station (ILRS)

Against this geopolitical backdrop, another space bloc has emerged: the ILRS. A year after the livestreamed signing of the *Artemis Accords* by the first signatories, China and Russia released a joint statement in April 2021 announcing cooperation to construct a lunar research station, likewise inviting other nations to join.[133] As of March 2026, thirteen nations are participating (inclusive of the two leads), mostly from the Global South.[134] Notably, Thailand and Senegal hold membership with the *Artemis Accords* too—a development examined in the final part.

Similar to Artemis, China has outlined multiple missions through its Chang'e programme: the uncrewed Chang'e-7 (scheduled for late 2026) will survey the lunar south pole for the existence and distribution of water ice in the PSRs and 'test cutting-edge technologies critical for sustainable human activities on the moon'.[135] In 2029,

---

[128] Reuters Staff, 'Russian Space Agency Says Trump Paving Way to Seize Other Planets' (*Reuters*, 7 April 2020) <https://www.reuters.com/article/world/russian-space-agency-says-trump-paving-way-to-seize-other-planets-idUSKBN21P1T7>; Jeff Foust, 'Russia Sceptical About Participating in Lunar Gateway' (*SpaceNews*, 14 October 2020) <https://spacenews.com/russia-skeptical-about-participating-in-lunar-gateway/>.

[129] See, for example, Johnson (n 30); Athar ud Din, 'The Artemis Accords: The End of Multilateralism in the Management of Outer Space?' (2022) 20(2–3) *Astropolitics* 135, 144.

[130] Andrew Jones, 'China Outlines Position on Use of Space Resources' (*Space News*, 6 March 2024) <https://spacenews.com/china-outlines-position-on-use-of-space-resources>.

[131] China, *Submission to the Working Group on Legal Aspects of Space Resource Activities* (Statement, 2024) 2–3 <https://www.unoosa.org/documents/pdf/copuos/lsc/space-resources/LSC2024/English_Chinas_submission_to_the_working_group_on_space_resources.pdf>.

[132] ibid. 3.

[133] CNSA and Roscosmos (n 3).

[134] This number is correct as of March 2026. Victoria Samson and Ian Christensen, 'Lunar Space Cooperation Initiatives' (Secure World Foundation, 27 January 2026). <https://www.swfound.org/publications-and-reports/lunar-space-cooperation-initiatives>.

[135] 'China Unveils International Lunar Research Station Details' (*State Council Information Office of the PRC*, 26 April 2024) <http://english.scio.gov.cn/m/chinavoices/2024-04/26/content_117152685.htm>.

the Chang'e-8 mission is scheduled to test and validate ISRU technologies, laying the foundation for the ILRS.[136]

The lunar research station itself is scheduled for establishment by 2035. Its first phase will see the construction of a 'basic model' of the ILRS on the lunar south pole, enabling scientific operations within a 100-kilometre radius and featuring 'an integrated information network between the Earth and the [M]oon to enable interconnectivity and interoperability among various missions, including unmanned [sic] lunar exploration, manned [sic] lunar landings, and international cooperation'.[137] Given ISRU's role in sustaining the ILRS, standards and interoperable technologies are expected to remain a priority.

In contrast to the relative transparency of the US-led *Artemis Accords*, the ILRS lacks 'a clear arrangement for a cooperative pattern and a legal framework'.[138] While the joint announcement declared that the ILRS would be '[g]uided by legislation of their respective countries, generally recognized rules and principles of international law, international treaties to which their States are participants and based on the principles of equality, openness and integrity',[139] there remains no comparable governing document. The *Declaration on the International Lunar Research Station*—announced in September 2021 as a joint effort between the lead partners—to set out 'General Principles, Ways of Participation and Guidelines for Activities' and to support 'follow-up cooperation' among the international community has yet to be published, despite intentions to do so in October 2021.[140] It is unknown why. This suggests that negotiations over a foundational document may have been, at some point, underway.

Similarly, no updates have been provided on plans to establish an ILRS Organization, which were first announced in May 2023.[141] Nonetheless, workshops among ILRS partners continue to be convened to 'discuss collaborative frameworks and technological roadmaps for the ILRS... to establish a sustainable robotic and human-involved lunar base by 2030'.[142] This ongoing activity indicates that, despite

---

[136] 'China to Launch Moon Base Mission as Early as 2026, Expected to be Equipped with Wireless Network' (*Global Times*, 24 September 2024) <https://www.globaltimes.cn/page/202409/1320295.shtml>.

[137] ibid.

[138] Xiaodan Wu, 'The International Lunar Research Station: China's New Era of Space Cooperation and Its New Role in the Space Legal Order' (2023) 65 *Space Policy* 101537, 2.

[139] CNSA and Roscosmos (n 3).

[140] CNSA, 'Russian–Chinese Joint Seminar on Cooperation in International Lunar Research Stations' (Web Page, 28 September 2021) <https://www.cnsa.gov.cn/english/n6465668/n6465670/c6812568/content.html>.

[141] Deep Space Exploration Laboratory of CNSA, *International Lunar Research Station* (Presentation, May 2023) 5 <https://www.unoosa.org/documents/pdf/copuos/2023/TPs/ILRS_presentation20230529_.pdf>; Samson and Christensen (n 134).

[142] 'APSCO Delegation Attended the Meeting of International Lunar Research Station Developers' (*Asia-Pacific Space Cooperation Organization*, 6 May 2025) <https://www.apsco.int/html/comp1/content/NewsHighlights/2025-05-06/54-732-1.shtml>; 'APSCO Co-Organized the Third Workshop on International Lunar Research Station ILRS' (*Asia-Pacific Space Cooperation* Organization, 19 September 2024) <http://www.apsco.int/html/comp1/content/NewsHighlights/2024-09-19/54-686-1.shtml>; Zhongmin Wang and Ruihong Yang, *International Lunar Research Station*

the absence of publicly available governance documents, international cooperation among the partners remains active.

Of note, what has been released is the *ILRS Guide for Partnership (V1.0)* (the Guide), published shortly after the initial ILRS announcement in June 2021. It is in this document that the emphasis on standards and interoperability becomes apparent.[143] Jointly developed by the space agencies of China and Russia, it acknowledges the necessity of a 'broad international partnership with an attraction of other countries, international organizations and international partners'.[144] It offers tiered levels of cooperation—spanning space mission cooperation, space system cooperation, and subsystem, ground and application cooperation—explicitly allowing nations to participate in accordance with their capabilities and resources.[145] The Guide states:

> All Partners are encouraged to join ILRS project based on their own situation. Any Partner willing to contribute to the ILRS, through a jointly coordinated negotiation with China and Russia, can participate including co-lead status in any part of the project. The objective, plan, interface, standards, interoperability and scientific application of the participating project(s) or missions shall be aligned with general architecture and functions of ILRS.[146]

This reflects an expectation that contributions align with ILRS-wide standards and interoperability of the broader ILRS systems. The Guide further commits to: 'Review[ing] existing standards in the field of launch vehicle and spacecraft technology development and potential definition of future standards that may be used during the creation of ILRS', with a dedicated engineering standards subcommittee to be established within the ILRS organisational structure.[147] This signals an openness to using current standards, while reserving the option to develop future standards. In this respect, it is not dissimilar to the approach of the *Artemis Accords* to leverage existing and develop new standards. Yet, the scope of ILRS standardisation—primarily on launch vehicles and spacecraft—is narrower than the *Artemis Accords*, which addresses surface infrastructure such as landing pads, power and communication systems.

(Deep Space Exploration Laboratory of the China National Space Administration, May 2023) <https://www.unoosa.org/documents/pdf/copuos/2023/TPs/ILRS_presentation20230529_.pdf>.

[143] CNSA and Roscosmos, *International Lunar Research Station Guide for Partnership (V1.0)* (Report, June 2021) <https://www.cnsa.gov.cn/english/n6465652/n6465653/c6812150/content.html>.

[144] ibid. 2.

[145] ibid. 12.

[146] ibid.

[147] ibid. 12, 14.

4.3 United Nations Committee on the Peaceful Uses of Outer Space: Working Group on Legal Aspects of Space Resources

The need for standards and interoperability has been identified by nations and civil society observers within multilateral fora. In 2021, UN COPUOS established the Working Group on Legal Aspects of Space Resources under its Legal Subcommittee, with a 5-year workplan and mandate to: 'Assess the benefits of further development of a framework for such activities, including by way of additional international governance instruments'.[148] Its primary focus has been negotiating an initial draft set of recommended principles for space resource activities,[149] continuing the governance trend from hard to soft law.

The initial draft, released in May 2025, significantly includes principle 7:

> (B) States shall in a timely manner share relevant scientific and technical data that they have identified as well as their experiences gained in conducting space resource activities [, to the greatest extent feasible and practicable], so as to contribute to multilateral efforts to:
>
> (a) develop interoperable and common space-based infrastructure and standards for space resource activities, as appropriate; and
>
> (b) ensure that space resource activities are conducted in a safe, sustainable, rational and peaceful manner for the benefit and in the interests of all States irrespective of their degree of economic or scientific development.[150]

It also flags a possible provision on '[procedures for] development of interoperable and common infrastructure and standards for space resource activities, as appropriate', noting there is not yet 'in general terms, a common understanding in the Working Group'.[151] Such procedures could tangibly progress benefit-sharing in space mining.

This was unsurprising, as interoperability had been repeatedly identified as important: the US proposed it as a principle drawing on the *Artemis Accords*,[152] while Australia did so both individually and in a joint submission with Austria and the Netherlands, identifying it as a common theme across the *Artemis Accords* and *Hague Building Blocks*.[153]

---

[148] United Nations, *Report of the Committee on the Peaceful Uses of Outer Space, Sixty-Fourth Session* UN Doc A/76/20 (2021) 53.

[149] United Nations Committee on the Peaceful Uses of Outer Space, *Initial draft set of recommended principles for space resource activities: Note by the Vice-Chair of the Working Group on Legal Aspects of Space Resource Activities* UN Doc A/AC.105/C.2/L.339 (27 March 2025)

[150] ibid. principle 7.

[151] ibid. annex, part 2, principle 10(c)(c).

[152] United Nations Committee on the Peaceful Uses of Outer Space, *Draft report: Addendum VII. General exchange of views on potential legal models for activities in the exploration, exploitation and utilization of space resources* (19 April 2024) UN Doc A/AC.105/C.2/L.329/Add.7, para 18.

[153] United Nations Committee on the Peaceful Uses of Outer Space, *Relevant considerations for developing a set of initial recommended principles for the exploration, exploitation and utilisation of space resources* (20 March 2023) UN Doc A/AC.105/C.2/2023/CRP.6, 22.

At the first UN Conference on Sustainable Lunar Activities, organised by UNOOSA to promote 'constructive, insightful and inclusive exchanges', nations, industry and civil society stakeholders identified 'the existence of common understanding' on issues, including interoperability.[154] Brazil and Luxembourg had separately canvassed it during conference planning.[155] For example, Luxembourg shared a white paper from its 2022 Space Resources Week, which included a session on 'balanc[ing] the environmental, economic, operational, and technological impacts that ensure the availability of resources for future generations and the long-term success of space resources companies'.[156] This white paper emphasised:

> International cooperation is also key to creating standards to ensure the best practices preventing the space resources sector from repeating the mistakes that terrestrial mining companies did, including harmful environmental impacts.[157]

Public interest concerns animated submissions from civil society. NGOs with observer status at UN COPUOS cannot vote, but may submit documents, deliver statements and engage informally with nations in-the-margins. The 'informal codification' that such submissions 'perform can subsequently serve as an important start for the formation of 'soft law'' through UN COPUOS.[158] As a case in point, the HISRG Working Group was active in UN COPUOS until its work programme concluded in 2019, but its *Hague Building Blocks* remain influential.

Additionally, SGAC previously submitted its call for a Lunar Governance Charter as a conference room paper to UN COPUOS.[159] Its Effective and Adaptive Governance for a Lunar Ecosystem (EAGLE) team reviewed the *Artemis Accords*, *Hague Building Blocks* and Moon Village Association proposals to 'identify shared ground and contentious issues around lunar policy'. Multilateralism, benefit-sharing, interoperability and the role of space resources were identified as consistent themes.[160] SGAC recommended:

---

[154] United Nations, *Report of the Committee on the Peaceful Uses of Outer Space, Sixty-Seventh Session* UN Doc A/79/20 (2024) para 362.

[155] United Nations Committee on the Peaceful Uses of Outer Space, *Input to the Working Group on Legal Aspects of Space Resource Activities on Scope and Topics to be Addressed at the International Conference to be Held in 2024: Contributions from Algeria, Australia, Brazil, Cuba, Ecuador, Russian Federation, Slovak Republic, and Türkiye* UN Doc A/AC.105/2023/CRP.7 (31 May 2023) 3.

[156] United Nations Committee on the Peaceful Uses of Outer Space, *Input to the Working Group on Legal Aspects of Space Resource Activities on Scope and Topics to be Addressed at the International Conference to be Held in 2024: Contribution of Luxembourg* UN Doc A/AC.105/2023/CRP.18 (31 May 2023) annex 1, 8.

[157] ibid.

[158] Irina A Chernykh, Denis A Gugunskiy and Alexander M Solntsev, 'The Impact of Epistemic Communities on the Development of Future International Legal Regulation of Lunar Activities' (2024) 28(1) *RUDN Journal of Law* 215, 228.

[159] United Nations Committee on the Peaceful Uses of Outer Space, *Effective and Adaptive Governance for a Lunar Ecosystem: Lunar Governance Report: Paper submitted by the Space Generation Advisory Council* UN Doc A/AC.105/C.2/2021/CRP.13 (27 May 2021).

[160] ibid. 5.

> A Lunar Governance Charter should recognise the fundamental importance of interoperability for the sustainable development of the Moon… [and] should invite States to identify a global platform for the multi-stakeholder development (and subsequent regular updates) of multiple open international standards for lunar activities.[161]

The open-versus-proprietary standards debate is also imperative. Open standards can expand access to the benefits of space technologies, whereas proprietary standards risk creating technology enclaves.[162] Relatedly, the proposed charter encourages 'the development and use of interoperable space systems and equipment' and calls for 'a global dialogue for the development and use of shared infrastructures, avoiding the duplication of infrastructure and enhancing the reusability and repurposing of lunar infrastructure'.[163] Thus, interoperability and sustainability emerge as deeply interlinked principles in the lunar context.

5 Concluding Analysis: Synergies, Schisms and Next Steps

This chapter has argued that standardisation and interoperability can be seen as advancing the benefit-sharing principle of the OST in a way that helps ensure space mining activities, particularly ISRU, are available to all nations, irrespective of their degree of economic or scientific development, not only some. Across the two major space blocs and the UN COPUOS Working Group on Legal Aspects of Space Resources, synergies—not necessarily schisms—emerge on the importance of standardisation and interoperability, at least as high-level principles. As the current Director of UNOOSA, Aarti Holla-Maini, reflected in October 2024: 'There is more that unites us than divides us', highlighting shared perspectives expressed by *Artemis Accords* and ILRS signatories at the first UN Conference on Sustainable Lunar Activities.[164]

While the draft set of recommended principles from UN COPUOS includes a commitment to 'develop interoperable and common space-based infrastructure and standards for space resource activities', it concedes that there does not yet appear to be 'a common understanding in the Working Group' on an additional provision detailing procedures to operationalise this commitment.[165] Moving from principle to practice will present with challenges—but equally, opportunities—for advancing benefit-sharing and peaceful coexistence in space.

While debates on benefit-sharing have often centred on the distribution of profits from lunar resources, particularly if returned to Earth for sale by private actors,

---

[161] ibid. 7.

[162] Li (n 88).

[163] United Nations Committee on the Peaceful Uses of Outer Space (n 159) 35.

[164] Aarti Holla-Maini, 'Today is United Nations Day & to Mark This, I'm Sharing Some Reflections of Year 1 as Director of United Nations Office for Outer Space Affairs (UNOOSA)' (LinkedIn, 24 October 2024) <https://www.linkedin.com/feed/update/urn:li:activity:7255110784811257857/>.

[165] United Nations (n 150) principle 7; annex, part 2, principle 10(c)(c).

attention must also turn to the benefit-sharing of enabling space technologies. This reflects long-standing, principled requests from Global South nations—including during negotiations for the *Benefits Declaration*—to safeguard space for all and reduce the gap between the space technological haves and have-nots. As discussed, open technical standards can foster interoperability and, in turn, potentially expand possibilities for more equitable participation in space mining. Yet, without global coordination, there is a risk of fragmented, incompatible standards; for example, a patchwork could emerge in which Artemis and ILRS missions utilise different standards, reducing the scope for shared infrastructure and consequently narrowing opportunities for benefit-sharing.

The *Artemis Accords* and ILRS are the primary internationally cooperative endeavours through which humanity is planning to return to the Moon, each with growing memberships, and comprising nations with vastly different space heritage and capabilities. If benefit-sharing is to extend to space mining technologies, these space blocs—alongside the work underway in UN COPUOS—must be leveraged in complementary ways. Given that operationalisation of benefit-sharing depends on voluntary cooperation of nations and the contemporary complex geopolitical environment, an important question arises: which aspects of space mining could be of mutual significance to all nations, particularly in relation to standardisation and interoperability, and serve as a foundation for cooperation despite the framing of a contemporary 'space race'?

Here, history offers lessons. With the spectre of a second 'space race', this time between the US and China over lunar sites and resources, aspects of the first 'space race' during the Cold War seem to be repeating, including its rhetoric. Yet, history can repeat positively. As discussed, the Apollo–Soyuz Test Project and the ongoing operation of the ISS are but two notable examples that demonstrate how a commitment to interoperable technologies can foster dialogue and 'more-than-human diplomacy' through technical cooperation, even between rival great powers.[166] With the space domain often described as 'congested, contested, and competitive', it is imperative to remember that opportunities concurrently exist for 'cooperation, collaboration and communication'[167]—and must be actively seized whenever they arise.

Until recently, there was an absence of overlap between the *Artemis Accords* and ILRS memberships, hence them being conceived of as separate membership blocs. This changed when Thailand became the first signatory of both the ILRS in April 2024 and the *Artemis Accords* in December 2024, with Senegal now becoming the second

[166] Davis Cross (n 56); Stewart and Dittmer (n 59).

[167] United States Department of Defense and Office of the Director of National Intelligence, *National Space Security Strategy* (January 2011) <https://www.dni.gov/files/documents/Newsroom/Reports%20and%20Pubs/2011_nationalsecurityspacestrategy.pdf>; Cassandra Steer and Matthew Hersch, 'Conclusion: Cooperation, Collaboration, and Communication in Space' in Cassandra Steer and Matthew Hersch (eds), *War and Peace in Outer Space: Law, Policy and Ethics* (Oxford University Press 2021) 301.

nation to bridge the two space blocs.[168] Whether such overlapping membership grows or acts as an impetus at a future date for cross-bloc cooperation remains to be seen.

And, while it is unknown whether the principles for ILRS will align with the *Artemis Accords* in the absence of a commensurate document, the Secure World Foundation does recognise: ‘the similarities between the activities planned under Artemis and ILRS (permanent installations, extraction and use of lunar water and mineral resources, and manufacturing on the lunar surface) suggests they will be similar’.[169] Common ground exists—and must be fostered—alongside a rhetorical shift away from framing this solely as a ‘space race’. Moving away from these silos is inseparable from advancing benefit-sharing, given the power concentrated in these blocs and how they will invariably shape the future of lunar governance, particularly over space resources.

With this background, a joint dialogue between the *Artemis Accords* and ILRS nation leads and signatories could be convened, one focusing on open standards and interoperable technologies. Such a space bloc-to-bloc dialogue could run in parallel with each bloc’s internal discussions, while informing and supplementing the work in UN COPUOS. Of course, the challenges are considerable (and perhaps insurmountable), not least because of the *Wolf Amendment* and the current improbability that this will be revisited soon. If direct US–China cooperation on space remains politically impossible, signatories or intermediaries such as academic institutions, industry bodies and NGOs could begin the technical groundwork. These non-state actors could develop shared standards for adoption across lunar initiatives.

Early priorities for such diplomatic dialogue could include agreement on standards and interoperable technologies to enable rescue missions for humans engaged in space mining, a first-order priority. Drawing upon the *Hague Building Blocks*, agreement between the *Artemis Accords* and ILRS on technical standards for safety zones around space resource activities could reduce tensions. Developing power standards will likewise be necessary, given the growing recognition of their importance in enabling shared infrastructure on the lunar surface. Power will be required for the extraction, processing and utilisation of lunar resources. There could be progress towards the development of ISRU-specific standards. These potential priorities, to return to Aganaba’s exhortation referenced earlier, are suggested as they seek to ‘imagine a cycle of positive feedback to build increasing cooperation between those on the margins of space activity and those gaining the greatest benefit’.[170]

Yet, while this chapter has focused on the benefits of standardisation and interoperability, it is essential to avoid treating them as techno-utopian quick fixes for benefit-sharing. The true extent to which they can meaningfully expand access to the use and exploration of space—particularly in relation to mining lunar resources and to enabling technologies and infrastructure—remains to be seen. Such principles

---

168 See for dates: Secure World Foundation, *Artemis Accords and International Lunar Research Station Signatories Tracking Sheet* (25 July 2025) <https://docs.google.com/spreadsheets/d/1dFNVSQeTto4DiDH9u2ds4ZAjzGje2Z0-XSiWyiPc8lo/edit?gid=1343837889#gid=1343837889>.

169 Samson and Christensen (n 134).

170 Aganaba-Jeanty (n 40).

must not be deployed merely as rhetorical commitments, reduced to hollow tokenism that fails to translate into meaningful implementation. It must be said that, even if standardisation and interoperability can advance the benefit-sharing principle, they do not, on their own, address the systemic and structural forces that have led to global inequalities and inequities—from the ongoing legacies of colonialism to the growth of unchecked capitalism—both of which pre-date any human activities in space, but have been reproduced in this domain from the outset. Onwudiwe and Newton capture the duality of the promise and the pitfalls:

> [I]nteroperability could be a double-edged sword in the African context. On the one hand, there would be clear guidelines for the development of space technology… On the other hand, having one set of development guidelines could make it easier for existing players to use their economies of scale to dominate the markets for these technologies. In this case, interoperability has the potential to cement the status of dominant players in the space economy.[171]

The sharper side of this 'double-edged sword' for emerging and aspirational space-faring nations would be particularly felt if technical, interoperable standards depend on SEPs owned and controlled by industry in dominant space-faring states, thereby consolidating their pre-existing advantage and constraining benefit-sharing for newer entrants. Indeed, if an open, technical standard requires payment of licensing fees, even on FRAND terms, for a patent deemed a SEP, this raises the question of how open the standard is ultimately.

The not-insignificant risk of entrenching dominance extends to the standard-setting process itself. As Dennerley notes, dominant space-faring powers may develop domestic standards to serve their own industries, then regionalise and globalise them through leading participation in international SSO/SDO.[172] Responding to this observation, Aganaba stresses 'the impetus of Africa to ensure that it can create its own standards, and develop interoperable solutions' as essential to asserting greater agency, relevance and influence within international standard-setting fora.[173] This observation has wider importance beyond Africa, too. Overall, any approaches to space law and governance for lunar resources that promote interoperability and standardisation—whether through the space blocs, multilateralism or other fora—must give equal attention to designing inclusive, transparent, consensus-based standard-setting processes that actively involve nations from the Global North and Global South. Perhaps the ultimate test for humanity, as we prepare to return to and remain on the lunar surface, will be whether the 'earthly ways'[174] we take with us are *not* those of exclusion and competition, but of inclusion and cooperation—so that outer space may be for the benefit of all nations.

---

[171] Onwudiwe and Newton (n 108) 812.

[172] Joel A Dennerley, 'Emerging Space Nations and the Development of International Regulatory Regimes' (2016) 35 *Space Policy* 27, 29.

[173] Timiebi Aganaba-Jeanty, 'Realizing a Regional African Space Program' (67th International Astronautical Congress, Mexico, 2016) 1, 7–8.

[174] To return to the phrasing introduced at the start of this chapter by McDougal and Lipson (n 1).

Art Cotterell LLB(Hons), BSW, GradDipLegalPrac, PhD. Art is a Postdoctoral Fellow at the University of New South Wales, Faculty of Law and Justice. He has previously served as an executive-level adviser in the Australian Government on regulatory reform, intellectual property and international cooperation, and in civil society on access to justice policy.

Space Resources in Markets and Politics: Considerations on Property in the Interests of Social and Environmental Welfare

Scott Schneider

Abstract Law significantly impacts how humans use and control resources in or coming from outer space. This chapter examines how the property is treated under international and domestic space law frameworks, beginning with an economic account of the nature of property, the nature of property rights and the influence of politics on property rights. The chapter frames the political dynamic of property rights against property rights confined exclusively to economics, or markets, in analysing to what extent each is able to yield the most just and environmentally sound use of space resources. The consideration of incentives and consequences in markets and in politics invites discussion on alternatives or complementary avenues in private law to guide space resource management in a manner consistent with existing international space law and in the interests of humanity's well-being.

Keywords Property rights · Space resources · International space law · Domestic space law · Economics · Environment

1 The Potential of Property Rights and Space Resources

A longstanding discussion on the subject of outer space resources is to what extent international law facilitates those resources being used for the benefit of humankind. The central issue, yet to be resolved, concerns the non-appropriation principle under Article II of the widely ratified 1967 Outer Space Treaty,[1] alongside the ability, or desire, for states and non-state actors to obtain, use and exchange space resources. To unwrap this central issue requires consideration of property, property rights and the treatment of property rights in space law.

[1] Treaty on Principles Governing the Activities of States in the Exploration and Use of Outer Space, including the Moon and Other Celestial Bodies (adopted 27 January 1967, entered into force 10 October 1967) 610 UNTS 205.

S. Schneider (✉)
Special Counsel, International Aerospace Law & Policy Group, Brisbane, Australia
e-mail: sschneider@ialpg.com

S. Bhat B and A. Variath (eds.), *Legal Responses to Space Resource Utilisation*, Space Law and Policy, https://doi.org/10.1007/978-981-95-8864-0_6

Much of the literature discussing the intersection of space resources, property and law tends to either examine how legal frameworks ought to manage property or presume property is a construct of law. The present chapter begins one step back by considering the economics and the social context of property rights, irrespective of law or of political objectives (part 2). A short history of how legal institutions came to influence property rights (part 3) and how they currently treat property rights specifically in outer space (part 4) lends insight into how politics impacts the management of space resources. The opportunities and risks presented by political and non-political approaches to property rights over space resources are discussed (part 5) with various avenues considered for humans to fairly, legally and practically manage space resources in the interests of social welfare and environmental sustainability (part 6). The chapter concludes by offering areas for further discussion based on its limitations and relatively narrow scope, not least due to the discussion in this chapter being primarily drawn from a Western perspective on the subject matter.

Emphasising the economic underpinnings of property rights in the context of space resource governance seeks to complement the majority of academic and political discourse on the matter. The present chapter begins by isolating the social imperatives of property rights, demonstrating how legal frameworks respecting people's right to peacefully acquire and associate with one another are in the interests of international space law principles, societal well-being and the environment. This perspective is aided by distinguishing between two types of property rights, market-based property rights and politically-based property rights. From a starting point of acknowledging the social relevance of property rights and identifying vulnerabilities in the political management of property rights, the present chapter aims to encourage further discussion among space law experts and, particularly, those with influence over space resources law and policy, all in the interests of benefiting humankind.

2 Property and Property Rights

2.1 Excludability and Scarcity

To accept that everyone has the inherent authority to exclusively control their body, so long as they do not involuntarily initiate harm upon the body of another, is to accept individual sovereignty as a natural right.[2] More simply, the natural right of self-autonomy means society respects individual A's authority to prevent individual B from interfering with individual A's body. Applying this principle to that which is external to the human body enables people to peacefully interact with each other

[2] See, eg, Epicurus, *Letter to Menoeceus*; Abu Hamid al-Ghazali, *The Revival of the Religious Sciences* (n.d.); Thomas Aquinas, *Summa Theologica*, Part II; John Locke, *Second Treatise of Government*, ch V. "Initiating harm" encompasses posing a real threat to harm, and includes acting recklessly. The qualifier of "involuntary", or "non-consensual", is needed as one can, and often does, permit another person to initiate harm upon them, such is the case in sporting matches.

in a way that goes beyond merely communicating.[3] When someone is able to make a claim to exclusively appropriate something external to their body, that thing it is denoted as "property".[4] To refrain from interfering upon something which has been peacefully appropriated by someone, provided that the one who appropriated it is not using that thing to initiate harm upon others, is to treat property as a natural right.[5] A natural rights approach to property rights contrasts with the positivist-oriented positions which hold that property is not self-evident but, rather, created through law.[6] In following the natural rights approach to property, the chapter necessarily assumes the Scholastic view on justice in that it is immoral when natural rights are not equally respected.[7]

The manner in which property is managed significantly influences how humans govern their affairs and how an individual may increase one's own welfare. Access to, restriction from, use of and disposing or acquiring something are examples of the types of relationships humans have with property. In particular, property's quality of "excludability", that someone may prevent others from interfering with what that person has appropriated,[8] grants people the opportunity to make decisions about how to use property in a way that they believe will best satisfy their needs and desires. When the qualities of appropriation, alienability and excludability are recognised by others, we arrive at the concept of "property rights".[9] Property rights bring natural rights into the field of economics,[10] and are most relevant to society in the context of economic goods.[11]

An "economic good" is determined by something's scarcity relative to how much it is in demand. Air (or oxygen) here on Earth is rarely an economic good due to its abundance. As there is more air available than the amount demanded by humans, it is not scarce and is, therefore, a "non-economic" good. It follows that property rights to air on Earth are, in most cases, inconsequential. On the other hand, the scarcity of bread, steel and medicine brings in a greater relevance of property rights. As there is currently an insufficient amount of these items to satisfy humans' needs and desires,

---

[3] See part 2.2 below for the basis of this reasoning.

[4] Albeit, more than a claim is needed for one to "appropriate" property, as to which see part 2.3 below. Though, depending on the context, property can also refer to one's body (see, e.g., Aristotle, *Politics*) the present chapter distinguishes one's self from one's property.

[5] This follow from the Dominican tradition, as to which, see the works of Francisco de Vittoria.

[6] In the context of space resources in particular, see HR Hertzfeld and FG von der Dunk, 'Bringing Space Law into the Commercial World: Property Rights without Sovereignty' (2005) 6(1) *Chicago Journal of International Law* 81, 94.

[7] Contrary to the utilitarian or social contractualist approaches, as to which see part 3.1 below.

[8] But see, James Y Stern, "What is the Right to Exclude and Why Does it Matter?" in James Penner and Michael Otsuka (eds), *Property Theory: Legal and Political Perspectives* (Cambridge University Press 2018).

[9] Noting a right for one person is always accompanied by an obligation of another to do or to refrain from doing.

[10] Discussed in part 3.1 below.

[11] See, e.g., Lionel Robbins, *An Essay on the Nature and Significance of Economic Science* (MacMillan & Co 1932) 8, 45.

their nature as economic goods calls attention to how the property rights of those goods are managed. Context is important in determining whether something is an economic good—what may be abundant in one situation can be scarce in another. Would you place a higher value on air when walking in a forest or when performing a spacewalk?

2.2 Incentivising Cooperation

Resource scarcity with human demand determines what is an economic good. Economic goods give social significance to property, but do not define what is or is not property. Part 2.1 above introduced the qualities necessary to constitute property, appropriation, alienability and excludability. The quality of alienability helps demonstrate why respecting property rights is in the interests of peace and welfare. Alienability includes transferability. When property rights are transferred on a voluntary basis—that is, with mutual consent of the transacting parties—individuals have the opportunity to peacefully satisfy their needs (which can directly or indirectly satisfy the needs of others).[12] This situation establishes a market—a forum accommodating voluntary exchanges of property for property (buying a good) or property for labour (buying a service).[13] Markets are the home of "private property", meaning property rights are universally respected.[14]

To what extent does property exchanged in markets, that being with the mutual consent of all transacting parties, benefit society? Assuming there is consensus on the meaning of "benefit", the opposing sides on this question can be most simply represented with the classical views of Thomas Hobbes and Adam Smith. Hobbes in the seventeenth century reasoned that civilisation is unsustainable without a centralised monopoly on property rights and, thereby, called for political institutions to restrict private property.[15] Writing a century later, Smith argued property rights through markets, that is private property, best facilitates prosperity and avoids aggressive conflict.[16] The present chapter refers to these forms of property rights as politically-based property rights and market-based property rights respectively.

Private property's social significance is by way of market-based property rights enabling cooperation between multiple parties competing for the same economic

[12] Technically, this capability to transfer property is "alienability", as it does not require a transfer in as much as a disposal.

[13] A "market economy", sometimes referred to as a "free market" distinguishes a situation of mutual consent from a situations where third party intervenes without consent, in which case a "mixed economy" or, depending on the degree of intervention, a "command economy" is established.

[14] The term "public property" refers to the goods controlled by a political institution.

[15] Thomas Hobbes, *Leviathan* (1651) ch XIII para 9.

[16] See, Adam Smith, *An Inquiry into the Nature and Causes of the Wealth of Nations* (1776) bk IV, chs V and IX, however, did posit the state held a responsibility to protect property rights: XX.

goods.[17] This is achieved by markets incentivising people to resolve their conflicting interests without the need of violence or theft, which includes any non-consensual appropriation of another's property. By limiting the presence of aggression in society, individuals enjoy not only greater physical security but also greater confidence to invest, produce and innovate.[18] Critiques of market-based property rights include, but are not limited to, the unequal outcomes that stem from transactions despite their voluntary nature, such as the tendency for power imbalances between transacting parties to favour the more capable or more ambitious.[19] Another criticism is the market's inability to provide adequate goods and services for certain needs and desires. However, such criticisms do not necessarily dispute the benefits of private property inasmuch as they present opportunities for entrepreneurs in the market or for policies by political actors. Moreover, as will be laid out in part 5 of this chapter, such challenges are often created, or made worse, due to politically-based property rights.

2.3 *Property and Commons*

Property's feature of excludability means the appropriator of a resource is able to prevent others from accessing that same resource. When someone is in possession of or makes a claim to a resource which is not subject to the exclusion of others, that resource is referred to as "submarginal".[20] It is submarginal because the satisfaction, or return, from owning the resource is not worth the cost of owning it. This is often due to the technical implausibility of excluding others, whereby there is no means to enforce any exclusion. Clouds, the ozone layer and sunlight cannot at present be exclusively appropriated because it is presently unfeasible to turn a profit from owning them, or there is too high a risk to make owning them worthwhile. As such, a submarginal good is defined not by the feature of the good itself but by the constraints upon humans in acquiring and excluding it from others.[21] Space resources, in fact, are subject to appropriation at the exclusion of others. For instance, one can feasibly land a spacecraft on a celestial body and physically defend that resource from any other spacecraft seeking to later access it. Or, one can extract an artefact from an asteroid, return it to Earth and secure it in a manner so as to prevent others from accessing it.[22] Thus, unowned space resources are *not* submarginal goods. Rather, they are prone

[17] See, e.g., Armen A Alchian and Harold Demsetz, 'The Property Right Paradigm' (1973) 33(1) *The Journal of Economic History* 16, 19.

[18] See, e.g., Douglass North and Robert Thomas, *The Rise of the Western World: A New Economic History* (Cambridge University Press 1973), ch 1.

[19] This is one of the outcomes of disparities in bargaining power.

[20] Sometimes referred to as "unownable".

[21] Ideas, names and designs, for example, are submarginal but their use, or the products derived from their use, are treated by society as akin property. Although such usage cannot be exclusively appropriated, the law classifies ideas, names and designs are as "intellectual property".

[22] How the law views both cases is a different matter and discussed in part 4 below.

to being exclusively appropriated and, therefore, are capable of becoming property. This type of resource, that which is non-submarginal but yet to become property, is called a "natural commons".

Space resources are a particular type of natural commons in that they are "rivalrous". A rivalrous good means one's use of a resource necessarily restricts the ability of another to use that same resource. On Earth, one can consume as much sunlight as desired with absolutely no effect on anyone else's ability to do the same. This means sunlight is a "non-rivalrous" good. The same is true for open-source software or information transmitted over a radio broadcast. On the other hand, eating an apple or consuming electricity means others cannot eat that same apple or extract the power from those same electrical currents.[23] Rivalrous goods, then, can be either property (owned rivalrous goods) or a natural commons (rivalrous goods yet to be owned).

A natural commons is a resource with no barriers to access. This means anyone can use or make a claim of ownership over the resources or any part thereof. Certain fish in international waters or truffles in a wild forest are examples of natural commons. By contrast, a "managed commons", sometimes referred to as a "common pool" resource, is an economic good that is actively controlled and typically restricts certain access. While managed commons are also non-submarginal and rivalrous, their added feature of active control necessarily means the resource is subject to excludability and, therefore, the resource has effectively been appropriated. As the controlling group also has the ability to, if desired, transfer rights to some of or to the entire resource pool to other parties, a managed commons fulfils the three qualities of property, despite the rights to that property being held collectively.[24] The control of a managed commons, usually through voluntary partnerships or by government decree, is one means to avoid the "tragedy of the commons" whereby a rivalrous economic good that is non-excludable is exhausted to the extent it can no longer be used by others.[25] Unlike a natural commons, which is an economic phenomenon, a managed commons is a social construct. Therefore, whether space resources remain a natural commons or become a managed commons is not a question of economics but one of politics and, by consequence, law.[26]

---

[23] Electricity is a product of energy, which is why it is rivalrous when the sun's raw energy is non-rivalrous.

[24] Elinor Ostrom, *Governing the Commons: The Evolution of Institutions for Collective Action* (Cambridge University Press 1990), ch 2. See also, part 5.2 below.

[25] See, Garrett Hardin, 'The Tragedy of the Commons' (1968) 162 Science 1243.

[26] See, e.g. Matthew Johnson, *Property Rights in Outer Space: Mining, Techno-Utopian Imaginaries, and the Privatisation of the Off-World Frontier* (Routledge 2024) 143.

3 Property Rights and Law on Earth

3.1 Political-Based Rights

In addition to establishing rules limiting one's individual sovereignty,[27] political institutions such as kingdoms, empires and states have for millennia determined one's right to property, the most notable example being taxation.[28] Contrary to the natural rights approach that property rights are an extension of individual sovereignty, philosopher David Hume reasoned that such rights are a formalised social convention.[29] Following from Hume's proto-positivist position, jurist Jeremy Bentham,[30] and philosopher John Stuart Mill,[31] went one step further to conclude that property rights require legitimisation by political institutions.[32] Bentham and Mill's view represents a utilitarian position on property rights. The natural rights approach, in contrast, acknowledges the many times in history where humans recognised and respected, and thereby legitimised, property rights in the absence of any social code or political enforcement.[33] The natural rights framing of property, therefore, begins ever so slightly behind the utilitarian position, and in effect alongside Hume, by asking, "from where comes a political institution's authority over an individual's property"? This question requires the distinction to be acknowledged between property "rights" and property "law", the latter of which, indeed, does often require codified norms or political institutions.

While the utilitarian approach considers property rights valid only when they are endorsed or established through politics, a more common rationale in Western culture for political authority determining property rights is social contract theory. Here, individuals are considered to have accepted a political institution's determination of the extent of their property rights as compensation for that institution maintaining civility in society. Social contract theorists did not necessarily argue from a standpoint of natural rights, let alone considering property as a natural right.[34] Yet, the movement suggests that politically based property rights are legitimised based on some form of

[27] Considers conscription or other forms of slavery.

[28] Some of near countless examples of political determination of property rights include the polities of in what is now Sumaria from 4500 BCE, Egypt from 3100 BCE and China from 2100 BCE.

[29] David Hume, *A Treatise of Human Nature* (1739–40) bk III, pt. II, s II.

[30] Jeremy Bentham, *Principles of the Civil Code* (1802).

[31] John Stuart Mill, *Principles of Political Economy* (1848).

[32] Legal positivism lends from utilitarianism, also rejecting property rights as natural rights, but goes yet another step further in holding that private property cannot exist at all without a preceding legal framework, irrespective of their legitimacy: see, e.g. Liam Murphy and Thomas Nagel, *The Myth of Ownership: Taxes and Justice* (Oxford University Press 2002), 8.

[33] See, e.g., Bruce L Benson, 'Enforcement of Private Property Rights in Primitive Societies: Law without Government' (1989) 9(1) *Journal of Libertarian Studies* 1; Thráinn Eggertsson, *Economic Behavior and Institutions* (Cambridge University Press 2009) Ch. 9

[34] John Locke being an exception.

approval from society.[35] That approval may come from a presumption that everyone wants to survive,[36] or because it is a preferred way to ensure justice,[37] or because all affected individuals have, in effect, individually[38] or collectively as a societal unit, consented.[39]

A third explanation for a political institution's involvement in the property affairs of individuals is the market failure approach. In his history of property rights, *Property and Freedom,*[40] Richard Pipes considers the authority for the state (a political institution) to restrict market-based property rights based on that restriction being in the interests of regulating activities and sustaining communal welfare.[41] This position necessarily implies a failure of markets, such as inadequate provision of the desired goods and services or the causing of externalities.[42] These brief descriptions of the utilitarian, social contract and market failure approaches help frame the proceeding examination of property and law concerning space resources. Further context still comes by way of tracing a history of how property rights shifted from market-based to political-based.

3.2 Legal Frameworks and Property Rights

Many of the laws throughout history respecting property rights do so selectively. That is to say, many legal frameworks did not afford property rights in all contexts. Neither did they universally recognise the sovereignty of the individual.[43] Such laws, nevertheless, offer insight into how political institutions came to govern the private and voluntary transactions of non-state actors. One ancient instance of property rights taking on political significance occurred in Asia some 5000 years ago through the Code of Ur-Nammu. By this time, many societies had become less nomadic with settlement becoming more commonplace. This led to agricultural practices advancing, which, in turn, saw many resources that were traditionally submarginal becoming economic goods due to their relevance in food production. As introduced in part 2.2 above, economic goods encourage individuals to facilitate a system of property rights as a means to fairly secure one's own interests and to cooperate with others rather than engage in violent conflict. The Code of Ur-Nammu is often cited as

[35] For opposing perspectives, see: Johannes Althusius, *A Digest of Political Methods* (1603); Lysander Spooner, *No Treason* (1867); Robert Nozick, *Anarchy, State and Utopia* (1974).

[36] Hobbes, *Leviathan* (1651).

[37] Rawls, *A Theory of Justice* (1971).

[38] Locke, *Second Treatise of Government* (1689).

[39] Rousseau, *On the Social Contract* [*Du Contrat Social; ou Principes du Droit Politique*] (1764).

[40] Richard Pipes, *Property and Freedom* (Vintage Books 2002).

[41] Richard Pipes, *Property and Freedom* (Knopf Doubleday 1999), 287.

[42] Both which are discussed in part 5 below.

[43] For instance, taxes, asset forfeiture, slavery and prohibitions on peaceful voluntary interactions were enforced through historical laws, as many legal frameworks continue to do today.

one of the earliest instances of a political institution codifying an express enforcement of property rights. Its provisions prohibited the theft of and damage to private property, which was likely motivated by the dependency of farming for food security and economic activity.[44] In the first-century BCE, the codification of property rights came to Europe through the Twelve Tables. However, the declining economic conditions of the Roman Empire incentivised further development of property rights protections, and in 438 CE, the Theodosian Code became the first known public law enforcing voluntary price agreements. Like the Twelve Tables, the Theodosian Code prohibited fraudulent acts and taking another's property without consent. Indeed, such principles were carried through to civil law the many significant developments beginning from the sixth century[45] as well as in common law several centuries thereafter.[46]

While bilateral agreements between political institutions had concerned property rights for centuries,[47] one of several possible starting points to examine property rights under contemporary international law is the Hague Convention of 1899.[48] Article 10 of this multilateral framework obliges states to respect the property rights of foreign subjects during wartime, a principle mirrored in the 1919 League of Nations Covenant[49] and in the later Geneva Conventions.[50] A transition from international law respecting property rights absolutely to restricting the nature of such rights is evident in the soft law (non-binding) instrument, the Universal Declaration of Human Rights.[51] Article 17 of that 1949 declaration reads "everyone has the right to own property alone as well as in association with others". Article 29 goes on to suggest the extent of one's property rights are determined by law and may be restricted not only in the interests of the rights of others but also for morality, public order and societal well-being. This soft law limitation of property rights has made its way to becoming actual international law in several regions. In Europe, for instance, property rights exist to the extent that the state restricts them for the purposes of taxation, to exercise punishment or in the general interest.[52] This principle represents the general limitation to market-based property rights under international law today, whereby state sovereignty includes the authority to qualify or restrict the property rights of their citizens, or of other entities under its jurisdiction.[53] International law is one avenue for states to clarify certain restrictions, that is, through mutual obligations between

44 As evidenced by the code's emphasis on agricultural infrastructure, land and slaves.

45 See, the Justinian Code; art II of the The Declaration of the Rights of Man and of the Citizen following the French Revolution; the French Civil (Napoleonic) Code established of 1804.

46 E.g., Magna Carta 1215 (25 Edw 1, cc 1–63).

47 E.g. Treaty of Cateau-Cambrésis (1559); Treaty of Osnabrück (1648); The Jay Treaty (1794).

48 Convention IV, arts 46, 52.

49 Covenant of the League of Nations (adopted 28 June 1919, entered into force 10 January 1920) 112 BFSP 13.

50 See, e.g. article 46.

51 GA Res 217A (III), UN Doc A/810 (1948).

52 European Convention for the Protection of Human Rights, Protocol 1, Art. 1 (1952).

53 See, also, Convention on Biological Diversity (1992); Arab Charter on Human Rights (22 May 2004), entered into force 15 March 2008, art 32.

states. But ultimately, it is states which are the actors responsible for establishing and enforcing politically-based property rights.

3.3 Mining and Managed Commons on Earth

Politically-based property rights regarding natural resources are evident from at least the fourth millennium BCE with Egyptian Pharaohs limiting public access to the Nile and to much of the land on its banks. A more recent example is the managed commons of much of England's land for farming from the Medieval period through to the Victorian era. As farming became more profitable, landowners harnessed political intervention as a means to cease certain managed commons. This led to the enclosure law movement, beginning under Henry VIII's reign[54] with the laws of most lasting impact being enforced from the late eighteenth century well into the 1800s. It is largely acknowledged that the enclosure laws significantly increased poverty in both rural England and, as a consequence of labour migration to more populated areas, in urban England.[55] The enclosure laws are an example of how non-state actors can engage in what is called "regulatory capture" as a means of using political institutions for gaining a competitive edge against those with less political influence. This phenomenon an entirely distinct concept from the phenomenon that is disparities of bargaining power in voluntary market transactions. Regulatory capture is no less present in today's natural resource laws than it was in since the Late Middle Ages.[56]

In 1962, the United Nations General Assembly passed a resolution affirming state sovereignty over the natural resources in a state's territory.[57] Although a soft law declaration, this resolution validated the legal position that political authority through the state is what determines how property rights to natural resources are managed. But what of natural resources that to not fall under any one state's jurisdiction? Such international natural commons have been addressed by international agreements establishing those resources as managed commons. Examples of legal frameworks for managed commons include the Antarctic Treaty's 1991 Protocol on Environmental Protection,[58] the 1982 United Nations Convention on the Law of the Sea[59] and the Vienna Convention's 1987 Montreal Protocol.[60] These state-to-state arrangements form part of the present-day multi-layering of political institutions governing property rights. In his 2023 book, *The Multi-level Governance of Space*

---

[54] Statute of Fines 1535, 27 Hen 8, c 10.

[55] Karl Polanyi, *The Great Transformation* (Farrar & Rinehart 1944).

[56] See, part 5.3 below.

[57] Declaration on Permanent Sovereignty over Natural Resources (adopted by the General Assembly of the United Nations on 14 December 1962) UNGA Res 1803 (XVII).

[58] Opened for signature 4 October 1991, entered into force 14 January 1998) 2941 UNTS.

[59] Opened for signature 10 December 1982, entered into force 16 November 1994) 1833 UNTS 3.

[60] Montreal Protocol on Substances that Deplete the Ozone Layer (opened for signature 16 September 1987, entered into force 1 January 1989) 1522 UNTS 3.

Mining,[61] space resource law scholar Antonino Salmeri argues that the multi-level nature of international law is, generally, to be welcomed due to states alone being incapable of resolving global issues.[62]

Supporting this contention is Ana Elizabeth Bastida's 2020 book*The Law and Governance of Mining and Minerals: A Global Perspective*, in which she contends that mining law is overly dependent on domestic legal frameworks. These national frameworks do not accommodate the truly global nature of mining, let alone meet the demand for sustainable development.[63] Also published in 2020, Jeffrey Bone explores the adverse impacts on local communities caused by resource extraction companies. However, Bone gives no weight to any possible solutions that consider reducing politically-based property rights in favour of market-based property rights. Rather, while recognising that regulation fails to prevent such adversity, Bones' analysis stays within the paradigm that political institutions dictating the extent of property rights is the accepted starting point for overcoming the adversities imposed on local communities by mining activities.[64] The present chapter offers an alternative by asking to what extent market-based property rights can serve community interests. Posing this question regarding space resources first requires an overview of the politically-based property rights on both the international and domestic levels.

4 Property, Law and Space Resources

4.1 International Law and Space Resources: Limitations on Property Rights

Part 2.3 of this chapter explained how space resources are rivalrous, non-submarginal and, when unowned, are a natural commons. This combination means, in economic terms, space resources are able to be appropriated as property. There is no economic reason why anyone acquiring space resources cannot benefit from property rights. Any limitations on the property rights of space resources, then, are by way of social artefact, namely through political institutions establishing and enforcing law. The primary legal instrument governing human affairs in outer space, the 1967 Outer Space Treaty. This instrument impacts property rights to space resources in several ways. A first consideration is the Outer Space Treaty's context. This multilateral agreement is derived from a United Nations General Assembly resolution five years prior. That resolution, short titled the Declaration of Legal Principles, aimed to reduce the risk of military conflict in outer space by ensuring space was to be explored and

[61] (Wolters Kluwer, 2023).

[62] Antonino Salmeri, *The Multi-level Governance of Space Mining* (Wolters Kluwer, 2023), ch 1.

[63] Ana Elizabeth Bastida, *The Law and Governance of Mining and Minerals: A Global Perspective* (Hart Publishing 2020).

[64] Jeffrey Bone, *Governing the Extractive Sector: Regulating the Foreign Conduct of International Mining Firms* (Hart Publishing 2020).

used for peaceful purposes and to benefit mankind.[65] This aim was carried over into the enforceable provisions of the Outer Space Treaty.[66] Notably, the first paragraph of article I of the Outer Space Treaty requires the exploration and use of outer space be for the benefit of "countries" rather than "mankind", a slight variation from the preceding declaration.

As for the second paragraph of article I:

> [o]uter space, including the moon and other celestial bodies, shall be free for exploration and use by all States without discrimination of any kind, on a basis of equality and in accordance with international law, and there shall be free access to all areas of celestial bodies.

These first two paragraphs of article I of the Outer Space Treaty show no express prohibition on space resources activities, including whether exploring or using space resources. However, from these same two paragraphs do establish several limitations to any potential property rights related to such activities. Such limitations include that space resource activities, and their uses, must:

1. benefit all countries;
2. comply with international law;
3. be carried out on a basis of equality and non-discrimination; and,
4. no prevent others from accessing any area of a celestial body.

Perhaps more directly relevant to property rights is Article II of the Outer Space Treaty, the full provision of which reads:

> [o]uter space, including the moon and other celestial bodies, is not subject to national appropriation by claim of sovereignty, by means of use or occupation, or by any other means.

While the primary intention of this "non-appropriation principle" of article II was to prevent territorial expansion of states in outer space,[67] Hobe et al's analysis of the drafting history behind article II, particularly regarding the records during negotiations, "clearly shows that most of the delegates participating in those discussions were of the opinion that Article II prohibits both public and private appropriation" and "neither State sovereignty nor private property claims apply to outer space".[68] This view is reinforced by article VI of the Outer Space Treaty, requiring all outer space affairs by any non-state actor to be authorised and undergo continued supervision by a state.[69] Salmeri summarises one of article VI's legal consequences in that "a State could not validly authorise a private entity to engage in an activity that the

---

[65] Declaration of Legal Principles Governing the Activities of States in the Exploration and Use of Outer Space (1962) GA Res 1962 (XVIII).

[66] Articles I, III and IV.

[67] International Institute of Space Law, 'Statement by the Board of Directors of the IISL on Claims to Property Rights regarding The Moon and Other Celestial Bodies' (2004) https://web.archive.org/web/ accessed 21 August 2025.

[68] *Cologne Commentary on Space Law* (vol 1, 2009), 237; But see, White, WN, 'Real Property Rights in Outer Space' (Proceedings of the 40th Colloquium on the Law of Outer Space 1997), 4.

[69] The obligation rests with the state to authorise the actor, not on the actor seeking authorisation.

State itself could not undertake."[70] By this reasoning, the non-appropriation principle under article II would apply to non-state actors.[71] As appropriation is one of the three core elements of property, if this interpretation of the Outer Space Treaty is accepted, there is no possibility of facilitating market-based property rights in the domain of space resources and, accordingly, space resource activities depend on politically-based property rights.

4.2 *Avenues for Property Rights in the Outer Space Treaty*

Yet another challenge of the Outer Space Treaty against market-based property rights to space resources is its article VIII whereby states must maintain a registry to determine which objects launched into space fall under their jurisdiction and control. In fact, despite being recorded on any registry, a spacecraft launched from Earth by a non-state actor, whether used for space resource activities or otherwise, cannot escape state jurisdiction or control.[72] By extension, this provision may even apply to equipment and spacecraft produced in outer space,[73] meaning the state would retain control over goods used along supply chains in space. As the Outer Space Treaty establishes politically based property rights, is there any pathway enabling space resources to be exchanged on a voluntary basis and, thereby, allowing non-state actors to participate in at least a quasi-market? One avenue is for states to develop a new multilateral agreement, in addition to the Outer Space Treaty, which specifically addresses space resources.[74] However, a much of the literature on this question suggests there may an avenues other than the relying on the further layering of international space law instruments.

One starting point is to consider what, exactly, constitutes state appropriation as prohibited under Article II of the Outer Space Treaty. Distinguishing state appropriation from the rights of non-state actors to engage in space resource activities provides insight. Space lawyer Virgiliu Pop considered how the non-appropriation principle prevents states from regulating space resources, while articles VI and VIII mean states can, nevertheless, regulate the space resource activities of non-state actors under their jurisdiction, noting "[t]the object of jurisdiction is in this case the conduct of nationals

[70] Antonino Salmeri, *The Multi-level Governance of Space Mining* (Wolters Kluwer, 2023), 59.

[71] See, also, Wayne White, 'The Legal Regime for Private Activities in Outer Space' in Edward Lee Hudgins (ed), Space: The Free-Market Frontier (Cato Institute 2002), 84.

[72] See, also, Convention on International Liability for Damage Caused by Space Objects (adopted 29 March 1972, entered into force 1 September 1972) 961 UNTS 187.

[73] Antonino Salmeri, 'Space Objects as a New Concept of International Space Law' (2021) 46(2) *Air and Space Law* 203.

[74] See, e.g. Imre Anthony Csabafi, *The Concept of State Jurisdiction in International Space Law: A Study in the Progressive Development of Space law in the United Nations* (Springer 1971); Ranjana Kaul, 'Outer Space: Is It a Global Common?' (2024) 5 *Frontiers in Space Technologies* 4.

and not a spatial extension."[75] Because space resources are non-submarignal and are able to be appropriated and exclusively controlled, the more relevant question is what incentives are there for non-state actors to engage in space resource activities. To incentivise non-state actors to engage in space resources activities, as non-submarginal resources, avenues have been proposed which focus on the returns derived from space resource activities without consideration as to whether the rights to those returns are market-based or politically based. In their 2005 paper, Hertzfeld and von der Dunk draw on examples on Earth where non-state actors hold limited property rights in factors of production but, nevertheless, enjoy returns from that which is produced by those factors of production.[76] In a presentation delivered in 1998, Wayne White Jr. made a case for functional property rights to space resources.[77] Developing on prior scholarly work, White argued property rights in the context of real property would not depend on state appropriation but, rather, on the state's control of the equipment and people, on any part of a celestial body or area of outer space otherwise.[78] Such an approach, argues White, poses no inherent contradiction to articles II, VI or VIII of the Outer Space Treaty. Rather, it is up to the states to establish laws ensuring those international provisions are met.

White's presentation was given at the International Institute of Space Law ("IISL")'s 40th Colloquium on the Law of Outer Space. The IISL is a non-governmental association of world-leading space law professionals seeking to facilitate the development of the laws governing outer space affairs. While the IISL's view on space resources is that the Outer Space Treaty does not preclude non-state actors carrying out space resource activities, it also follows the reasoning presented above in that articles II and VI of the treaty prevent non-state actors from appropriating space resources.[79] More recently, the IISL has confirmed this position while suggesting a new framework for a dedicated legal regime should be established in addition to the Outer Space Treaty to address space resources.[80] To escape the tendency of seeking states and international organisations to spend efforts to negotiate, draft, agree and ensure the enforcement of new international agreements for space resources and property, soft law measures can be proposed as an alternative.

---

[75] Virgiliu Pop, 'Appropriation in Outer Space: The Relationship between Land Ownership and Sovereignty on Celestial Bodies' (2000) 16(4) Space Policy 275, 279.

[76] H. R. Hertzfeld and F. G. von der Dunk, 'Bringing Space Law into the Commercial World: Property Rights without Sovereignty' (2005) 6(1) *Chicago Journal of International Law* 81, 98.

[77] White, WN, 'Real Property Rights in Outer Space' (Proceedings of the 40th Colloquium on the Law of Outer Space 1997).

[78] White, WN, 'Real Property Rights in Outer Space' (Proceedings of the 40th Colloquium on the Law of Outer Space 1997), 7.

[79] International Institute of Space Law, 'Statement by the Board of Directors of the International Institute of Space Law (IISL) on Claims to Property Rights Regarding the Moon and Other Celestial Bodies' (2004) https://web.archive.org/web/ accessed 1 August 2025.

[80] International Institute of Space Law, 'Statement of the Board of Directors' (2004) https://web.archive.org/web/ accessed 3 September 2025.

4.3 *Soft Law Proposals for Property Rights alongside the Outer Space Treaty*

In addition to the limitations on market-based property rights under the Outer Space Treaty outlined so far, one further provision has a direct impact on the actual operations when conducting space resource activities. Article IX of the Outer Space Treaty requires states exploring and using outer space to afford due regard to other states in their exploration and use. An interpretation of this article is that a state conducting mining operations on a celestial body cannot carry out such operations in a manner that is against the interests of another state. By virtue of Article VIII requiring states to control the activities of their nationals engaged in space resource activities, it is legally plausible that the due regard principle applies equally to non-state actors. Thus, articles I, II, III, VI, VIII and IX of the Outer Space Treaty form a somewhat complex web of ambiguities, possibilities and prohibitions to the extent property rights may be harnessed in the domain of space resources. Because of these ambiguitires, several proposals have been put forward in an attempt to provide a greater degree of certainty while remaining within the bounds of all aforementioned provisions of the Outer Space Treaty.

One such proposal is from the Hague Space Resources Governance Working Group, established in 2014, to fill the gaps in the legal framework concerning the use of space resources. Its final report, published in January 2020, presented the proposal as "Building Blocks" for the development of an international framework on space resource activities.[81] The initial principles of the Building Blocks do not list property rights in any fashion but do address the prevention of disputes arising out of space resource activities and take into account the contributions of pioneer operators.[82] Section 8 does expressly suggest that domestic legislation should permit resource rights over raw and volatile minerals. However, section 7 suggests "priority rights", rather than being absolute, should be determined by law on a case-by-case basis. Thus, sections 7 and 8 of the Hague Space Resources Governance Working Group's Building Blocks evoke a positivist approach to space resources. Section 13 offers clarification to the obligation under Article I of the Outer Space Treaty by suggesting that space resource activities should benefit all countries not in any compulsory revenue-sharing but, rather, countries should promote benefit-sharing measures.[83] Notably, section 9 restates, but does not proposes any equivalent clarification to, the due regard principle of Article IX of the Outer Space Treaty.

Less than one year after the Building Blocks were published, the United States released a multilateral agency-level initiative for cooperation in lunar and related space resource activities. In October 2020, the National Aeronautics and Space Administration released the Artemis Accords as a set of soft law principles to

[81] The Hague International Space Resources Governance Working Group, 'Final Report 2018–2019' (27 January 2020) Ref no HSRGWG/FR/1/27012020 ("Building Blocks").

[82] Building Blocks, 4.2 d), 4.20 k).

[83] Section 13.1 offers a non-exhaustive definition of what "benefit" means.

"enhance the governance of the civil exploration and use of outer space" in line with the United States' Artemis Program.[84] As of January 2026, government representatives of 61 states have joined the collaboration, notably with no participation from China and Russia. The Artemis Accords do not attempt to provide any clarity on how property rights are managed. Nonetheless, they do confirm the position of the signatory state representatives that space resource activities, including extracting and using space resources, are not inconsistent with the Outer Space Treaty.[85] The significance of this is in the context of the relatively unpopular 1979 Moon Agreement which sought to establish lunar resources as the "common heritage of mankind".[86] While the provisions of the Moon Agreement arguably limit the possibility of market-based property rights to nearly nought, no major space-faring states ratified that treaty and its provisions. Its status as customary international law, therefore, is questionable.[87] While the Artemis Accords suggest space resource activities are lawful, several states have gone the extra mile and established domestic laws to afford some extent of property rights to national actors engaging in space resource activities.

4.4 Property Rights in Domestic Law

Complementary soft law is one avenue to clarify the Outer Space Treaty's provisions concerning space resources and any associated property rights. Another option is for individual states to take advantage of Article VI's obligation to authorise non-state actors' affairs in outer space. While a state's domestic law must comply with all provisions of the Outer Space Treaty, the scope of interpretation in the treaty's rules governing space resources and property lends to flexibility in how states implement the treaty's obligations on a national level. As of January 2026, four states have established legislation specifically addressing the rights of non-state actors with regard to space resources. All four of them, the United States, Luxembourg, the United Arab Emirates and Japan, expressly afford non-state actors a certain degree of property rights over space resources.

The first nation to clarify the extent of such property rights was the United States. In 2015, it enacted legislation referred to as the *Commercial Space Launch Act,*[88] part IV of which is dedicated to space resource activities. Under this law, a citizen of the United States is entitled to property rights over space resources at least to the extent of possession, ownership, transportation, usage and sale in accordance with

[84] United States Department of State and National Aeronautics and Space Administration, 'Artemis Accords' (13 October 2020).

[85] Section 10.

[86] Agreement Governing the Activities of States on the Moon and Other Celestial Bodies (adopted 5 December 1979, entered into force 11 July 1984) 1363 UNTS 3.

[87] However, the government agencies of several states parties to the Moon Agreement have signed the Artemis Accords.

[88] *Commercial Space Launch Act of 1984*, Pub L No 98–575, 98 Stat 3055 (codified as amended at 51 USC ch 509).

other United States law and international law. This is all the while the legislation expressly states the United States does not assert any claim of sovereignty over a celestial body. Two years later, Luxembourg followed the lead of the United States with the Luxembourg Space Resources Law.[89] From the outset, the Luxembourg law clearly states that "space resources are capable of being owned,"[90] and provides a framework for authorising companies registered in Luxembourg to carry out space resources activities.

In 2019, the United Arab Emirates enacted a law establishing space resource exploration and extraction as regulated activities.[91] Four years later, a dedicated Space Resources Regulation was passed stating space resources may be "explored, exploited or used" and that operators exploring, exploiting or using space resources are entitled to exercise ownership rights to the extent afforded by law.[92] Finally, Japan's space resources legislation of 2021 enables ownership of space resources so long as the individual conducts their activities in accordance with an state-approved business plan.[93] While these four instances of domestic legislation bring a greater emphasis on property rights into the domain of space resources, in each case, such rights are entirely subject to political decree. This begs the question of what are the effects of political intervention in economic activity, particularly when law and regulation are enforced under the banner of safety, environmental protection or equality.

5 Incentives and Consequences of Markets and of Politics

5.1 The Social Consequences of Property Rights

Disparities in bargaining power mean markets can facilitate inequalities of wealth. However, when property rights remain outside of politics, those with lower bargaining power have a far greater chance of fairly improving that power. The alternative is relying on political institutions, which means non-state actors who are connected to or are direct beneficiaries of political institutions maintain an unfair advantage and usually at to increasing degrees.[94] In Europe, this is perhaps most evident with the European Union's Common Agriculture Policy, where larger corporations with ties to the state are favoured over smaller, or "inconvenient", operators. This phenomenon is more apparent in states where political corruption is accepted or tolerated and is, therefore, in plain sight. Markets, in contrast, provide an even ground for non-state

[89] *Law of 20 July 2017 on the Exploration and Use of Space Resources* (Luxembourg)

[90] Article 1, *Law of 20 July 2017 on the Exploration and Use of Space Resources* (Luxembourg).

[91] Federal Law No. (12) of 2019 On the Regulation of the Space Sector, issues 19 December 2019.

[92] Resolution No (19) of 2023 regarding the Space Resources, art (7).

[93] Act on the Promotion of Business Activities for the Exploration and Development of Space Resources (Act No. 83 of 2021), art 5.

[94] Part 5.3 below discusses this concept in the context of transfer payments.

actors to compete with one another without political favouritism and, by so doing, equality is inherent in their interactions.[95]

When news media and politicians refer to societal development, the term economic growth is employed. Economic growth is often measured by a state's gross domestic product ("GDP"), the sum of internal expenditures of both non-state actors and political institutions. While GDP is a helpful indicator to monitor the activity within any one state, it gives little insight into the standards of living of the individuals living in that state. More reliable metrics of an individual's standard of living are based on identifying the range, or variety, of what a person has access to. For instance, the greater options one has to access shelter, healthcare, education, nutrition, security, the greater one's standard of living can increase. Distinct from economic "growth", access to greater options in one's life determines the degree of one's economic "prosperity". Political institutions can directly influence economic growth to the extent they contribute to or influence GDP. Increasing economic prosperity, however, requires more options in society accessible to all citizens. This is a feat best enabled through entrepreneurship, investment and trade, including in the domain of capital goods. Space resources are, first and foremost, a capital good, meaning their value is in the production of other goods or the provision of services. As such, market-based property rights enable space resources to contribute to economic prosperity.

A primary critique of market-based property rights in space resources is that wealth will remain concentrated in first movers. This position is laid out in Mathew Johnson's 2024 book *Property Rights in Outer Space* in which he argues that once non-state actors come into ownership of space resources, "the benefits of off-world minerals would flow to the owners rather than to the other parties".[96] This is the same as saying that the benefits from a company that explores or extracts iron ore are limited only to that company. In fact, beneficiaries of iron ore include anyone whose life has been made better off due to the products developed from iron ore, such as steel (and, in turn, housing, roads and manufacturing), water treatment and ammonia-based products. This is alongside the employment opportunities made available from the associated production process in each product's supply chain. Financial returns on mineral operations in space will, indeed, travel to those who endured the uncertainty while investing or making the efforts in the entrepreneurial endeavours of space resources activities, as is the case with any industry, mining or otherwise. Yet, the flow of benefits derived from that activity is widespread and, indeed, is the reason the financial returns arrive for those investors and entrepreneurs. In the absence of politics,[97] fraud or coercion, the only way first movers can earn returns on space resources is by using or transferring the rights to those resources in a way which yields benefit to at least some demographic in society.

---

[95] See e.g., Hernando de Soto, *The Mystery of Capital*, NB. arguing from a utilitarian position.

[96] Matthew Johnson, *Property Rights in Outer Space: Mining, Techno-Utopian Imaginaries, and the Privatisation of the Off-World Frontier* (Routledge 2024), 72.

[97] See part 5.3 below.

5.2 An Environmental Consequence of Property Rights

A 2025 article by Pelton, Mishra and Vitoloni argues that the lack of regulation managing space resource activities compromises the sustainability of the space environment.[98] This is a common position generally regarding the intersection of environmental protection, markets and politics, whereby, so the argument goes, without strict regulation controlling non-state actors, the environment will face a more rapid rate of deterioration. It should first be pointed out that Pelton, Mishra and Vitolni suggest a global commons would be the ideal situation for space resources yet they claim "disparities in capabilities and interests often lead human beings from making use of these resources towards exploiting these resources. This phenomenon has sometimes also been known as the 'tragedy of the commons'". In fact, this is an incomplete representation of the tragedy of the commons or, more particularly, the causes of the tragedy of the commons. Part 2.3, which introduced the concepts of both a natural commons and a managed commons, explained why noted that the tragedy of the commons is the exhaustion of a rivalrous, non-excludable good. The exhaustion of such a resource is preventable by making the resources excludable and, thereby, turning a natural commons into property.

As referred to in part 2.3, a collective controlling access to a common-pool resource exercises property rights to those resources in the form of a managed commons. This control allows for the safeguarding of the resources, which can ameliorate the risk of arriving at the tragedy of the commons. The second means to avoid the tragedy of the commons is to allow the resources to be acquired as private property. In fact, both a managed commons and private property, all else equal, are market phenomena. There is nothing at all political about a group of individuals, whether within one or among several communities, forming an agreement or following a social custom to establish a managed commons. The common-pooled land in England used for agriculture prior to the enclosure laws mentioned in part 3.3 above is one example of a managed commons established through voluntary participation of non-state actors. When one or more states, or another form of political institutions, uses force through law to establish a managed commons, the situation becomes political. Examples of politically managed commons are resources contained in the deep seabed,[99] Antarctica,[100] and the Rhine River Basin.[101] Enforcing the Moon Agreement would have likely established a politically managed commons for space resources. That political institutions are veering away from this consequence of the Moon Agreement, not least by way of the Artemis Accords and

---

[98] Joseph N Pelton, Nishith Mishra and Martina Elia Vitoloni, 'Mining the Ocean Floor vs Mining the Moon: What Can We Learn from Our Past Experiences?' (2025) *Frontiers*.

[99] United Nations Convention on the Law of the Sea (adopted 10 December 1982, entered into force 16 November 1994) 1833 UNTS 3.

[100] Antarctic Treaty (adopted 1 December 1959, entered into force 23 June 1961) 402 UNTS 71.

[101] Berne Treaty, (1963)—Revised Convention for the Navigation of the Rhine ("Mannheim Convention" 1868).

domestic legislation, allows markets to contribute to environmental sustainability where political institutions may fall short.

Throughout history, it is clear that non-state actors have caused severe environmental degradation. The pollution caused by the industrial revolution and the deforestation in several regions of South America and Southeast Asia are two examples. However, it is no less true that political institutions also cause severe environmental degradation. A prominent example is the European Union's Common Fisheries Policy, which has been criticised for decades due to its skewed incentives and structural inadequacies leading to overfishing.[102] Aside from such immediate adversities, political institutions distort markets and facilitate certain non-state actors in actually causing environmental harm. In the United States, there are several industries in which non-state actors benefit from government assistance despite harming the environment, with fossil fuels, agriculture and military among those domain causing the most environmental degradation.

Moreover, where non-state actors harm the environment in the absence of political influence, it is often because the immediate stakeholders of the environment tend to place a lower value on sustainability than on the benefits derived from the activity that causing the environmental harm. During the Industrial Revolution, for example, environmental protection was not valued by Western society as much as in present times. People were more concerned with immediate needs such as health, security and earning an income. Accordingly, private firms faced no market pressure to spend resources to prevent pollution of waterways and the atmosphere. More contemporary examples, such as the lumber industry in many forests in South America and Asia, have a similar situation, where the relatively low standards of living compared to many European or Northern American communities mean individuals can either accept adversities to the environment in return for greater opportunities or security in their livelihoods.[103] Even where people do place a high value on environmental protection, they may nevertheless be unaware of the harms caused by certain ventures carried out by non-state actors. When enough individuals reach a level of welfare in which they can afford to worry about negative externalities, property rights through markets incentivise non-state actors to reduce those negative externalities, including environmental degradation. This principle is applicable in space just as it is on Earth, suggesting a proper cost-benefit analysis between the costs and effects of market-based property rights to space resources compared to politically-based property rights to those resources.

---

[102] Regulation (EU) No 1380/2013 of the European Parliament and of the Council of 11 December 2013 on the Common Fisheries Policy [2013] OJ L354/22.

[103] See, e.g., World Bank, *Sustainability Review 2019* (World Bank 2020).

5.3 The Politics of Law

Johnson's *Property Rights in Outer Space* explores the role of the mining of space resources in the context of the domestic legislation governing commercial activities. His contends that the NewSpace phenomenon[104] shows how private property compromises any humanitarian goals which may be espoused by the leaders of NewSpace. Johnson ultimately argues NewSpace's neoliberal approach will likely lead to the benefits of space resources favouring an elite cohort rather than all of humankind. Part 5.1's discussion on the social consequences of property rights mentioned Johnson's rather narrow interpretation of the meaning of "benefit". A broader, and more practical, approach is available, however. Just as we can trace the origin of harms caused to people, immediate or distant, so, too, can one benefit from an activity despite being several links away from that originating activity.[105] This is the entire basis of why capital goods have value. Beyond this, Johnson articulates another explanation for his concern that NewSpace will channel wealth to an elite few. This concern is not dependent on definitions but on the activity of political institutions.

In Chapter 6 of *Property Rights in Outer Space*, Johnson demonstrates how the prominent non-state actors of NewSpace frequently espouse that government needs to step away from controlling the space sector, letting private actors take the lead with minimal constraint. However, at the same time, those NewSpace actors are dependent on the government for their survival: "the 'rebel alliance' of Han Solo-style entrepreneurs turn out to be part and parcel of the Empire."[106] This paradox is in large part due to a majority of space technologies being tied to military priorities or other security-related applications. Yet, even a majority of civilian space endeavours undertaken by non-state actors exist because of the involvement of political institutions. This political involvement is not necessarily by way of law but through policy, particularly in the form of transfer payments. A transfer payment occurs when a political institution takes resources away from certain non-state actors and provides them to other non-state actors. Examples include grants, subsidies and procurements. In each case, a non-state actor is receiving resources from a political institution that obtained those resources either through taxation, inflation or public debt. Therefore, it is always the case that when, for example, a non-state actor receives a monetary grant from the state to develop a space technology, that money was previously held by another non-state actor (taxation), obtained by reducing the wealth of other non-state actors (inflation) or obtained by placing an obligation on non-state actors to pay in the future (public debt).

It is one thing for transfer payments to be made on a sound basis with genuine and transparent effort in determining who should get what resources. It is another

---

[104] "NewSpace" is the term for the trend of an increasing market for space activities outside of politics.

[105] For a practical and consistent meaning of definitions, see, e.g., Scott Schneider, "Beneficiaries of the New Space Age: Following the Benefit to all Humankind" *Acta Astronautica* (2024).

[106] Matthew Johnson, *Property Rights in Outer Space: Mining, Techno-Utopian Imaginaries, and the Privatisation of the Off-World Frontier* (Routledge 2024), 179.

matter entirely when transfer payments are made recklessly, dishonestly or maliciously. This is more common than what may be easily observable. Political institutions often engage in boondoggling,[107] cronyism[108] and bailouts[109] and are susceptible to rent seeking,[110] including regulatory capture.[111] One consequence of such transfer payments is the creation of monopolies or barriers to entry for new actors to participate in the space industries.

Separate from transfer payments are the vulnerabilities in political decision-making generally. Public choice is the discipline that examines the actions and consequences of politics using economic principles. An example of the space sector seen through the lens of public choice is considering the incentives behind elected and appointed decision-makers. A politician seeks to please his or her most vocal constituents, while an appointee or delegate must appeal to those responsible for their appointment. Bureaucrats are neither free from the mechanics of public choice as their incentives may include budget maximisation, job security and maintaining or increasing control. Essentially, public choice treats political decision-makers as humans rather than as humanitarian, flawless and all-wise agents. The reason this part 5.3 considered politics is not to claim always and everywhere law and policy is harmful. Rather, this part 5 in general aims to present a more realistic account of the drawbacks of politically-based property rights against some of the lesser-known positive possibilities of market-based property rights. In short, hidden incentives and unintended consequences can render otherwise well-intentioned political decisions as, or more, harmful for equality and sustainability than private property rights.[112]

---

[107] See the Space Launch System of the United States.

[108] See, e.g., J Yunfan Yu, Saskia de Klerk and Michael Hess, 'The Influence of Cronyism on Entrepreneurial Resource Acquisition' (2021) 40 *Asia Pacific Journal of Management* 121.

[109] Bailouts can come in the form of continued payments despite poor performance or programmes ending, as in the case of several of Boeing's contracts with the United States government.

[110] See, e.g., Robert T Deacon and Ashwin Rode, "Rent Seeking and the Resource Curse" in Roger D Congleton and Ayre L Hillman (eds), *Companion to the Political Economy of Rent Seeking* (Elgar 2014).

[111] See, e.g., Andrea Saltelli et al., "Science, the Endless Frontier of Regulator Capture" (2022) 135 *Futures* 102,860.

[112] The caveat here that if property rights are to be market-based, they must be entirely market based without any political interference. Otherwise the inequalities caused by political involvement compromise the positive aspect of private property outlined in parts 5.1 and 5.2.

6 Bridging a Gap or Paving a New Path?

6.1 Alternatives to the Political Management of Property Rights

To rely on political institutions in solving societal and environmental issues necessarily means relying on them to also avoid irresponsible transfer payments while simultaneously overcoming public choice incentives. While subject to their own flaws, market-based property rights can encourage non-state actors to invest in preventing, or seeking resolutions of, societal challenges without the vulnerabilities that come with politically-based property rights. At the very least, markets can complement political means of managing property rights. Among the most common non-political mechanisms for enforcing property rights is through private law. While public law is the product of politics, private law empowers those with the most direct interest to manage their property rights. The many merchant guilds in Europe during the Medieval period demonstrate how voluntary associations were formed to protect the common interests of their members. A primary example is the Hanseatic League of Northern Europe, which dominated trade from the thirteenth to the sixteenth centuries CE and was the product of non-state (at that time, non-empire) actors rather than political institutions. A contemporary example of a private law mechanism to enforce property rights is the American Arbitration Association, a forum where non-state actors have been voluntarily resolving contractual disputes since 1926. The primary arguments against a private law system for managing space resources include the inequality argument and the externality argument, both of which have been addressed in part 5 above.[113]

When investors and entrepreneurs engage in space resource activities, they experience two states of flux. On the one hand, they endure uncertainty upon speculating on how to address a need that may not even be known by consumers. On the other hand, they hold a certain amount of risk whereby certain activities they undertake during the entrepreneurial venture carry a certain likelihood of going wrong, such as a rocket exploding prior to lift-off or a satellite veering off course. Uncertainty is inherent in entrepreneurship and is borne by the entrepreneur and the entrepreneur's investors. Risk, however, is a quantifiable concept and, as such, the liability for risk can be transferred from the entrepreneur to another party. The transfer of risk is facilitated through the insurance market, whereby certain non-state actors—underwriters—accept the risk of the non-state actors performing business ventures. Insurance markets have contributed to both equality and sustainability, with examples including minimising the impact of agricultural losses in African and Asian regions,[114] thereby giving

[113] Parts 5.1 and 5.2 respectively.

[114] See, e.g., 'Micro-insurance plan for Kenya's farmers', 37(2) *Appropriate Technology* (June 2010) 35.

greater security to local farmers. On the sustainability side, industry standards and firm reputation incentivise companies to obtain environmental insurance.[115]

Under politically-based property rights, a legal claim can only be made where a regulation has been broken. An alternative to regulation, however, is already semi-present in many societies today—tort law. Regulation is a prescribed set of rules restricting how people can act. Tort law traditionally sat outside of regulation, enabling non-state actors, including individuals, to seek remedies for harm caused outside of what the state deemed suitable. This form of decentralisation provides for greater access as individuals and businesses do not need to rely on the state judicial systems to pursue a claim for suffering. The reason tort is semi-present today is because, in most states, a claim of tort can only be made in accordance with state-sanctioned law. Tort in most Western societies has become quasi-regulation. Tort law is also in the interests of sustainability. Under regulation, a non-state actor can still harm the environment at least to the extent of the law. In some cases, it may do so beyond the legislation with the cost of a penalty or, if the non-state actor is politically connected, without any penalty. In contrast, managing externalities through tort would enable stakeholders to bring claims against bad actors, whether or not there is any breach of public law.

6.2 *The Role of Politics in the Market of Space Resources*

Can the state play a role in facilitating property rights, rather than preventing them? Consider the state-coordinated Geodata for Agriculture and Water, which, among other outcomes, enabled small farmers in several regions of Africa to obtain insurance.[116] This is one example of transfer payments that yield positive results. Such results, then, are able to be placed against the costs of how that money could otherwise have been used. Political actors can also actively enforce certain property rights. Ensuring non-state actors have effective means of recourse upon experiencing instances of fraud, coercion or other forms of involuntary force in the context of space resources will disincentivise bad actors.

States ensuring they meet their international obligations, such as those under the Outer Space Treaty, without further restricting property rights, is perhaps the most effective action in the interests of equality and sustainability. An easy example is for a state to maintain its space object registry, clearly delineating who has ownership of which equipment in space. More sensitive issues, however, concern domestic licensing for space resource activities and associated activities. Regarding article VI of the Outer Space treaty, a streamlined, unambiguous approvals framework

---

[115] See, e.g., Erskin Blunck, 'Germany BMW's Sustainability Strategy of Evolution and Revolution towards a Circular Economy' in V Anbumozhi and J Kim (eds), Towards a Circular Economy: Corporate Management and Policy Pathways (ERIA Research Project Report 2014–44, ERIA 2016) 75–92.

[116] G4AW, *"G4AW: Global Monitoring for Agricultural Productivity — Space Office (NL)"* (website) https://g4aw.spaceoffice.nl/en/ accessed 3 September 2025.

proportional to the capabilities and maturity of the domestic industry will best allow the development of those capabilities. States can also facilitate the enforcement of contracts and promote international cooperation and peace.

6.3 Concluding Remarks

Acknowledging the distinction between market-based property rights and politically based property rights allows for conducting a cost-benefit assessment. Such assessments provide vital information when determining how to address particular issues of space resources. Particularly in the interests of equality and sustainability, political decision-making is not always the answer which yields the most fair or environmentally sound outcomes. This chapter began with a natural rights perspective on property rights in establishing the moral basis for property. However, even if one takes a utilitarian, social contractarian or legal positivist perspective on property rights, the social and environmental consequences of market-based property rights remain evident.

In presenting an overview of space resources in the context of law, property rights, markets and politics, the discussion presented in the present chapter may be developed upon, challenged or clarified through further research in one or more of several areas not considered herein. Among its most apparent limitations is that the present analysis was made predominantly from a Western perspective. Assessing the approaches to property rights, equality and environmental protection of non-Western cultures would complement this analysis and provide insight into how international frameworks can best accommodate variations in the treatment of property rights and desired outcomes. In addition, the present chapter is concerned with the law and economics of property rights in the context of resources in outer space. As such, the analysis omits comparatives, most notably, between the relevance of market-based and political-based property rights concerning maritime, airspace and land-mining activities. As to the latter, a dedicated examination tracing the developments of politically based property rights, and the challenges caused by market-based property rights, concerning natural resources on earth would likely inform decision-makers in the domain of space resource activities on how to achieve an appropriate balance between both forms of property rights in the interests of achieving effectiveness and fairness.

In the absence of intercultural viewpoints and interdisciplinary comparatives, the present chapter nevertheless sought to lay a coherent lineage between the economics of property rights and the potential for markets to contribute to equality and environmental sustainability. Whether market-based property rights are the answer to any of the issues in the domain of space resources management is questionable due to the overwhelming consensus that political institutions are best suited to determine who holds the right to what. Nevertheless, the analysis herein seeks to stimulate discussion among the international and domestic space law community before commissioning development of, and enforcing, of further public legal frameworks on space resources should they be pursued. Through such targeted discussions, it may be found that contributions from apolitical systems, which might have otherwise gone unconsidered, can be incorporated into the global systems for increasing the welfare of humankind and protecting the environment of Earth and outer space.

Scott Schneider is Special Counsel at International Aerospace Law & Policy Group, where his expertise includes approvals and liability management for space activities, law reform and public international law. He holds a Master of Laws in Air and Space Law and teaches law and economics to undergraduates, MBA students and professionals around the world. In 2020, Scott led the team obtaining Australia's first commercial launch and spaceport approvals and has contributed to several amendments in legal and policy frameworks to facilitate safe and competitive aviation and rocketry activities. Scott is the Managing Director of Raker, providing specialist assistance and training in risk management, corporate governance and commercial affairs.

Space Diplomacy Toward a Legal Regime for Space Resource Activities on the Moon

Nancy Riordan, Miloslav Machoň, and Lucia Kozák Csajková

Abstract The development of national policy and legal outputs toward space resource utilization across several countries accelerated after the introduction of the Artemis Accords in 2020. The announcement also further intensified the debate about the necessity of specifying a regime for using space resources on the international level. In 2021, the Legal Subcommittee of the UN Committee on the Peaceful Uses of Outer Space finally approved the establishment of its Working Group on Legal Aspects of Space Resource Activities. Therefore, this chapter builds on diplomacy under conditions of polycentric governance and aims to map the diversity of actors' ideas and interests about governance for space resource utilization on the level of the Working Group. The research process follows a qualitative analysis methodology and assigns the results to the matrix of prospective alternatives for governing space resources. The available materials of the Working Group, including reports and recordings, constitute the core empirical data used in the analysis.

Keywords Artemis Accords · Space resource utilization · International governance · Polycentric diplomacy · UN COPUOS working group · Qualitative analysis

N. Riordan (✉)
University of Massachusetts, Boston, MA, United States
e-mail: Nancy.Riordan@umb.edu

M. Machoň
The Ambis University, Prague, Czech Republic
e-mail: miloslav.machon@ambis.cz

L. K. Csajková
Prague University of Economics and Business, Prague, Czech Republic
e-mail: csal00@vse.cz

S. Bhat B and A. Variath (eds.), *Legal Responses to Space Resource Utilisation*, Space Law and Policy, https://doi.org/10.1007/978-981-95-8864-0_7

1 Introduction

The Working Group on the Legal Aspects of Space Resource Activities (hereafter as the Working Group) was established under the UN Committee on the Peaceful Uses of Outer Space (UNCOPUOS) Legal Subcommittee agenda item "General Exchange of Views on Potential Legal Models for Activities in the Exploration, Exploitation and Utilization of Space Resources" at the sixtieth session of the Legal Subcommittee in 2021 and later renamed in 2022.[1] The establishment of the formal working group was preceded by years of informal international discussions and permissive regulatory measures taken at the national level by a few states, beginning with authorizing commercial activities in U.S. legislation.

While commercial efforts in space had existed, they were significantly supported with the 2010 NASA Authorization Act. It called for an extension of space exploration for which space resource utilization would be needed.[2] Also, it anticipated active U.S. involvement and leadership in commercial space activities, presuming to support commercial endeavors. Within this Act, the United States strived for research, development, space exploration, and international cooperation in the field of space exploration,[3] as well as maintaining safe and affordable space exploration capabilities throughout the next century.[4] These assumptions for space resource utilization were formalized in the U.S. Commercial Space Launch Competitiveness Act of 2015. It elevated commercial activities in space to "promote the right of United States citizens to engage in commercial exploration for and commercial recovery of space resources free from harmful interference, in accordance with the international obligations of the United States and subject to authorization and continuing supervision by the Federal Government."[5] However, the U.S. Act ruled out asserting sovereignty, ownership, or other exclusive rights over any celestial body.[6]

In addition to the United States, other countries expressed a strong interest in the commercial exploitation of space resources. The Luxembourg Space Resources Initiative was formed and later supported by the Law of 20 July 2017 on the exploration and use of space resources, even before Luxembourg established its own

---

[1] UNOOSA, 'Working Group on Space Resources' https://www.unoosa.org/oosa/en/ourwork/copuos/lsc/space-resources/index.html accessed 7 May 2025.

[2] *The Space Review*, 'Putting the White House Executive Order on Space Resources in an International Context' https://thespacereview.com/article/3939/1 accessed 7 May 2025.

[3] Elizabeth M Robinson, 'Statement before the Committee on Commerce, Science and Transportation, United States Senate' (National Aeronautics and Space Administration, 2010) https://www.commerce.senate.gov/services/files/FC2E6930-DCB2-44C0-87E3-F776884B6C10 accessed 7 May 2025.

[4] Elizabeth M Robinson, 'Statement before the Committee on Commerce, Science and Transportation, United States Senate' (National Aeronautics and Space Administration, 2010) https://www.commerce.senate.gov/services/files/FC2E6930-DCB2-44C0-87E3-F776884B6C10 accessed 7 May 2025.

[5] Sen. Rockefeller, *S 3729—NASA Authorization Act of 2010*, 111th Congress (2009–10).

[6] *S 3729—NASA Authorization Act of 2010*, 111th Congress (2009–10).

Space Agency in 2018.[7] The Luxembourg initiative advocates that space resources should be used for peaceful purposes for the common benefit of humankind, and are gathered sustainably and in compliance with international law, such as the Outer Space Treaty.[8] According to this initiative, the resources from Near-Earth Objects should be explored and utilized commercially and indicates that an appropriate legal and regulatory framework for space resource utilization activities should be established.[9] Moreover, the program supported the relevance of polycentric governance[10] and connected diverse public and private organizations to aid in space exploration—including the Luxembourg Government and Ministry of Economy, as well as U.S. and Japanese private companies.

In response to these developments and the intensifying debate on the commercial use of space resources, in 2017, upon a proposal by the Belgian delegation, the topic of space resources first appeared as a single-year agenda item of the UNCOPUOS Legal Subcommittee (LSC). This became agenda item 14, General Exchange of Views on Potential Legal Models for Activities in Exploration, Exploitation and Utilization of Space Resources, and received ever-increasing attention by the legal subcommittee.[11]

Furthermore, in 2016, the Hague International Space Resources Governance Working Group was formed, following ongoing informal discussions and a Roundtable on the Governance of Space Resources, convened by The Hague Institute for Global Justice on 1 December 2014. The group's responsibility was to ensure that the discussion about space resource activities was meeting the existing treaty obligations, regarding on-orbit operations and space resource rights. In November of 2019, the Hague Working Group adopted the Building Blocks for the Development of an International Framework on Space Resource Activities (hereafter as the Building Blocks). The Building Blocks have since served as a key reference for negotiations in the Working Group.[12]

The commercial use of space resources became a more significant topic and could be observed during the ongoing sessions of the UNCOPUOS Legal Subcommittee. By the fifty-seventh session of the Legal Subcommittee held in Vienna from 9 to

---

[7] *Journal Officiel du Grand-Duché de Luxembourg: Loi du 20 juillet 2017 sur l'exploration et l'utilisation des ressources de l'espace* (Legilux, 2017) https://legilux.public.lu/eli/etat/leg/loi/2017/07/20/a674/jo accessed 7 May 2025.

[8] SpaceResources.lu, 'The SpaceResources.Lu Initiative Positions Luxembourg as a Pioneer in the Exploration and Utilisation of Space Resources' https://www.spaceresources.lu accessed 7 May 2025.

[9] SpaceResources.lu, 'The SpaceResources.Lu Initiative Positions Luxembourg as a Pioneer in the Exploration and Utilisation of Space Resources' https://www.spaceresources.lu accessed 7 May 2025.

[10] Sung-Wook Kim, 'Is Global Governance Fragmented, Polycentric, or Complex?' (2017) 24 *Global Governance* 905, 905–6.

[11] Gouvernement.lu, 'SpaceResources.Lu: The Luxembourg Government Becomes a Key Shareholder of Planetary Resources, Inc.' (2016) https://gouvernement.lu accessed 7 May 2025.

[12] Universiteit Leiden, 'The Hague International Space Resources Governance Working Group' https://www.universiteitleiden.nl/en/law/institute-of-public-law/institute-of-air-space-law/the-hague-space-resources-governance-working-group accessed 7 May 2025.

20 April 2018, the need to ensure sustainability of space activities was stressed by several delegations, and that the LSC would require a new impulse to keep up with current trends of commercialization and privatization of outer space.[13]

Linking to the previous discussions, in the fifty-eighth session, Belgium and Greece introduced a working paper containing a proposal for working methods and a work plan for a new working group on legal aspects of the exploration, utilization and exploitation of space resources.[14] The initiative of Belgium and Greece and the following discussion on this topic were supported by various delegations. However, not all delegations supported the formation of such a group. According to the opposing views, the legal framework was sufficient in regard to space activities and the exploration, exploitation, and utilization of space objects were still being innovated and forming a working group on legal aspects could hinder these innovations. If the working group were to be established, it should encourage public and private investment in this area, as well as respect individual rights and collective interests in the commercial exploitation of space objects. According to the members, it was essential that the discussion, in the potential working group, should also include scientific, economic, technical, and legal experts.[15]

Reflecting on these developments in the international arena, by April 2020, the United States issued Executive Order 13,914 "Encouraging International Support for the Recovery and Use of Space Resources." The commercial activities and utilization of lunar resources were backed by this new executive order that encouraged commercial partners to participate in the U.S. "innovative and sustainable program,"[16] in order to return humans to the Moon and launch a mission to Mars. Although, according to this executive order, the United States would not view outer space as a global common, the Secretary of State should negotiate and arrange bilateral and multilateral agreements with foreign states.[17] This executive order would constitute a crucial part in achieving the objectives of the Artemis program and is one of the foundations for the future of a U.S. commercial space economy.[18] Later that year, in October

---

[13] UNOOSA, *Report of the Legal Subcommittee on Its Fifty-Seventh Session, Held in Vienna from 9 to 20 April 2018*, UN Doc A/AC.105/1177 (2018).

[14] UNOOSA, *Proposal for Working Methods and Work Plan of the Working Group on Legal Aspects of the Exploration, the Utilization and the Exploitation of Space Resources, with Reference to Document A/AC.105/C.2/L.311: Working Paper by Belgium and Greece*, UN Doc A/AC.105/C.2/2019/CRP.26 (2019) 1–2.

[15] UNOOSA, *Report of the Legal Subcommittee on Its Fifty-Eighth Session, Held in Vienna from 1 to 12 April 2019*, UN Doc A/AC.105/1203 (2019).

[16] *Executive Order 13914: Encouraging International Support for the Recovery and Use of Space Resources* (2020) Federal Register.

[17] *Executive Order 13914: Encouraging International Support for the Recovery and Use of Space Resources* (2020) Federal Register.

[18] Space Review, "The Space Review: Putting the White House Executive Order on Space Resources in an International Context."

2020, the U.S.-led Artemis Accords were announced.[19] The Artemis Accords reinforce that space resource extraction and utilization can and will be conducted under the auspices of the Outer Space Treaty, with specific emphasis on Articles II, VI, and XI.

After introduction of the Artemis Accords, the variety of national and international debates and initiatives accelerated and deepened.[20] Therefore, the aim of our research is to understand a diversity of actors' interests and emerging ideas about governing space resource utilization with an emphasis on the Moon in the Working Group. Understanding diplomatic negotiations under conditions of polycentrism[21] has the potential to deepen knowledge about the requirements for achieving a compromise between diverse actors with unique political demands, socio-economics, and cultures. Also, it opens avenues for future theorizing about whose ideas about governance goals under a schema for common pool resources that are as yet not in use as such (future) but are evolving and how.

The research methodology had three steps. First, we began with a narrative literature review approach intended for topics that have been conceptualized differently and studied by various groups of researchers within diverse disciplines[22] in order to detect relevant emerging themes, theoretical perspectives, and common issues.[23] A Web of Science advanced bibliometric search and analysis for 1984–2024 yielded 1,578 documents published in the Web of Science Core Collection using topic search terms "lunar resources." Of these, 998 sources were published between 2015 and 2024 with the most articles, book chapters, and other sources published in 2023. Further refining the process, we also used the HeinOnline database (a collection of legal and other journals) using search terms "Lunar resource governance," yielding 448 results, with the most research published on Lunar resource governance between 2020 and 2024.

Second, the results of the literature review highlighted the space diplomacy mechanisms conceptualized by Cross and Pekkanen[24] and the model for governing space resources introduced by Yap et al.[25] to systematize trends in starting diplomatic negotiations at the Working Group. The space diplomacy conceptualization outlined

[19] Christian Riordan, František Machoň and Katarína Csajková, 'Space Diplomacy and the Artemis Accords' (2021) 19 *Space Policy* 1, 19.

[20] Christian Riordan, František Machoň and Katarína Csajková, 'Space Diplomacy and the Artemis Accords' (2021) 19 *Space Policy* 1, 19.

[21] Christian Dorsch and Christian Flachsland, 'A Polycentric Approach to Global Climate Governance' (2017) 53 *Global Environmental Politics* 53; Jan Scholte, 'From Government to Governance' (2001) *Global Governance* 56.

[22] Gillian Wong and others, 'RAMESES Publication Standards' (2013) 2 *BMC Medical Research Methodology* 2.

[23] Hannah Snyder, 'Literature Review as a Research Methodology' (2019) 57 *Journal of Business Research* 335.

[24] Allan Cross and Saadia Pekkanen, 'Introduction: Space Diplomacy' (2020) 2 *Space Diplomacy Review* 202–6.

[25] Wendy Yap and others, 'Four Alternative Scenarios of Commons in Space' (2018) 12 *Futures* 397–402.

communication, persuasion, and bargaining as three common mechanisms for diplomatic activities of governmental and non-governmental actors. The matrix model for governing space commons by Yap et al. recognized two dimensions. The first dimension covered the distribution and resource access; the second one dealt with Earth-bound sustainability. It stemmed from global governance studies elaborating aspects of non-rivalry and non-excludability in IR,[26] future studies conceptualized and promoted scenario models for predictions,[27] and science and technology studies connected the scenario analysis with a socio-technological nexus.[28]

Third, a qualitative analysis strategy constituted the core of methodologies used. Applying the rules of systematic sampling,[29] we compiled submitted materials from countries and observer organizations 2022–2023 publicly available from the Working Group, including reports and recordings. In the first phase, qualitative content analysis followed the rules of structured coding to sort the collected data into two categories: government inputs and observer (non-government) inputs to the Working Group.[30] The pattern coding, which is the analogy for cluster analysis used in quantitative fields, constituted the second cycle of the qualitative analysis. We then combined the summaries from the previous cycle into the four categories developed by Yap et al.[31] This process revealed the meaning within the data and integrated this into higher-order key themes to explain initial codes in a more general and thematic way.[32]

The chapter has six sections. It begins with a theoretical discussion about polycentricity followed by its manifestations in space diplomacy. The second section reviews a selection of existing literature on space resource utilization and governance. The third section provides operationalization of the Yap et al., model including four scenarios for global common governance in the context of space diplomacy strategies. The fourth section provides the results of the qualitative analysis. Fifth, the discussion confronts the results with the latest findings on space resource utilization identified in the literature review. The conclusion offers suggestions for future research.

[26] Inge Kaul and Ronald Mendoza, 'Advancing the Concept of Public Goods' (2003) 9 *Global Public Goods Review* 82–83.

[27] Oliver Schwarz, 'Assessing the Future of Futures Studies in Management' (2010) 45 *Futures* 238–39.

[28] Sheila Jasanoff, *Future Imperfect: Science, Technology, and the Imaginations of Modernity* (University of Chicago Press 2015) 7–9.

[29] Klaus Krippendorff, *Content Analysis: An Introduction to Its Methodology* (3rd edn, SAGE 2013) 118–19, 213.

[30] Johnny Saldaña, *The Coding Manual for Qualitative Researchers* (3rd edn, SAGE 2016) 10, 238.

[31] Johnny Saldaña, *The Coding Manual for Qualitative Researchers* (3rd edn, SAGE 2016) 10, 238.

[32] Jane Ritchie and others, *Qualitative Research Practice: A Guide for Social Science Students and Researchers* (SAGE 2013) 271.Karl Polanyi, *The Logic of Liberty* (Routledge 1951) 184.

2 Toward Polycentric Governance

In the 1950s, Michael Polanyi dealt with managing polycentric tasks for which he emphasized an element of mutual harmonization of individual activities.[33] Following his works, Elinor Ostrom elaborated his concepts and labeled polycentricity a useful analytical approach for understanding climate change governance, because it can capture diverse actors' actions participating in natural resource, environmental, and climate governance at different scales.[34] In the 1990s, polycentricity was adopted in International Relations (IR) for studying the sustainability of common pool resources under conditions of global interdependence,[35] because the globalization processes significantly blurred the boundaries between the foreign actions of governmental and external activities of non-governmental actors and deepened interdependence between them.[36] Polycentricity admits adjusting and aligning toward a collective goal in conditions of decentralism.[37] Therefore, the polycentric approach presumes the fragmentation of state and competitive growth of public–private modes of governance with larger and smaller jurisdictions than a country domain.[38]

In 2012, Aligica and Tarko conceptualized monocentrism as the opposite of the polycentric mode of governance.[39] These modes of governance differ in two fundamental aspects.[40] First, centralization is a prerequisite of monocentric policy processes, whereas polycentrism allows a far more dispersed formulation, implementation, and enforcement of social rules to reduce an individual failure.[41] Second, the monocentric mode of governance includes only government as the relevant actor, while polycentric governance invites the empirical diversity of global governance. It broadens the analytical scope to capture various kinds of actors and their diverse interactions. Theorizing about the diversity of interactions distinguishes local governments, national governments, regional institutions, transnational companies, and

---

[33] Michael Polanyi, *The Logic of Liberty* (Routledge 1951) 184.

[34] Dominik Dorsch and Christian Flachsland, 'A Polycentric Approach to Global Climate Governance' (2017) 17(1) *Global Environmental Politics* 45, 47.

[35] Robert O Keohane and Elinor Ostrom (eds), *Local Commons and Global Interdependence: Heterogeneity and Cooperation in Two Domains* (Sage 1995) 3–5.

[36] Wilfried Bolewski, 'Compass for Public/Private Management in Turbulent Times' (2012) 9(1) *Public Organization Review* 1, 8.

[37] Sungjoon Kim, 'Is Global Governance Fragmented, Polycentric, or Complex?' (2013) 16(4) *International Studies Review* 903, 905–6.

[38] Jan Aart Scholte, 'From Government to Governance' (2004) 50 *International Political Science Review* 57, 67.

[39] Paul Dragos Aligica and Vlad Tarko, 'Polycentricity: From Polanyi to Ostrom, and Beyond' (2012) 25(2) *Governance* 237, 244.

[40] Jan Aart Scholte, 'From Government to Governance' (2004) 50 *International Political Science Review 49.*

[41] Ahmed Alshamy, Denys Kiriukhin and Oleksandr Yukhymenko, 'Polycentric Defense, Ukraine Style' (2023) 64(2) *Survival* 37, 40.

distinguished individuals.[42] These diverse actors with local, national, regional, and global jurisdictions carry out societal regulation across public and private sectors, including their hybrid public–private forms.[43]

From an analytical perspective, monocentric governance primarily focuses on a country level, whereas polycentric governance recognizes several spatial scales.[44] While monocentrism emphasizes neatly separated territorial jurisdictions, polycentrism often involves overlapping jurisdictions. Following empirical results, considering the variety of spatial configurations allows tackling diverse problems on spatial scales effectively, supporting the legitimacy of decision-making processes and their outputs.[45] A distinction also lies in allowable modes of regulatory control. While supreme, comprehensive, and exclusive regulatory control is acceptable for monocentrism, polycentrism labels this form of control as undesirable and prioritizes conditions in which no actor of governance can aspire for a predominant position. Research on polycentricity focuses on autonomous external policies and activities of sub-governmental actors to highlight their foreign policy-making competencies.[46]

In the context of IR, polycentrism opens avenues for studying the self-organization and coordinated networking of these diverse actors in policy arenas in specific contexts.[47] The self-organization principle claims the nondominant position of an individual or a group of actors and secures the freedom to set up their own rules. The site-specific context—i.e., a multilateral institution such as the UN Committee on the Peaceful Uses of Outer Space (COPUOS) and resources in areas beyond national jurisdiction (ABNJ)—implies a polycentric task for an individual or a group of actors. These tendencies highlight the relevance of global governance networks in which institutional ties and collaboration accelerate negotiations among diverse actors in international policy arenas.[48] Understanding diplomatic negotiations under polycentrism has the potential to deepen knowledge about the requirements for achieving a compromise between diverse actors with unique socio-economic social bases, political interests, and cultures.[49] It may also clarify conditions for achieving governance

---

[42] David A Baldwin, Thomas A Börzel and Adrienne Héritier, 'Empirical Research on Polycentric Governance' (2019) 28(2) *Perspectives on Politics* 319, 322–23.

[43] Dominik Dorsch and Christian Flachsland, 'A Polycentric Approach to Global Climate Governance' (2017) 17(1) *Global Environmental Politics*.

[44] Jan Aart Scholte, 'From Government to Governance' (2004) 50 *International Political Science Review* 49.

[45] Björn Newig, Daniel Schulz and Frauke Jager, 'Disentangling Puzzles of Spatial Scales and Participation in Environmental Governance—The Case of Governance Re-Scaling Through the European Water Framework Directive' (2016) 46 *Environmental Policy and Governance* 999, 1000–1001.

[46] Bernhard Reinsberg and Sebastian Dellepiane, 'The Domestic Sources of Sub-State Foreign Policymaking' (2020) 25(6) *Journal of European Public Policy* 922, 927–28.

[47] Tiffany Morrison, John Dryzek, Robyn Eckersley and Hayley Stevenson, 'Building Blocks of Polycentric Governance' (2019) 24(3) *Global Environmental Politics* 476, 486–89.

[48] Sungjoon Kim, 'Is Global Governance Fragmented, Polycentric, or Complex?' (2013) 16(4) *International Studies Review* 903, 905–6.

[49] Graham Smith, 'Innovating for the Global Commons' (2019) 25(2) *Environmental Politics* 52, 56.

goals under schema for common pool resources, which are as yet not in use as such (future).[50]

2.1 *Specifics of Diplomacy Under Conditions of Polycentrism*

Whereas, under conditions of monocentrism, diplomacy has always involved a diplomat who exclusively represented one nation-state in the territorial jurisdictions of another state, in the case of public diplomacy, polycentrism recognizes separate governance agencies with diversity in their missions and the responsibility to represent a government in foreign territories.[51] The prerequisite for a diversity of actors under conditions of polycentrism stems from democratization trends.[52] In democratic regimes, national governments usually support a variety of sub-governmental and transgovernmental actors for greater political autonomy to supplement governmental roles, covering para-diplomatic and external relations agendas.[53]

These diverse actors with ambitions to participate in space resource utilization have acquired access to evolving multi stakeholder diplomatic arenas to provide public goods and generate soft power. Therefore, they became diplomatic co-actors under conditions of polycentrism.[54] In doing so, they have developed specialized departments in their organizational structures to deal with external relations, which are equivalent to governmental foreign affairs ministries. The newly created departments acquired internal responsibility for negotiations with external actors to obtain access to policy arenas and markets and intensify relations with governments and other actors in global civic society through partnerships and strategic coordination or acting as independent actors.[55]

The integration of diplomatic practices into the portfolio of their strategic planning activities helped to increase the predictability of their operations and interact with external actors in the complex and volatile global environment.[56] These new actors in diplomatic negotiations gained an opportunity to shape human rights, improve

---

[50] Dominik Dorsch and Christian Flachsland, 'A Polycentric Approach to Global Climate Governance' (2017) 17(1) *Global Environmental Politics* 53.

[51] Jan Aart Scholte, 'From Government to Governance' (2004) 50 *International Political Science Review* 56.

[52] Megan Bloodgood, 'New Multilateralism: The United Nations and Governance in the Era of Nonstate Actors' (2009) 16(1) *Global Governance* 59, 611.

[53] Benjamin Leffel, 'Principles of Modern City Diplomacy and the Expanding Role of Cities in Foreign Policy' (2021) 22(2) *Journal of Urban Affairs* 179, 183–84.

[54] Bolewski, "Compass for Public/Private Management in Turbulent Times," 8.

[55] Vincent Guilbaud, 'Diplomacy by Non-State Actors' (2017) 14(2) *Place Branding and Public Diplomacy* 187, 189; Dirk P White and Wilfried Bolewski, 'Corporate Diplomacy' (2017) 6(2) *Journal of International Business Policy* 61, 62–63.

[56] Huub Ruël, 'Multinational Corporations as Diplomatic Actors' (2013) 9(1) *The Hague Journal of Diplomacy* 1, 2.

global economic stability, promote global peace, and resolve local social conflicts.[57] They represent diverse overlapping and interweaving collective identities, including substate authorities, communities, and distinguished individuals.[58] Accompanying governmental officials at their diplomatic visits enables forming partnerships and coordinating activities with the officials.[59] These trends spill over into communication styles. Under polycentric governance conditions, a more diverse group of actors has the ability to communicate messages externally.[60] Therefore, diplomatic negotiations became more complex due to the larger number of diverse actors involved in diplomacy, which also increases aspects of cultural diversity.

Regarding the Working Group participants in particular, these new actors can provide consultations in preparing multilateral negotiations or specialized conferences to articulate their interests and protect the existing legal environment. Moreover, if their activities in the global environment aim to provide economically and socially sustainable solutions, they have the potential to increase the legitimacy and trust of their activities for global society and enhance the image of the country of their origin.[61] Ensuring social and economic responsibility became an internally organizational form of governance for addressing sustainability issues. These new actors developed internal mechanisms for monitoring compliance with sustainability principles themselves, like codes of conduct.[62]

According to Cashore et al., interactions between traditional diplomats and new actors can take three forms: complementarity, competition, and coexistence. They act complementarily when they combine resources and talents to minimize transaction costs and mutually reinforce the same demand in negotiations.[63] The situation could be classified as competition when an antagonism for limited resources exists between traditional and new diplomatic actors aiming to monopolize power over these resources. Coexistence exists when traditional and new diplomatic actors cohabitate in an issue area, but traditional diplomats apply divergent goals and modes of governance.[64] Even if the actors in co-existent settings can act simultaneously,

---

[57] Huub Ruël, 'Multinational Corporations as Diplomatic Actors' (2013) 9(1) *The Hague Journal of Diplomacy* 1, 2.

[58] Richard Jackson, 'Paradiplomacy and Political Geography' (2008) 32(3) *Political Geography* 1, 5.

[59] White and Bolewski, "Corporate Diplomacy," 64–67.

[60] Jan Aart Scholte, 'From Government to Governance' (2004) 50 *International Political Science Review* 56.

[61] Wilfried Bolewski, 'Compass for Public/Private Management in Turbulent Times' (2012) 9(1) *Public Organization Review* 144.

[62] Steffen Renckens, *Private Governance and Public Authority: Regulating Sustainability in a Global Economy* (Cambridge University Press 2020) 1–3.

[63] Milton Milstein, 'Capitalism and Democracy' (2015) 33(1) *Critical Sociology* 7, 9.

[64] Benjamin Cashore, Jette Steen Knudsen, Jeremy Moon and Hamish van der Ven, 'Private Authority and Public Policy Interactions in Global Context: Governance Spheres for Problem Solving' (2021) 15(4) *Regulation & Governance* 1166, 8–10.

the coexistence may not be classified as harmonious.[65] In cases of the unpopularity of governments in foreign territories, these new actors have the potential to serve as balancers in public diplomacy.[66]

2.2 Literature Review

One recurring theme across the last two decades of research on lunar governance involves complementary and contrasting responsibilities of governments and private actors in using space resources, usually resulting in policy recommendations for adaptive governance.[67] A previous study introduced structuring data on public–private interactions through scenarios to effectively display the advantages and disadvantages of each form of space governance effectively.[68] These authors also raised a dilemma between naturalist and positivist legal perspectives on whether existing international legal principles for property rights could be applied to space resources or not. Theorizing on historical experiences with human settlements of remote regions to build alternative predictions for lunar governance has gained popularity in scholarly journals across social science disciplines.[69]

Also, common among the literature is a growing recognition of the need for international cooperation to manage space resources effectively and sustainably.[70] Some authors provided comprehensive interpretations of the common heritage principle,[71] reflecting the recent needs of actors interested in lunar activities. The theorizing

65 Brian Milstein, 'Capitalism and Democracy: Complementarity, Complicity, Conflict, Compatibility' in Albena Azmanova and James Chamberlain (eds), *Capitalism, Democracy, Socialism: Critical Debates* (Springer 2022) 19.

66 Candace L White and Wilfried Bolewski, 'Corporate Diplomacy' in A Research Agenda for Public Diplomacy (Edward Elgar 2023) 59–73, 63–67.

67 Lotta Viikari, 'Natural Resources of the Moon and Legal Regulation' in Viorel Badescu (ed), *Moon: Prospective Energy and Material Resources* (Springer 2012) 519–52; Tanja Masson-Zwaan and Mark J Sundahl, 'National and International Norms towards the Governance of Commercial Space Resource Activity' in *Routledge Handbook of Commercial Space Law* (Routledge 2023).

68 Ram Jakhu and Maria Buzdugan, 'Development of the Natural Resources of the Moon and Other Celestial Bodies: Economic and Legal Aspects' (2008) 6(3) *Astropolitics* 201, 206, 228.

69 James Clay Moltz, 'Toward Cooperation or Conflict on the Moon? Considering Lunar Governance in Historical Perspective' (2009) 3(3) *Strategic Studies Quarterly* 82, 89–90; Clare M Mouat, Erika J E Techera, Lies Notebaert, Meredith Blake and Renae Barker, '(Un)Earthly Governance: Beyond Functional Frameworks to Flourishing Spacescapes' (2021) 13(2) *Journal of Property, Planning and Environmental Law* 122, 152–53.

70 Moltz (n above); Gershon Hasin, 'From "Space Law" to "Space Governance": A Policy-Oriented Perspective on International Law and Outer Space Activities' (SSRN Scholarly Paper, 3 August 2022) 413; Masson-Zwaan and Sundahl (n above); Melissa de Zwart, Stacey Henderson and Michelle Neumann, 'Space Resource Activities and the Evolution of International Space Law' (2023) 211 *Acta Astronautica* 155.

71 Following recent discussion on preference of using gender-neutral terms, this chapter uses "common heritage of humankind" term to replace a well-established, but gender-biased terms "common heritage of humankind." "province of all mankind": Cassandra Steer, "The Province of

included correlation of the non-appropriation and common heritage principles used in the Outer Space Treaty and the Moon Agreement as legal concepts for setting relationships between spacefaring and non-space-faring countries.[72] Studies also used the similarities and differences in international legal regimes for evaluation and comparison of domestic legislation regulating the use of outer space natural resources.[73] Moreover, some authors used the current state of the international legal regime for proposing prospective legal innovations for lunar resource utilization mining like environmental impact assessments[74] and priority rights for natural resources in space.[75]

Stakeholders and experts are increasingly advocating space governance approaches that provide flexible, policy-oriented frameworks to adapt to evolving technological capabilities.[76] They especially underline many unresolved issues in commercial lunar mining activities, as well as in the context of potential environmental degradation of these resources, for which the flexibility of governance mechanisms would be desirable. Besides the aspect of uncertainty of uses of natural space resources, Hasin[77] considered a space governance model, applicable also for lunar activities, based on the needs and desired outcomes of diverse actors as the most promising solution for achieving equal access to outer space and benefits from resource extraction.[78] Following the practical experiences in previous international policy and legal negotiations, some authors admitted that creating a multilateral regime for future space resource activities could be "a complex and lengthy process"[79] that will need considerable activity of non-governmental organizations to prepare an environment for reaching a consensus among countries. This process needs to include the diverse interests of different stakeholders, commercial actors in

All Humankind' – A Feminist Analysis of Space Law' in Melissa de Zwart and Stacey Henderson (eds), *Commercial and Military Uses of Outer Space* (Springer 2021) 169–88, 182–83.

[72] Guoyu Wang and Xinyi Huang, 'On the Common Heritage of Mankind Principle in Space' (2023) 211 *Acta Astronautica* 926, 230–32.

[73] Melissa de Zwart, Stacey Henderson and Michelle Neumann, 'Space Resource Activities and the Evolution of International Space Law' (2023) 211 *Acta Astronautica* 157–59; Mahulena Hofmann, 'Luxembourg Completing Its Space Legislation' (Elsevier, 2022) 524–26.

[74] Anne-Sophie Martin and Steven Freeland, 'Back to the Moon and Beyond: Strengthening the Legal Framework for Protection of the Space Environment' (2021) 46(3) *Air and Space Law* 423, 428–29.

[75] Guoyu Wang and Xinyi Huang, 'The Establishment of the Priority Right in Space Resources Activities' (2024) 73(1) *Advances in Space Research* 918, 920–23.

[76] Clive R Neal and others, 'The Moon Needs an International Lunar Resource Prospecting Campaign' (2024) 214 *Acta Astronautica* 737, 743–44.

[77] Hasin, "From 'Space Law' to 'Space Governance,'" 413.

[78] Maya Nasr, Jeffrey Hoffman, Tanja Masson-Zwaan, Donald Rapp and Dava Newman, 'A Policy Framework for Sustainable and Equitable Space Resource Utilization' (SSRN Scholarly Paper, 2 March 2023).

[79] Masson-Zwaan and Sundahl, "National and International Norms towards the Governance of Commercial Space Resource Activity," 402.

particular.[80] Experts close to the negotiations also found that "the future lunar legal landscape may well comprise international and national law evolving in parallel, at least for the near future."[81]

3 Governing Space Resources

The trends in diplomacy under polycentrism and literature review helped to highlight the space diplomacy mechanisms outlined by Cross and Pekknanen[82] and four scenarios for space governance by Yap et al.[83] as relevant approaches for operationalization of the stated research aim and further qualitative sorting of the collected data.

The model for governing commons in space introduced by Yap et. al, in 2023, proposes four alternative scenarios. The specific scenarios include Earth-space sustainability, Open space, Earth-centric gold rush, and Space cartel.[84] Using space diplomatic mechanisms to operationalize these scenarios led to an underlining of the activities of scientific research networks. The networks aim to collaborate and reach shared goals. This tendency contrasts with governmental actors who pursue the fulfillment of national self-interests.[85] On the level of diplomatic negotiations, these dichotomies open space for operationalizing the four alternative scenarios for predicting possible negotiation outcomes.

---

[80] Jakhu and Buzdugan, "Development of the Natural Resources of the Moon and Other Celestial Bodies," 228–32; Moltz, "Toward Cooperation or Conflict on the Moon?," 100; Hasin, "From 'Space Law' to 'Space Governance,'" 430; Masson-Zwaan and Sundahl, "National and International Norms towards the Governance of Commercial Space Resource Activity," 399–402.

[81] Mark J Sundahl and Tanja Masson-Zwaan, 'The Lunar Legal Landscape: Challenges and Opportunities' (2021) 46(1) *Air and Space Law* 1, 54.

[82] Mai'a K Davis Cross and Saadia M Pekkanen, 'Introduction. Space Diplomacy: The Final Frontier of Theory and Practice' (2023) 18(2–3) *The Hague Journal of Diplomacy* 193, 202–06.

[83] Xiao-Shan Yap and others, 'Four Alternative Scenarios of Commons in Space: Prospects and Challenges' (2023) 17(1) *International Journal of the Commons* 390, 397–402.

[84] Xiao-Shan Yap and others, 'Four Alternative Scenarios of Commons in Space: Prospects and Challenges' (2023) 17(1) *International Journal of the Commons* 390, 397–402.

[85] Cross and Pekkanen, "Introduction. Space Diplomacy," 204.

3.1 Earth-Space Sustainability

In the Earth-space sustainability scenario, the ideas of a space ecological epistemic community promoting environmental protection of outer space are the most relevant.[86] Its concepts shape and facilitate an information exchange within groups of diplomats and epistemic communities. Also, it frames outreach initiatives for the general public toward accomplishing a shared sustainability goal for the Earth and outer space.[87] The scenario introduces sustainability as a complex and relevant issue and the Earth as a non-isolated space object. The sustainability concept is embedded in "diplomacy for science"[88] policies to shape the top-down research priorities and bottom-up collaboration between individual researchers and their interactions with political and diplomatic officials. Implementing diplomacy for science policies involves negotiating and reaching agreements supporting participation in multinational programs and building major research infrastructures.[89] In these activities, the business sector emphasizes the philanthropist role as it shares benefits from outer space across regions on Earth. Therefore, science communicators frame space activities as opportunities for solutions to present-day challenges such as mitigating the effects of climate change and minimizing environmental degradation. As this scenario emphasizes a fair distribution of space resources among governmental and non-governmental actors across countries at different developmental stages, it stimulates diversity and inclusion of global audiences toward space activities. Moreover, the competition among countries is less intense, so it lowers the developmental gap between developed and developing countries, improves coordination, and provides equal access to space resources. Countries with a status of "followers"[90] have guaranteed access to the means for space activities and space resources, while achieving democracy in their political systems represents the top priority.

3.2 Open Space (Space Utopia)

In the open space model, actors operate in multiple multi-level governance settings. It encompasses global, regional, national, and local levels. A mixture of larger and smaller business ventures replaces the exclusivity of multinational corporations

---

[86] Aurélie Trur, 'Governance Aspects of Space Sustainability: The Role of Epistemic Actors as Enablers of Progress' (2021) 180 *Acta Astronautica* 451, 453; Mai'a K Davis Cross, 'Outer Space and the Idea of the Global Commons' (2021) 35(3) *International Relations* 384, 386.

[87] Cross and Pekkanen, "Introduction. Space Diplomacy," 202–3.

[88] The Royal Society, 'New Frontiers in Science Diplomacy—Navigating the Changing Balance of Power' (2020) 11(54) *Revista Conjuntura Austral* 9, 9.

[89] Pierre-Bruno Ruffini, *Science and Diplomacy: A New Dimension of International Relations* (Springer 2017) 12.

[90] Yap et al., "Four Alternative Scenarios of Commons in Space," 401–2.

and implies double regulatory requirements, hindering innovation and commercialization. The model aligns with recent trends in environmental governance, which assume decentralization and localization of agendas. Therefore, diverse space actors consider various environmental and societal factors, including location, economic development, and specific knowledge discourse. The open space governance setting itself indicates fragmentation from an external perspective. Internally, the diversity of actors and their needs establishes cross-scale communication as a priority to use a market-driven rational choice for avoiding the tragedy of space commons to secure the sustainability of their space activities.[91,92] Although the open space model provides opportunities for high competition among actors, sustainability as a prevalent scientific paradigm provides a basis for building trust among the diverse actors.[93,94] It makes their actions confidential. Together with a need to solve global challenges being out of traditional governmental control and a prevalent narrative highlighting science's ability to rationalize and pacify egoistic self-interest in politics via providing complex information, the trust among diverse actors in the open space model allows spillover of confidentiality into non-scientific dimensions of politics to provide a foundation for building or improving diplomatic relations, and fostering cross-border policy integration and international policy collaboration.[95,96]

3.3 Earth-centric Gold Rush

The Earth-centric gold rush scenario and the space cartel scenario share an assumption. In both, large multinational companies are the primary actors responsible for utilizing outer space and have strong financial, political, and legal support from the governments of space-faring countries. The difference between scenarios lies in the framing associated with persuasion targeting, changing, and uniforming the

---

[91] Yap et al., 400–401.

[92] Robert G Patman and Lloyd S Davis, 'Science Diplomacy in the Indo-Pacific Region: A Mixed but Promising Experience' (2017) 45(5) *Politics & Policy* 862, 865.

[93] Tim Flink, 'Taking the Pulse of Science Diplomacy and Developing Practices of Valuation' (2022) 49(2) *Science and Public Policy* 191, 192–94.

[94] Tim Flink, 'Taking the Pulse of Science Diplomacy and Developing Practices of Valuation' (2022) 49(2) *Science and Public Policy* 191, 192–94.

[95] Meredith L Gore, Elizabeth S Nichols and Karen R Lips, 'Preparing Scientists for Science Diplomacy Requires New Science Policy Bridges' (2020) 15(3) *The Hague Journal of Diplomacy* 424, 430.

[96] Ping Su and Maximilian Mayer, 'Science Diplomacy and Trust Building: "Science China" in the Arctic' (2018) 9(S3) *Global Policy* 23, 24, 26; Charlotte Rungius and Tim Flink, 'Romancing Science for Global Solutions: On Narratives and Interpretative Schemas of Science Diplomacy' (2020) 7 *Humanities and Social Sciences Communications* 1, 2, 4.

attitudes of involved actors.[97,98] Therefore, this diplomatic strategy has the ability to provide more profound knowledge for a more persuasive argument to change the frame of understanding of space exploration for facilitating diplomatic negotiations and reaching a consensus on a new set of global values.[99] While the space cartel scenario frames the activities of multinational private actors as the most crucial aspect to accelerate accomplishing scientific and technological progress and narrows their regulation to non-binding best practices, the Earth-centric gold rush scenario uses the mitigation of environmental degradation within national territories as shared norms, principles, history, and motivation to appropriate use of space resources. So, mitigation of transboundary environmental degradation provides new aspects for a preferred way of thinking, casual interpretation, and moral evaluation to serve as minilateral agreements, which raises the predictability and manageability of policy consequences stemming from the implementation of this scenario.[100] However, from a future perspective, minilateralism indicates a high probability of tragedy for the outer space commons.[101]

3.4 Space Cartel

The Working Group on Legal Aspects of Space Resource Activities has existed since 2021 under the COPUOS Legal Subcommittee. As of 30 August 2024, nineteen state members and seven permanent observer organizations submitted proposals (see Appendix 1). The set formed the core of the qualitative analysis under the four scenarios of space governance.

In the space cartel scenario, national governments in leading space-faring countries delegate responsibilities for implementing outer space activities to technologically advanced private actors due to their innovative potential. Governments prioritize establishing public–private partnerships, prioritizing large private companies as the most relevant actors for outer space's use and exploration. Public–private partnerships respect the conceptualized national interest. So, their activities aim to avoid costly domestic or international conflicts and protect resources within national

[97] Christopher Provis, *Negotiation, Persuasion and Argument* (Ashgate 2004) 99; Cecilia J Morales and Kent C Berridge, "Liking' and 'Wanting' in Eating and Food Reward' (2020) *Philosophical Transactions of the Royal Society B: Biological Sciences* 375 (1797).

[98] Bettina Bosse, Martin Höpner and Marco Vieira, 'Persuading the Giant? Explaining the European Union's Position on Investor-State Dispute Settlement' (2019) 26 *Journal of European Public Policy* 289.

[99] Christopher Provis, *Negotiation, Persuasion and Argument* (Ashgate 2004).

[100] Rianne Dekker and Peter Scholten, 'Framing the Immigration Policy Agenda: A Qualitative Content Analysis of Media and Political Debates in the Netherlands' (2017) 65 *International Journal of Public Administration* 197.

[101] Yap Jason, et al., 'Four Alternative Scenarios of Commons in Space: Towards a Governance Model' (2021) 47 *Space Policy* 396.

borders from internal and external threats. The exclusiveness of public–private partnerships opens doors for private actors who represent governmental sovereigns and provide a basis for their hegemonic aspirations.[102] "*Strategic negotiations result in tit-for-tat trade-offs among space diplomats (state or non-state) who seek to maximize their state's self-interest through bargaining with other states.*"[103] Private actors' activities follow the "first-come-first-served"[104] principle for using and occupying space resources. The growing role of private actors would lead to fragmentation of international coordination and slow down negotiations about a new comprehensive regulative instrument for global space activities. Instead, international diplomatic negotiations will aim for specific, self-coordinated, and non-binding best practices aligning with private actors' interests, which would marginalize principles for global justice and sustainable development and their enforcement.[105] Private actors lead diplomatic bargaining aiming to promote national policy objectives in international arenas.[106] The range of policy implications and benefits stemming from promoting a state's foreign policy interests may increase if governmental actors use epistemic motivation and social value justification, having reciprocal fairness at its core.[107]

4 The Alternative Scenarios Applied to the Working Group Negotiations

The Working Group on Legal Aspects of Space Resource Activities has existed since 2021 under the COPUOS Legal Subcommittee. As of 30 August 2024, nineteen state members and seven permanent observer organizations submitted proposals (see Appendix 1). The set formed the core of the qualitative analysis under the four scenarios of space governance.

4.1 Space Cartel

In the space cartel scenario context, the delegations of Russia, Belarus, and China recognized the innovative potential and financial resources of technologically

---

[102] Mario Silva Pasquali, 'The Role of Diplomatic Culture in International Disputes' (2018) 70 *International Journal* 267.

[103] Mai'a K Davis Cross and Saadia M Pekkanen, 'Introduction. Space Diplomacy: Theory and Practice' (2022) 30 *Cambridge Review of International Affairs* 199.

[104] Mark McKeown, Adam Dempster and Mustafa Saydam, 'The Artemis Accords: Cooperative or Competitive?' (2022) 48 *Space Policy* 13.

[105] Yap Jason, et al., 'Four Alternative Scenarios of Commons in Space: Towards a Governance Model' (2021) 47 Space Policy 397–99.

[106] Victoria Hart and Asaf Siniver, 'The Meaning of Diplomacy' (2020) 96 *International Affairs* 1.

[107] Edward C Rathbun, Joshua D Kertzer and Brian C Paradis, 'Homo Diplomaticus: Mixed-Method Evidence of Variation in Diplomatic Styles' (2017) 71 *International Organization* 40-41.

advanced actors in outer space research. Concurrently, these delegations called for substantive attention to elaborate the non-appropriation and benefit-sharing principles within the future legal regime for space resources activities. The prospective debate in the Working Group should prevent the first-come-first-served principle for using and occupying space resources and maintain the international status of space resources.[108] The delegations outlined the consequences of the limited volume and location of space resources as the limitations can stimulate economic rationality and the self-centered behavior of technologically advanced space actors. Especially China set on the agenda a demand for an effective governmental supervision over space resources activities by non-governmental entities, especially by private actors. The supervision should prevent hegemonic aspirations of these non-governmental actors.[109]

These undesirable tendencies in the context of the Outer Space Treaty would slow negotiations for subsequent diplomatic talks about a new comprehensive regulatory instrument for global space activities and contribute to the undesirable fragmentation of the international community of policymakers. So, especially the Russian submission highlighted a need for maintaining the international status of space resources after their extraction and their subsequent transformation to consider the level of economic and scientific development to ensure the fair distribution of benefits from space mining activities.[110]

Regarding the level of observer inputs to the Working Group, of the seven observer organizations, Open Lunar Foundation's (OLF) contribution to the debate exhibits the most concern about the known challenges of common pool resource issues, such as scarcity, rivalry, and developing sustainable polycentric governance mechanisms.[111] Input from Space Generation Advisory Council (SGAC) differs in significant ways from OLF, where OLF is defining and categorizing types of resources and predetermining scarcity and rivalry of Lunar resource systems with a potential for private property (policy permitting),[112] the SGAC, otherwise advised that currently the "extent to which off-Earth resources can contribute to public and private space activities cannot be exactly determined due to the lack of sufficient and reliable data

[108] United Nations Office for Outer Space Affairs (UNOOSA), *A/AC.105/C.2/2023/CRP.20—Russian Federation: Input to the Working Group on Legal Aspects of Space Resource Activities* (2023).

[109] United Nations Office for Outer Space Affairs (UNOOSA), *A/AC.105/C.2/2024/CRP.5—China: Input to the Working Group on Legal Aspects of Space Resource Activities* (2024).

[110] United Nations Office for Outer Space Affairs (UNOOSA), "A/AC.105/C.2/2023/CRP.20 - Russian Federation – Input to the Working Group on Legal Aspects of Space Resource Activities," 7.

[111] United Nations Office for Outer Space Affairs (UNOOSA), *A/AC.105/C.2/2023/CRP.25—Open Lunar Foundation: Input to the Working Group on Legal Aspects of Space Resource Activities* (2023).

[112] United Nations Office for Outer Space Affairs (UNOOSA), *A/AC.105/C.2/2023/CRP.25—Open Lunar Foundation: Input to the Working Group on Legal Aspects of Space Resource Activities* (2023) 8.

on their availability and the economic viability of their extraction and use."[113] In this regard, SGAC asserted that "what constitutes a highly valuable or limited resource is likely to evolve as more data becomes available, and so will the related need to regulate their access and use." Moreover, SGAC contends that "the concept of what constitutes a *lunar resource* and the legal conditions for their access and use should evolve side by side."[114]

4.2 Open Space (Space Utopia)

Regarding the open space model, Luxembourg, the Netherlands, Belgium, and other Western countries submitted proposals to spark the debate about multiple multi-level governance settings. The Building Blocks for the Development of an International Framework on Space Resource Activities for which the outputs of the achievements of the International Seabed Authority may serve as a good practice constituted the core of their positions.[115] Considering the interests and benefits of all countries, the countries' proposals recognized the diversity of space actors with various socio-economic and environmental needs. Therefore, the Building Blocks should be supportive of legal certainty, cross-scale communication, and mitigate the risk of international conflicts. These aspects stimulate the roles and needs of commercial actors for whom market-driven mechanisms help to avoid the tragedy of space commons.

Moreover, the benefit-sharing principle for utilizing space resources ensures access and exchange of information and might result in the establishment of joint ventures. The introduction of the benefit-sharing principle may require the establishment of an international fund for voluntary contributions, especially from the space operators responsible for conducting space resource activities. The safety zones introduced in the Artemis Accords have potential to reduce uncertainty and specify the critical support for safe and sustainable operations.[116] In sum, the submitted proposals from Western countries equally predicted an emergence of decentralized and localized agendas on space resources that may result in a form of adaptive governance.

---

[113] United Nations Office for Outer Space Affairs (UNOOSA), "A/AC.105/C.2/2023/CRP.26 - Space Generation Advisory Council – Input to the Working Group on Legal Aspects of Space Resource Activities," 49.

[114] Id.

[115] United Nations Office for Outer Space Affairs (UNOOSA), "A/AC.105/C.2/L.315 - Building Blocks for the Development of an International Framework on Space Resource Activities: Working Paper Submitted by Luxembourg and the Netherlands," 4; UNOOSA, "A/AC.105/C.2/L.325 - Contribution by Belgium to the General Exchange of Views on Potential Legal Models for Activities in the Exploration, Exploitation and Utilization of Space Resources: Working Paper Submitted by Belgium," 3.

[116] United Nations Office for Outer Space Affairs (UNOOSA), "A/AC.105/C.2/2023/CRP.7 - Australia – Input to the Working Group on Legal Aspects of Space Resource Activities," 6.

At the observer level, Western ideas predominated (see: Appendix 1.), we evaluated that at this stage of the Working Group, all observer organizations are based in the United States and Europe. Of these, Moon Village Association (MVA), SGAC, and OLF, were most vocal about aspects of adaptive governance. MVA views the challenge for COPUOS as not only to adopt a legal model for space resources but to develop an adaptive legal model with sustainable rules for future generations.[117] Similarly, OLF views the development of adaptive governance mechanisms as necessary to understanding how to recover, process, and use space resources safely and sustainably.[118] In this regard, both MVA and OLF prioritized development enabled by adaptive governance mechanisms.

Of the seven observers, concerns over benefit sharing were most pronounced in SGAC, MVA, National Space Society (NSS) and European Space Agency (ESA) inputs. The NSS and ESA emphasized the importance of clarifying benefit sharing, especially in the legal field, and what it means for incentivizing investment, the development of technology, capabilities, and education, including for developing countries.[119] MVA called for a "broad and inclusive debate on measures to mitigate future inequalities that may result from lack of sufficient consideration of access and benefit sharing issues" with the view that this theme should be "mandatory at relevant legal and operational levels," and this must be preceded by resolving access issues in "close concertation with key operators and strongly invested stakeholders."[120] The SGAC, with regard to a future Lunar Governance Charter, should include an Annex protocol for internationally agreed benefit sharing and capacity building based on principles of adaptive governance.[121] Still with observer views regarding adaptive governance and benefit-sharing priorities in the Working Group's deliberations, it does not guarantee any level of equal access or share of future benefits but could potentially evolve toward this scenario.

---

[117] United Nations Office for Outer Space Affairs (UNOOSA), "A/AC.105/C.2/2023/CRP.23 - Moon Village Association – Input to the Working Group on Legal Aspects of Space Resource Activities," 2.

[118] United Nations Office for Outer Space Affairs (UNOOSA), "A/AC.105/C.2/2023/CRP.25 - Open Lunar Foundation – Input to the Working Group on Legal Aspects of Space Resource Activities," 5.

[119] United Nations Office for Outer Space Affairs (UNOOSA), "A/AC.105/C.2/2023/CRP.24 - National Space Society – Input to the Working Group on Legal Aspects of Space Resource Activities," 3; UNOOSA, "A/AC.105/C.2/2023/CRP.22 - European Space Agency – Input to the Working Group on Legal Aspects of Space Resource Activities," 3.

[120] United Nations Office for Outer Space Affairs (UNOOSA), "A/AC.105/C.2/2023/CRP.23 - Moon Village Association – Input to the Working Group on Legal Aspects of Space Resource Activities," 3–4.

[121] United Nations Office for Outer Space Affairs (UNOOSA), "A/AC.105/C.2/2023/CRP.26 - Space Generation Advisory Council – Input to the Working Group on Legal Aspects of Space Resource Activities," 7.

4.3 Earth-Space Sustainability

Although the aspects of Earth-space sustainability were marginal in the submitted proposals, some European countries and New Zealand considered space resource utilization a novel domain for space exploration with uncertain consequences for the Earth and outer space environments. They expressed their interest in establishing an epistemic community to evaluate various aspects of accomplishing shared sustainability for Earth and outer space.[122] Also, future negotiations in the Working Group should build a consensus on responsibilities for damage to the environment, including legal instruments for the prevention of adverse changes in the Earth's environment.[123]

Notably, the UK delegation highlighted taking into account COSPAR's recommendations for planetary protection policies.[124] The expert panel elaborated five categories to operationalize potential hazards from space exploration for the Earth's environment.[125] The categories range from cases requiring no protection up to missions having a status needing continuous control throughout the return to Earth phase.[126] Following the expected risks, COSPAR's panel called for reporting information on space missions that considered the concept of planetary protection. The content of these reports should provide detailed data on biological burdens.

Observer inputs to the working group have used the term "sustainability" with regard to space resource activities in multiple ways referring to a variety of issues whether this involves the sustainability of a legal framework, sustainability of projects and human missions, or building and sustaining economic aspects in space and on Earth. Yet few, if at all, involve sustainability in an environmental sense. The National Space Society prioritized protecting heritage sites in their statement but did not separate this from the distinct complexities of interference aspects.[127] For those

---

[122] United Nations Office for Outer Space Affairs (UNOOSA), "A/AC.105/C.2/2023/CRP.18 - New Zealand – Input to the Working Group on Legal Aspects of Space Resource Activities," 3; UNOOSA, "A/AC.105/C.2/2022/CRP.21* - Working Paper on the Endorsement of the Work Plan for the Working Group Established under Agenda Entitled 'General Exchange of Views on Potential Legal Models for Activities in the Exploration, Exploitation, and Utilization of Space Resources' and Proposals for a Dedicated International Conference: Working Paper Submitted by Austria, Belgium, Czech Republic, Finland, Germany, Luxembourg, Norway, Portugal and Romania," 3.

[123] United Nations Office for Outer Space Affairs (UNOOSA), "A/AC.105/C.2/L.328 - Updated Summary by the Chair and Vice-Chair of Views and Contributions Received on the Mandate and Purpose of the Working Group on Legal Aspects of Space Resource Activities," 2; UNOOSA, "A/AC.105/C.2/L.315 - Building Blocks for the Development of an International Framework on Space Resource Activities: Working Paper Submitted by Luxembourg and the Netherlands," 4; UNOOSA, "A/AC.105/C.2/2023/CRP.13 - Germany – Input to the Working Group on Legal Aspects of Space Resource Activities," 3; UNOOSA, "A/AC.105/C.2/2023/CRP.14 - Greece – Input to the Working Group on Legal Aspects of Space Resource Activities," 5.

[124] United Nations Office for Outer Space Affairs (UNOOSA), "A/AC.105/C.2/2023/CRP.21 - United Kingdom – Input to the Working Group on Legal Aspects of Space Resource Activities," 3.

[125] COSPAR, "Committee on Space Research (COSPAR)» Panel on Planetary Protection (PPP)."

[126] COSPAR, "COSPAR Policy on Planetary Protection," 1–3.

[127] UNOOSA, "A/AC.105/C.2/2023/CRP.24 - National Space Society – Input to the Working Group on Legal Aspects of Space Resource Activities," 3,5.

fitting aspects of an Earth-Space sustainability scenario, the NGO For All Moonkind viewed Space resource utilization as an absolute necessity and advocates for enabling resource use while dealing with protecting "cultural heritage" sites. Moreover, they view the resources of space as vast and have the potential to better the life of our human species both on Earth and in the Universe.[128] Open Lunar Foundation view "the Moon as part of a developing Earth–Moon system"[129] and the Space Generation Advisory Council asserted that governmental funding should be complemented by dedicated measures to promote deeper integration with Earth's economy and industries. Therefore, funding a private lunar economy would democratize access to the Moon and reduce pressure upon States.[130] Of all the observer inputs, the European Space Agency space resources strategy aligns most with this scenario declaring that the challenge of space resources will also drive innovation on Earth.[131] According to ESA, learning to use the resources of space creates opportunities for Earth through technology innovation and new approaches to managing the resources of Earth. In this regard, ESA views the challenges and environments of space as "driving more optimized technologies and systems with implications for sustainability and the Sustainable Development Goals and by addressing the increasing challenge of resource scarcity where the resources of space may even be used on Earth as the cost of access falls and some resources on Earth become increasingly scarce."[132]

4.4 *Earth-Centric Gold Rush*

In this scenario, countries' proposals indicated the lack of data needed to evaluate future business behavioral practices and business actors' internal motivations. At the declaratory level, the proposals underlined existing relevant international legal principles that should shape the subsequent utilization of outer space, such as benefits-sharing, considering outer space as the province of all humanity, and minimizing negative externalities for Earth's environment.[133] Moreover, the working group should elaborate on the responsibilities of strengthening conflict resolution mechanisms among actors participating in the extraction and utilization of space resources.[134]

[128] UNOOSA, "A/AC.105/C.2/2023/CRP.35 - For All Moonkind – Input to the Working Group on Legal Aspects of Space Resource Activities," 5.

[129] UNOOSA, "A/AC.105/C.2/2023/CRP.25 - Open Lunar Foundation – Input to the Working Group on Legal Aspects of Space Resource Activities," 4.

[130] UNOOSA, "A/AC.105/C.2/2023/CRP.26 - Space Generation Advisory Council – Input to the Working Group on Legal Aspects of Space Resource Activities," 37.

[131] ESA, "ESA Space Resources Strategy," 6.

[132] ESA, 8, 11.

[133] UNOOSA, "A/AC.105/C.2/2023/CRP.18 - New Zealand – Input to the Working Group on Legal Aspects of Space Resource Activities," 1–3.

[134] UNOOSA, "A/AC.105/C.2/2023/CRP.20 - Russian Federation – Input to the Working Group on Legal Aspects of Space Resource Activities," 2–4.

A Belgian contribution to the debate in the Working Group conceptualized and framed risks that could lead to the tragedy of outer space commons.[135] Although unilateral and club-based policy and legal instruments can potentially increase the efficiency of project-driven tasks, the approach is not in line with existing Space Treaties and may deepen economic inequalities worldwide. The Belgian delegation welcomed the future development of forecast models for space resource utilization. It also emphasized responsibilities for private actors involved in space mining activities to avoid resource exhaustion. The multilateral global dialogue could facilitate information exchange, create conditions for transparency, and build confidence to reduce the risk of conflicts.[136] Also, future negotiations should discuss establishing an international fund for efficient benefit sharing.[137]

All observer inputs to the Working Group in subtle ways portend the possibility of aspects of both the Earth-centric gold rush and space cartel scenarios in that private actors are critical to scientific and technological progress and require legal certainty. Input from the European Space Agency prioritized public–private partnerships[138] while the Hague Institute for Global Justice asserted that the private sector needs a "global framework that is transparent, uniform, simple, and predictable" and "achieved through a bottom-up approach that relies on the advice and experience of civil society."[139] Additionally, the SGAC highlighted the development of a lunar economy as an essential component, finding that lunar companies would benefit from support of the public sector where governments and space agencies should provide assistance for non-governmental lunar activities.[140] The NGO For all Moonkind asserted that to avoid excessively high costs of space exploration both States and private companies "must be able to extract, use and create commerce around space resources" in their view it is expected and must be "permitted that space resources will be able to be used to supplement dwindling natural resources on Earth."[141]

---

[135] UNOOSA, "A/AC.105/C.2/L.325 - Contribution by Belgium to the General Exchange of Views on Potential Legal Models for Activities in the Exploration, Exploitation and Utilization of Space Resources: Working Paper Submitted by Belgium," 2–3.

[136] UNOOSA, "A/AC.105/C.2/2023/CRP.13 - Germany – Input to the Working Group on Legal Aspects of Space Resource Activities," 3; UNOOSA, "A/AC.105/C.2/2023/CRP.37 - United States – Input to the Working Group on Legal Aspects of Space Resource Activities," 4–5.

[137] UNOOSA, "A/AC.105/C.2/L.315 - Building Blocks for the Development of an International Framework on Space Resource Activities: Working Paper Submitted by Luxembourg and the Netherlands," 5; UNOOSA, "A/AC.105/C.2/2023/CRP.12 - France – Input to the Working Group on Legal Aspects of Space Resource Activities," 5–6.

[138] ESA, "ESA Space Resources Strategy," 11.

[139] UNOOSA, "A/AC.105/C.2/2023/CRP.32 - The Hague Institute for Global Justice: The Washington Compact on Norms of Behavior for Commerical Space Operations," 4.

[140] UNOOSA, "A/AC.105/C.2/2023/CRP.26 - Space Generation Advisory Council – Input to the Working Group on Legal Aspects of Space Resource Activities," 37–39.

[141] UNOOSA, "A/AC.105/C.2/2023/CRP.35 - For All Moonkind – Input to the Working Group on Legal Aspects of Space Resource Activities," 5.

5 Discussion

The results of our qualitative analysis indicate a diversity of views about policy and legal aspects of space resource utilization within the Working Group. They highlight aspects of the discussions about future governance and legal frameworks for using space resources and exploring near-Earth outer space, emphasizing the lunar environment. In sum, general agreement about the need for a comprehensive clarification of hard and soft legal principles governing recent uses of outer space was visible across the submitted proposals by countries and observer organizations. In general, setting this topic in the legal subcommittee of COPUOS renewed scholarly discussion about how to continue the subsequent developments in regulatory aspects of lunar activities.

Our analysis captured a variety of demands on the country level, indicating that negotiations at the Working Group are still in the introductory phase. Developments at the Working Group should clarify whether and what historical international governance models, namely the Antarctic governance regime and the high seas and seabed international governance regimes, and what their particular elements are eligible to be included in desired lunar governance principles.[142] Although the international situation of the 1950s allowed the establishment of the Antarctic governance regime as an example of a temporary conflict prevention situation, in 2009, Moltz had noticed that freezing territorial claims in outer space has minority support among policy and legal scholars.[143] Our results also confirmed these tendencies at the level of international political negotiations. In the space cartel scenario, where fears about resource exhaustion were the most noticeable, neither countries nor observers demanded freezing space mining activities. Instead, countries and observers called for fair sharing of benefits and maintaining their international status of the resources after extraction from outer space objects if necessary. At the level of observers, the NGO *For All Moonkind* outlined a hierarchy to support efficient space resource management.

Therefore, the debate on equitable uses of lunar resources, resulting in setting an environment without significant political and economic conflicts between spacefaring and non-space-faring countries became more profound instead. On the country level, actors' perspectives converge into two main scenarios. Most non-Western countries called for preventative measures against applicability of the first-come-first-served principle for the cases of space resource utilization. In their proposals, these countries warned against growing global inequalities and the chain of events leading toward a tragedy of the commons. Conversely, Western countries clustered their visions of lunar governance around the space utopia scenario in which emphasis on liberal values was visible. Overall, they prioritized power dispersion among actors with various socio-economic and environmental needs for whom principles of cross-scale communication and market-driven mechanisms became crucial principles.

---

[142] David Mouat, Tanja Schünemann, and Peter Martinez, '(Un)Earthly Governance: The Challenge of Implementing Earthly Norms in Outer Space' (2022) 37 *Space Policy* 133.

[143] James Clay Moltz, 'Toward Cooperation or Conflict on the Moon?' (2022) 74 *Bulletin of the Atomic Scientists* 87.

In line with Western countries' demands, the prevalent scholar discourse highlighted the inefficiency of the Outer Space Treaty for developing commercial aspects of space exploration, mainly in connection with lunar resources.[144] As space lunar resource governance remains undeveloped, the UN system should support academic and political events dealing with various aspects of excludability and rivalry.[145] These aspects should include development of advanced mechanisms for environmental impact assessments.[146] The scholarly demands were in line with the obtained results in our analysis in which countries and observers showed respect for the existing provisions of the Outer Space Treaty and further called for their future elaboration to enhance legal and technological certainty.

6 Conclusion

This analysis mapped the diversity of demands of countries and observers at the Working Group to provide preliminary results about international negotiations on future governance models for space resource utilization. The obtained qualitative data indicated the progress of the ongoing expert and diplomatic talks, whose complexity is expected to increase. Their future outcomes have the potential to provide an innovative variety of governance tools for exploiting lunar resources which the existing governance regimes for global climate and oceans may inspire.[147] Also, some complex proposals for regulating radio frequencies may contribute to increasing the efficiency of the governance of limited lunar resources.[148]

The presented results showed a need for more structure in political debates in COPUOS. Future negotiations in the Working Group should cover at least two essential aspects—feasibility and effectiveness.[149]In terms of feasibility, previous research has noted the rapidly changing global environment and the differing positions of Western democratic and non-Western countries. The differences in political regimes in these two blocks portend the mutually opposite values at the UN level.[150] Whereas space activities have been the domain of governments up until

[144] James Clay Moltz, 'Toward Cooperation or Conflict on the Moon?' (2022) 74 *Bulletin of the Atomic Scientists* 94.

[145] Nicoletta Krolikowski and Martin Elvis, 'Space Resources and Prospects for Contested Governance' (2021) 73 *Space Policy* 671.

[146] Paul Martin and Steven Freeland, 'Back to the Moon and Beyond: Legal Aspects of Space Resource Activities' (2018) 61 *Proceedings of the International Institute of Space Law* 428.

[147] Rephael Hasin, 'From "Space Law" to "Space Governance"' (2022) 37 *Space Policy* 398.

[148] Tanja Masson-Zwaan and Mark J Sundahl, 'National and International Norms towards the Governance of Commercial Space Resource Activity' (2020) 45 *Air and Space Law* 400; Christopher J Roberts and Juan F Linares, 'A Method for Assessing Satellite Operators' Compliance with Geosynchronous Orbital Assignments' (2019) 44 *Space Policy* 226.

[149] Rephael Hasin, 'From "Space Law" to "Space Governance"' (2022) 37 *Space Policy* 409.

[150] Rephael Hasin, 'From "Space Law" to "Space Governance"' (2022) 37 *Space Policy* 408.

recent times, current policy and legal environments in Western democratic countries open doors for private actors' involvement. These developments raise questions about the matrix of governmental, hybrid, and private responsibilities and their consequences in competing environments between Western and non-Western countries. These differences in values would be visible in framing private actors' activities in the lunar environment, especially in the structure of public–private partnerships enabling governmental incentives for these new space players.[151] Regarding the aspect of effectiveness, Neto highlighted managing space diplomacy negotiations toward widely respected solutions.[152] The ability to persuade reluctant actors to remain involved in negotiations increases relevance, especially in enabling a sustainable environment for space activities and securing the natural environment on Earth. In the context of the recognized diversity of actors existing in the space domain, building conditions toward adaptive space governance could contribute to finding a policy consensus in multilateral space governance fora.[153]

Future research could consider overcoming limitations that emerged during our research process. Dealing with a newly established working group at the UN and a rapidly evolving topic with limited data input thus far allowed us to provide preliminary outputs only. As the process continues and negotiations evolve, future research should consider conducting interviews with distinguished individuals and stakeholders. Future studies should also consider doing a cost–benefit analysis for particular governance models expected in the Working Group in the future.

Appendix 1: Submissions from state members and permanent observer organizations to the Working Group as of 30 August 2024

Actor type	Name	Submission year
State members	Australia	2022
	Azerbaijan	2022
	Bahrain	2022
	Belarus	2022
	Belgium	2023
	Canada	2023

(continued)

[151] Ram Jakhu and Joseph N Buzdugan, 'Development of the Natural Resources of the Moon and Other Celestial Bodies: Economic and Policy Issues' (2017) 60 *Proceedings of the International Institute of Space Law* 206.

[152] Miguel Gonçalves Neto, 'Outer Space as a Global Commons and the Role of Space Law' (2019) 31 *Revista de Direito Internacional* 13.

[153] Daniel L Oltrogge and Ian Christensen, 'Space Governance in the New Space Era' (2019) 35 *Space Policy* 434.

(continued)

Actor type	Name	Submission year
	China	2024
	France	2022
	Germany	2022
	Greece	2022
	Japan	2023
	Jordan	2022
	Luxembourg	2022
	Morocco	2022
	New Zealand	2023
	Norway	2022
	Russia	2022
	United Kingdom	2022
	United States	2022
Permanent observer organizations	European Space Agency[154]	2022
	For All Moonkind[155]	2022
	Hague Institute for Global Justice[156]	2023
	Moon Village Association[157]	2022
	National Space Society[158]	2022
	Open Lunar Foundation[159]	2022
	Space Generation Advisory Council[160]	2022

Source UNOOSA. Working Group on Legal Aspects of Space Resource Activities. available at: https://www.unoosa.org/oosa/en/ourwork/copuos/lsc/space-resources/index.html

Nancy Riordan is a Global Governance and Human Security PhD candidate at the University of Massachusetts Boston. She has also served as a part-time faculty member in the Department of Political Science, College of Social Science and Humanities at Northeastern University. She spent more than a decade working with cities actively experiencing and transitioning

[154] The European Space Agency is a governmental organization based in Paris, France.

[155] The For All Moonkind is a non-governmental organization based in New York, the U.S.

[156] The Hague Institute for Global Justice is a non-governmental organization based in the Hague, Netherlands.

[157] The Moon Village Association is a non-governmental organization based in Vienna, Austria.

[158] The National Space Society is a non-governmental organization based in Washington, D. C., the U.S.

[159] The Open Lunar Foundation is a non-governmental organization based in San Francisco, the U.S.

[160] Space Generation Advisory Council is a non-governmental organization based in Washington, D.C., the U.S. and Vienna, Austria.

from conflict (Baghdad, Belfast, Derry~Londonderry, Haifa, Jerusalem, Kirkuk, Kaduna, Mitrovica, Mitte, Mostar, Nicosia, Sarajevo, Srebrenica, Ramallah, and Tripoli) — she is building on those experiences to understand current and future conflict dynamics in outer space. Her research focuses on the political aspects of space sustainability and conflict over space resources.

Miloslav Machoň is a postdoctoral researcher and assistant professor in international relations at the AMBIS University, Prague. He specializes in various qualitative-based topics covering environmental governance and outer space issues. In 2020, he was appointed as a visiting postdoctoral scholar at Northeastern University and the University of Massachusetts Amherst, under the supervision of Professor Mai'a K. Davis Cross and Professor Peter M. Haas, where he focused on the bureaucratic and market aspects of epistemic authority. He obtained his PhD in 2017 at the Prague University of Economics and Business for processing qualitative data on the activities of Czechoslovak and Czech researchers in astrophysics and space policy under the Czechoslovak and Czech Ministries of Foreign Affairs at the UN Office for Outer Space Affairs.

Lucia Kozák Csajková is an International and Diplomatic Studies PhD student at the Prague University of Economics and Business with a research focus on environmental and city diplomacy. During her studies, she was awarded a Prize for Excellent Student Thesis at the Faculty of International Relations for a seminar paper, bachelor thesis and diploma thesis, which explored diplomacy and sustainability.

Exploitation of Celestial Mineral Resources: Challenges and Solutions

G. S. Sachdeva

Abstract Outer space is believed to be a treasure trove of mineral resources of different kinds, some of which are direly needed by mankind, having depleted its terrestrial reserves. It has been found that the dusty layer of Helium covering the Moon is rich in material that can be used to generate power. Geo-intelligence of celestial bodies has revealed many other precious metals available in different asteroids and heavenly bodies. As of today, the US and Japan have attempted to harness asteroids for the excavation of minerals, and success seems around the corner. India, with its start-ups, e.g., Agnikul and Pixxel, are following the lead. International Space Law does not permit commercial exploitation of celestial resources nor appropriation by state or private entities. It is, however, a different matter to bring small quantities as samples for scientific investigation or in-situ use of such indigenous resources. Nevertheless, natural resources of outer space are *res hominem* and need to be protected as a province of mankind for the collective good and common benefit. Some countries like the US, Luxembourg, et al are raring to go for this spoil, seemingly, in disregard of the restrictions of international space law. Thus, there is a need for legal expedients. Besides, there is a need for establishing a governing authority, say, Outer Space Trusteeship Council (OSTC), to create domination over the resources and grant leases on rentals with specific demise for controlled exploitation. This will also require inspection of compliance and other professional instrumentalities of confirmation. The challenge does not seem insurmountable, provided a sincere effort is made to find viable solutions.

Keywords Space resources · Asteroid mining · International Space Law · Commercial exploitation · Global governance · Outer Space Trusteeship Council

---

G. S. Sachdeva (✉)
Chandigarh University, Mohali, Punjab, India
e-mail: sach1938@gmail.com

NALSAR University of Law, Shameer Pet, Hyderabad, India

S. Bhat B and A. Variath (eds.), *Legal Responses to Space Resource Utilisation*, Space Law and Policy, https://doi.org/10.1007/978-981-95-8864-0_8

1 Introduction

The fact that there lies a limitless bounty of mineral resources in outer space has been proved by remote sensing, close observation through space probes and analysis of samples brought to the Earth. And that these are useful to humanity and can augment the depleting natural resources of the Earth is also established beyond doubt. Further, the technology to excavate supports viability and, to some extent, processing of the same in situ so that the end product of ores would make for a smaller quantum of high utility mineral resources or precious metals that can be transported to Earth in economical capsules. Possibility can also be envisaged where processed material can be, in some manner, used on the celestial body itself with 3D printing. Thus, the usefulness of celestial mineral resources to mankind is affirmative, but the analytics of comparative economics of excavation, processing and transportation are not yet clear to convince of commercial viability and reasonable profits. Prospects appear positive, and business sentiment is bullish.

Attention to space reserves has been attracted due to depleting natural resources on the Earth, the growing scarcity and consequently escalating prices of metallic ores and precious metals. The pinch has been felt rather acutely, which has diverted attention to considering prospects from other probable sources with viability of excavation, the quality of the excavated ore, and the cost parameters. One such source that zeroed in on asteroids was long known for their mineral richness and existence of precious metals, but technological viability for exploitability on a commercial scale was lacking. Technology has now progressed to challenge this opportunity, and viability knocks at our door. Some countries, for example, the US, Luxembourg, Japan, and Saudi Arabia, through non-governmental entities, are making frantic efforts in competition to actualise this possibility.

The Outer Space Treaty 1967 is regulatory in law and expressive of permissions and prohibitions on space activities. It also mandates certain principles for engaging in space activities but contains no procedures or modalities for ensuring implementation of the treaty provisions and a governance structure to facilitate commercial space activities. These lacunae have tempted and prompted some countries to encourage their juridical or natural citizens of private enterprise, almost in defiance of the treaty, to take advantage of the absence of a governing institution. The national interest and business advantage may be the prime considerations, but the majesty of international space law is at stake. This chapter seeks to find pragmatic solutions in the interim and proposes an institutional structure for the proper governance of celestial mineral resources.

2 Existing International Space Law

Considering the trend of state agencies selectively withdrawing from commercial space activities, except those of strategic importance or deep space exploration, it seems certain that such an activity as mining in space, being of a commercial nature and import, will not be undertaken by any governmental space agency, nor would the public exchequer allocate funds for this purpose.[1] Perforce, such activity will have to depend upon a business initiative from private enterprise, whether for commercial adventurism or entrepreneurial vanity or business aggrandisement. And given the present standing and competence of private enterprises, these can harness complex space technology, are willing to invest huge funds in space projects, and patiently wait during the gestation period. It would, therefore, not be presumptuous to believe that Space-X, Boeing, Lockheed Martin, etc. or a consortium of willing and enterprising astro-preneurs would venture into this nascent yet apparently promising segment for prestige or profit, irrespective of the attendant financial risks or hazards of space activities. Hope prevails with positivity.

Outer Space Treaty, in Article I, permits "The exploration and use of outer space, including the Moon and other celestial bodies, ...for the benefit and in the interest of all countries...and shall be the province of all mankind." The next paragraph of this article assures all states that "there shall be free access to all areas of celestial bodies." However, the freedom of access is subject to the principle of non-interference. Further, the Treaty ordains in Article II, "Outer space, including the Moon and other celestial bodies, is not subject to national appropriation by claim of sovereignty, by means of use or occupation, or by any other means." Thus, this clearly precludes the customary possibility of occupation under the maxim of *res nullius*.

The OST also mandates under Article VI, "States Parties to the Treaty shall bear international responsibility for national activities in outer space, including the Moon and other celestial bodies..." Hence, state liability arises and attaches for any or all unauthorised activity, by whomsoever, in the outer space domain. At the same time, the states, their agencies and juridical citizens are required to engage in space activities "with due regard to the corresponding interests of all other State Parties to the Treaty." (Article IX). This invokes the cardinal principle of non-interference, well established in international law that correlates mutual rights and obligations.

The legality of private enterprise undertaking activities relating to use and exploration in outer space can be validly derived from Article VI of the Outer Space Treaty

[1] It seemed likely that certain countries like China and India may not involve private enterprise in strategic or commercial space activities for a long time. Yet a small beginning has been made by China which permitted establishment of a private company in 2015 to undertake space activities. Nevertheless, state-owned companies will remain the main players in Chinese space activities for quite some time. In India also, start-up space entities, Agnikul, Pixxels and others have shown interest in this field while ISRO remains the major space agency.

(OST),[2] which permits non-governmental entities to undertake such ventures. There is, however, a rider in the same article of the Treaty that these activities "shall require authorization and continuing supervision by the appropriate State Party to the Treaty." Thus, whatever may be the nature of conditionality, a requirement of basic permission exists for private enterprise to enter this domain for legitimately permitted and *bonafide* activities within the ambit of the Treaty. In fact, this is happening, and satellites owned and operated by the private sector, like Starlink, are already in outer space. Even mining for in-situ utilisation of space mineral resources, raw or processed, can be justified under existing law. So is the US space enactment titled Commercial Space Launch Competitiveness Act, 2015,[3] valid and benign to this extent.

3 Activities Contra to the Treaty

The basic law on the point of exploitation of the uses of outer space and celestial bodies is loud and clear, but problems would arise at commercial recovery and private appropriation of celestial mineral resources. Perhaps, quantum of recovery, its commercial intent and appropriation of proceeds as private could be relevant here. Small quantities as samples have been obtained in the past from celestial bodies and distributed for scientific investigation as urged under the Treaty. Such soil and rock samples weighing 1.9 kg have now been brought back by China on Chang'e-6 lunar probe.[4] It is yet premature to surmise whether parts of it will be shared for scientific investigation with other desirous states. However, the US enactment of the Commercial Space Launch Competitiveness Act, 2015 is of a different genre and flags a wider import relating to commercial exploitation and private appropriation of space mineral resources. Its clause that "United States citizens engaged in commercial recovery of an asteroid resource or space resource obtained, including to possess, own, transport, use and sell the asteroid resources or space resources obtained in accordance with applicable law, including the international obligations of the United States" is murky and tricky. This provision of the Act seems to contain a syllogistic fallacy and lacks legal synchronicity with the Treaty mandates.

Scholars upholding and justifying the US stance maintain that space harvesting needs no more international regulatory permissions.[5] They seem to draw strength from the famous Lotus Principle, well established in international law and reinforced by state practice. This rule evolved by the Permanent Court of International Justice

[2] *Treaty on Principles Governing the Activities of States in the Exploration and Use of Outer Space, including the Moon and Other Celestial Bodies* (adopted 27 January 1967, entered into force 10 October 1967) 610 UNTS 205, 18 UST 2410, TIAS No 6347, 6 ILM 386 (Outer Space Treaty).

[3] *US Commercial Space Launch Competitiveness Act 2015* (HR 2262, signed into law 25 November 2015).

[4] China National Space Administration, statement reported in *South China Morning Post Plus* (28 June 2024).

[5] Chris Tolton, 'Space Mining Doesn't Need More International Regulation from the UN' *SpaceNews* (26 June 2024) Opinion https://spacenews.com accessed 7 May 2025.

(PCIJ) in a case between France and Turkey in 1927, which held that what is not expressly prohibited is deemed permitted. This rule is state-centric and was laid a century ago. In a fast-developing world of technology and innovation, such a prohibition cannot be "treatised" in advance for the oncoming potentialities and possibilities. Further, this rule of international law cannot bulldoze OST and needs to be applied mutatis mutandis with due regard to other provisions of the Treaty, such as province of all mankind, non-sovereignty, non-appropriation and equitable sharing of common benefits, among others.

Another argument is that the OST is binding on states that are party to the treaty and does obligate individual or legal citizens as these are not subjects of international law. Hence, their activities are beyond this implicit prohibition. But this nuance carries dubious character and a syllogistic fallacy. Firstly, states are obligated and the only ones competent to authorise activities in outer space and celestial bodies by governmental agencies and private entities. Secondly, the liability for any damage caused in outer space to any space object attaches to the state and in turn, may devolve on the offending entity. So, the state's responsibility to regulate its space activities gets multiplied. Thirdly, the state cannot bestow any better rights or powers on its citizens than it has under any agreement or treaty. It is common wisdom that a person cannot give more than it has, and states are no exception to this.

Another set of pro-advocates follows a line of argument eliciting support from the Lockean doctrine, which maintains that "mixing one's labour with property would create rights in it as everyone is entitled to the fruits of their own labour."[6] This doctrine validly applies to the terrestrial domain under the regime of sovereignty, property rights and res nullius. But while extending this generality to Outer Space and celestial bodies, it encounters a syllogistic fallacy. The outer space domain is neither open to claims of sovereignty nor private appropriation of celestial assets. The totality of the outer space domain is in the nature of res communis with no tenable claims to ownership and with a clear mandate to share the benefits accrued from the outer space domain. The maxim of mutatis mutandis becomes relevant and over-riding.

In another bold step, NASA has devised and enunciated the Artemis Accords prescribing almost similar measures as the OST, but in its enthusiasm ambles farther than that, beyond the permissions under the Treaty. Further, NASA is soliciting support for the Accords from different countries and space agencies to build a strong front parallel to or even supra to the UN treaties and resolutions. It has won many signatories for different reasons. India, too, approved of this document and signed it in 2023. Whatever be the merits and demerits of the Accords, the style and modus of the US agency in propagating this document and seeking confirmatory accessions is hardly commendable.

To take advantage as a follower of the US advances and initiatives, the Duchy of Luxembourg is encouraging foreign corporations to register in the country as juridical entities and is offering state patronage, fiscal concessions and liberal incentives for

---

[6] H M Kaushik, 'Sovereignty in Space: Changing Dimensions' (2023) 3(2) *Centre for Aviation and Space Laws Newsletter* (6 August 2023).

exploitative space activities. Saudi Arabia, in league, is also following a similar space policy for promoting space exploitation business with financial doles and other attractions. In a bandwagon effect, there are others taking the cue and mustering courage for defiance. At the same time, Japan is busy with its technology experiments at the excavation of celestial bodies and seems to be reaping success. Thus, the distance between this dream and the reality of obtaining and appropriating celestial mineral resources seems short and traversable, though not really straight and smooth.

4 Towards Plausible Solutions

To reiterate, Articles I and II of the OST clearly omit permission for such an attempt at commercialisation and appropriation. The US, in an act of "business populomacy," has made a domestic law and proffered Artemis Accords to compromise this obstacle that impinges on the OST. The germane clause in the statute guarantees enlarged rights to its citizens than those bestowed on the state by the Treaty. In a way, the US enactment has strains of negativity against the established legal position and is not as innocuous as is argued to appear. Be that as it may, mineral harvesting viability is almost a *fait accompli,* and there is no need to be dismissive of the motivated provisions of the enactment and the vested narrative of the Accords.

Under the circumstances, the matter deserves to be rationally debated to find the best amends to the situation within the gamut of international space law and explore viable solutions in the common interest of humanity. It would seem equally sensible not to shut out this possibility from the celestial bounty. In the ultimate, there appears to be a need for a revised and modern construct of law because the narrative must change with advancing technological feasibilities, economics of exploitation, tempting profitability scenarios, emerging new business alignments and commercial initiatives. Under this scenario, let us seriously search for hints of permissibility for commercial exploitation under the existing space law.

First, the Preamble of the OST permits "...use of outer space for peaceful purposes." And the same dictate is reiterated in Article IV that enjoins, "The Moon and other celestial bodies shall be used...exclusively for peaceful purposes." And exploitation of celestial resources is undeniably a peaceful activity. Secondly, the Treaty permits activities by governmental agencies or by non-governmental entities, subject to authorisation by the state.[7] Here too, permissibility stands valid. Thirdly, some jurists vindicate permission by arguing that what is not specifically prohibited in law is permitted. This, however, constitutes an adverse inference and "[T]his dichotomy may not always be exact and sacrosanct...[with] grey areas of doubt..."[8] Moreover, this reference should not be taken for granted under the scenario of fast business developments and advancing technology.

---

[7] Outer Space Treaty, art VI.

[8] G S Sachdeva, *Space Commercialisation: Prospects, Challenges and Way Forward* (Pentagon Press LLP 2019) 75–77.

4.1 Evaluation of Solutions

In my opinion, OST is a majestic *grundnorm* and a symbol of global trust. Therefore, any voluntary and intentional infringement of the Treaty would be inappropriate, internationally condemnable and set a bad precedent. In fact, such a violative action will maul the spirit of the Treaty and will turn highly embarrassing to a space-leader like the US, and in turn, its allies and followers. Hence, there is a need to find an unimpeachable legal solution because the existing regimen of OST offers no concession and treats the outer space domain as "province of all mankind."[9] This position defies all arguments and legal constructs.

Experience shows that amendments to treaties are painfully arduous in diplomacy and time-consuming in negotiations. Therefore, possible solutions could be to initiate amendments to the OST or conclude a *specialis* protocol to the Treaty or propose an amending resolution in the UN General Assembly for a voice vote. Whichever proposal is agreed upon, it can be nuanced for success by offering a bait of benefits. Nevertheless, long negotiations, diplomatic hurdles and possible technical objections may foil the effort in a rather disappointing manner. Success is not guaranteed in this option.

The Moon Agreement (MA),[10] however, is more permissive and liberal in this regard. In Article 8, the Agreement permits excavation on the Moon, on or below the surface, and allows for free movement of equipment and personnel, etc. Further, provisions of this Agreement "...shall also apply to other celestial bodies within the solar system...." So near-Earth planets and asteroids get covered by this affirmation. But Article 11(1) of the MA treats these natural resources of outer space as "Common Heritage of Mankind (CHM)." Under this concept, celestial natural resources are treated as *res communis omnium,* which defies proprietary individualisation and consequent private appropriation in any manner. As a possible solution, it may be more prudent to ratify this Agreement and initiate a suitably modified legal regime for the exploitation of natural resources under Article 11(5) and get this negotiated by the requisite number of state parties to the Agreement by a majority vote. Even otherwise, MA admits of additional regulatory provisions when circumstantially so need. Hence, the Agreement deserves to be mainstreamed in international space law and is ripe for a revision under Article 18. This offers hope to obviate crass competition, chaotic economic order and a possible conflict in space.

Another pragmatic and conforming solution can be that an agreed portion of accruing profits from such exploitative activities can be shared and reserved in trust to benefit all states in some ratio and weightage as Space Benefits Fund (SBF) on the lines of Corporate Social Responsibility (CSR). And this frozen fund in trust can be disbursed to the entitled states once the criteria on proportion and modalities

---

[9] Outer Space Treaty, art I.

[10] *Agreement Governing the Activities of States on the Moon and Other Celestial Bodies* (opened for signature 18 December 1979, entered into force 11 July 1984) 1363 UNTS 3 (Moon Agreement).

of sharing have been finalised.[11] Under such an all-round beneficial arrangement, a broad agreement may be easier, if not a true consensus. This would thus satisfy the inclusive spirit as well as abide by altruism enshrined in the Treaty to pacify and benefit non-stakeholders. This would need a change of capitalist mindset; although "freebies" is also not a laudable concept.

5 Claims to Celestial Bodies

An international association called "For All Moonkind" deems itself as the guardian of the Moon and strives for the interests of its moon-lover members and preservation of the Moon. It bears representative recognition from the United Nations. Again, a group of Asgardians, who claim to be a Space Nation, are also striving for a habitat in outer space that they can call as their own and are propagating their cult. Similarly, adherents of the Church of Scientology profess faith in the belief that they are descendants of a Space Confederacy of King Xenu and have the first right on planetary bodies of the confederacy.[12] Thus, competition or conflict of interest with excavational activities and commercial usage of celestial resources by global astropreneurs is possibly going to be fierce because global corporates will be governed by the terrestrial regulations, while the Asgardians and other space-claimants will have no such legal shackles to cause irritants and obstacles. Conflict is again on the horizon.

Apart from stakes from organisational entities, even Russia has proclaimed that the planet Venus is theirs for having landed there first.[13] Howsoever specious, an assertion has been made. Further, there are individuals who have also laid claim to certain celestial bodies. An Arkansas businessman founded Planet Mars Development Corporation in September 1954. He later offered to convey the title for a suitable remuneration.[14] In 1953, Janeiro Gajardo Vera, a Chilean lawyer, claimed the Moon and made a public proclamation of his right. Surprisingly, no objections were received by the authorities for the adverse claim, and the Moon was registered in his name.[15] Again, in a jocular vein, Virgiliu Pop proclaimed ownership of the Sun with an online register in 2002 and threatened cosmic property owners with legal notices for the use of Sun's light.[16] However, a serious claim on the asteroid Eros

[11] G S Sachdeva, 'Commercial Mining of Celestial Resources: Case Study of US Space Laws' (2018) *Astropolitics* 202.

[12] G S Sachdeva, *Space Commercialisation: Prospects, Challenges and Way Forward* (Pentagon Press LLP 2019) 63–67.

[13] Dmitry Rogozin (Chief of Roscosmos), statement at the HeliRussia Exhibition, reported in *TASS* (18 September 2020).

[14] G S Sachdeva, *Space Commercialisation: Prospects, Challenges and Way Forward* (Pentagon Press LLP 2019) 66.

[15] G S Sachdeva, *Space Commercialisation: Prospects, Challenges and Way Forward* (Pentagon Press LLP 2019) 66.

[16] Virgiliu Pop, *Who Owns the Moon?* (Springer 2009) 2.

was made by Gregory Nemitz, a space activist. In 2001, when NASA had made an automated robotic landing of the NEAR-Shoemaker probe on Eros, Nemitz billed NASA and the US State Department for parking fees. This charge of fees for landing and stay was rejected for payment. In consequence, Nemitz filed a lawsuit in a US Court of Appeals. The court dismissed the appeal because he could neither prove his ownership right with valid legal documents nor substantiate his actual and physical possession of the property with valid evidence.[17]

Law, logic and rationality may dismiss these frivolous claims on whatever reasoning or lack of scientific justification from the contesting side, yet seeds of an endemic conflict reside in the situation that may adversely affect the commercial ventures on certain celestial bodies. At present, the existing law of space treaties is silent on this aspect of commercial exploitation of space resources, and consequent localised occupation and controls may hamper freedom of access and investigation in all parts and areas as promised in the Treaty. Such misadventures must be denied and curbed through national statutes because "[I]deally, [the Treaty] secures outer space including celestial bodies as a unified expanse of 'Humankind Commons.'"[18]

Frankly, it could not have been anticipated half a century back that man would achieve progress on space technology in diverse applications and, that too, so fast. But this development is already upon us, and we cannot unwisely shut off a treasure trove of spatial natural resources for sheer lack of legal solutions. These have to be found and equally fast. Some short-term and long-term legal measures and policy expedients have already been proposed for the space-faring nations to take initiative and resolve issues on the legal canvas, keeping in view the principles of common heritage, common benefit, inclusiveness and equity.

5.1 Governance of Commercialisation of Celestial Resources

At the very outset, let us accept that celestial bodies and the natural resources embedded therein are the province of all mankind, whatever meaning is attached to it, whether as Common Heritage of Mankind or Global Commons or any other label. Same way, in legal doctrines, the celestial bodies as assets remain *res publica* as a trust or *res communis humanitatis* as common property of humanity, past, present and future. Thus, "space law acknowledges collective sovereignty and universal community ownership of all mankind…"[19] Hence, private and individualised appropriation of these resources is not conceivable without proper safeguards of common sharing of benefits and sustainability for future generations.

Commercialization of outer space activities, in particular exploitation of celestial natural resources, whether renewable or non-renewable, by private corporate

---

[17] *Nemitz v United States* (US District Court, Nevada, 2004).

[18] G S Sachdeva, *Space Commercialisation: Prospects, Challenges and Way Forward* (Pentagon Press LLP 2019) 70.

[19] G S Sachdeva, *Outer Space: Law, Policy and Governance* (KW Publishers 2014) 173.

sector, needs a new set of rules under Private Space Law, e.g., leasing for excavation or inspection of compliances, collection of lease rentals and exercising demise of leases on expiry in contra-distinction to Public Space Law that governs and regulates public order in the space domain. The differences can be broadly treated as analogous to International Law, Public and Private. Nevertheless, it necessitates governance of resources, their exploitation, appropriation by private entities, contractual nuances, realisation of mine royalties, resolution of conflicts of interest, settlement of disputes and other incidental issues. The listed narrative, however, is not complete and comprehensive of ominous challenges.

The existing infrastructure of the Office of Space Affairs (OOSA) seems too inadequate for compliance with the Treaty and the governance function because the new dispensation would have to ensure additional measures for public order, regulatory regimen and governance procedures. This calls for a bigger set-up, for example, World Space Authority (WSA)[20] under the aegis of the United Nations to manage the Common Heritage of Mankind in the outer space domain with executive, regulatory and quasi-judicial effectuations. But the word Authority may cause apprehension and be construed as a supra-national body in an international context, whereas its relevance and writ would operate only for governance and administration in the outer space domain, in particular, for the management of mineral deposits in the celestial bodies and commercial utilisation of astro-estate in rental leases and trust holdings. It is wisely said that without a governing Authority, there is no enjoyment of liberty nor an assurance of rights. However, on a second thought, it has been proposed that the defunct Trusteeship Council may be politically revived and functionally resuscitated for the governance of the outer space domain in all aspects and purposeful management of celestial resources. It could be rechristened as the Outer Space Trusteeship Council (OSTC). Thus, the tasks of the Council would be holistic domain governance, policing of compliance, regulation of commercial race for resources, professional management of contractual contingencies and other specialised functions. "Hence, the Council would have to be vested with a chartered mandate of governance, a roll of authority, a suitable organisational structure, professional staffing, and adequate funding with a befitting office edifice."[21]

It may be hoped that the Trusteeship Council, with its earlier experience of governance, administration, policing, infrastructure construction, economic development and social engineering, would be best suited to its new role. Further, it could partly generate its own sources of revenue from cess, fees, royalties, rentals and other charges levied on permitted mining leases, space traffic regulation and space situational information, among provision for other services. Additionally, the Council could be authorised to levy a charge on rocket launches based on stages shed in space, payload carried, number of satellites to be launched, their size, purpose and life and whether having an inbuilt possibility for de-orbit and removal from the designated orbit. The accruals from the launch activity may constitute Space Sustenance Fund (SSF) for sustenance of the outer space environment and remediation of space debris

[20] G S Sachdeva, *Outer Space: Law, Policy and Governance* (KW Publishers 2014) 198–200.

[21] G S Sachdeva, *Crimes in Outer Space: Perspectives from Law and Justice* (Springer 2023) 237.

by outsourcing of scavenging and cleansing of dense and congested space orbits, viz., low-Earth-orbit and Geo-stationery-orbit.

5.2 *Ethics and Altruism in Commercial Space Activities*

An interest in the ethical dimension of space activities emerged in the 1990s of the last century. Rolston, Callicott, Berry and Lovelock were pioneers of this theme,[22] and Hargrove formalised the study of this topic.[23] However, a more cohesive discourse on this emerged in the current century when technology had achieved diverse viability and business showed eagerness to invest in space activities.[24] This was also the stage where key lines of arguments in favour and against became clear and crystallised with the emergence of a school of thought expressing "concern that ethics may be a source of unhelpful constraints which could stand in the way of the emerging space economy…".[25]

Considering the divergent views, it can be maintained that ethics does not exist in a vacuum or a static mould and is verily influenced by the progressive revelation of truths as well as compulsion of attendant change in circumstances because there is no referrable text of right ethics for everywhere and for all times. And certainly, ethics for the outer space domain has to be "cosmocentric," taking cognisance of relatively correct and intrinsic values pertaining to the existential concerns, environmental vulnerabilities and cosmic sustainability of its own domain. Thus, space ethics has to bear relational values to cosmic realities and progressive discoveries.[26]

Be that as it may, space exploitation devoid of ethical values and elementary morality, as well as altruism and inclusivity, is bound to be iniquitous and unjust. History is a witness that humanity has not covered itself with glory in its economic activities. No doubt, the motivation for any commercial or industrial activity is profit, but the business agenda has been to squeeze profit by means legal, pseudo-legal, or even illegal. This does not behove well and has created inequity, acrimony and discord among the participants for the production of wealth. Management-labour disputes have existed for centuries, and the sharing of wealth has been disproportionate to the contribution to production. Thus, inequities have writ large, and it needs no citations or statistics because the economic milieu is open to empirical observation and daily comment.

---

[22] James Schwartz and Tony Milligan (eds), *The Ethics of Space Exploration* (Springer 2016) 4.

[23] Eugene Hargrove (ed), *Beyond Spaceship Earth: Environmental Ethics and the Solar System* (1986).

[24] Constance Bertka (ed), *Exploring the Origins, Extent and Future of Life: Philosophical, Ethical and Theological Perspectives* (2009).

[25] James Schwartz and Tony Milligan (eds), *The Ethics of Space Exploration* (Springer 2016) 6.

[26] Mark Lupisella, 'Cosmological Theories of Value: Relationalism and Correctedness as Foundation for Cosmic Creativity' in Schwartz and Milligan, *The Ethics of Space Exploration* (Springer 2016) 75–91.

It will, thus, seem desirable not to carry such a baggage of business mindset and commercial culture to the outer space activities, even with a corrective dissect of foul beliefs and agonal competition for space ventures. In fact, the principles of ethics and altruism ooze out of the Treaty at different levels and have been epitomised in its doctrinal exhortations. It assures freedom of access to all states to all parts of outer space and celestial bodies for scientific investigation and peaceful activities subject to the principle of non-interference. Mutual cooperation is advocated in rescue and return of astronauts and providing shelter during accidents, disasters and emergencies in outer space. It advocates consultations when in doubt for missions with unknown risks or possible contamination of the environment or disturbance to space ecology. These recommendations are, indeed, worthy and laudable.

In a similar stance, the Treaty ordains the tenets of altruism in the best interest of humankind and inclusivity of all states to the benefits accruing from space activities. This flags incumbent responsibility on space-faring states to help fledgling nations in their flight to outer space. Ergo, it does not envisage self-serving nationalism, business populism, an isolationist attitude or crass competition for self-enrichment. On the other hand, it implicitly upholds that outer space is a province of mankind or global commons or common heritage of mankind; and space activities shall be carried out in the interest of all countries and the benefits shared. Thus, this carries a strand of humanitarianism and philanthropy that ordains mutuality of help and sharing with the needy.

The principles of egalitarianism, altruism, humanism and inclusiveness are espoused for the commonality of interest and universality of their acceptance. Hence, in a nutshell, these are the principles of natural justice, equity and fair play, sharing of benefits, corporate humanitarian responsibility, and sensitivity towards space workforce, else it may lead to a social catastrophe, even an existential threat. In view of the above, the commercially exploited resources cannot be privately appropriated or commercially disposed of, nor accrued profits fully pocketed by the corporates. Everyone among humanity has a common interest in *res*, in the astro-estate, its resources and deserves to get an equitable share as compensation for exploitation and utilisation by private commercial entities.

On the other hand, inequality and the need for help in global growth are emotive issues and draw different inferences. At the same time, there is a theory of unearned profits and unfair enrichment that discredits this stand on sharing. But OST still constitutes law on the subject, and its germane provisions stand valid. Of course, the modalities of sharing and proportions of benefit are yet to be determined and promulgated, but this inadequacy does not allow for violation of the basic law. There are, however, temporary expedients and pragmatic solutions, ad interim, that have been discussed earlier and so proposed, that the dignity of extant law is respected and its integrity maintained.

This bifurcation of opinions leads us to explore ethical values for space business and suss out basic tenets of altruism from international space law for adoption in activities in outer space. It is further emphasised that diversification of space activities and commercial exploitation of space resources bring humanity to a different threshold and usher in a new era of prosperity. No operator in space is fully self-sufficient in all respects to operate independently, and their dependence on other space players is manifest in normal operations as well as in emergencies. Therefore, the new space age will be an "Era of Cooperation and Responsibility"[27] among space players. The law of international responsibility, and consequent liability, is well established and is applicable to space activities; it only needs to be amplified mutatis mutandis. More specifics on this aspect have been stipulated in the Liability Convention, 1972.[28] Similarly, international cooperation between space players has been repeatedly urged in the OST and more specifically enunciated in the Declaration on International Cooperation, 1996.[29] Civilised space-faring nations do not need a nudge to abide by this advice. It will be in the mutual and reciprocal interests of all space actors to abide by this Declaration and respect its recommendations. Sagacity lies in choosing this prudent option rather than regretting the holocaust, post-facto.

6 Conclusion

In conclusion, the era of commercial exploitation of celestial bodies is right upon us, with technological capabilities, economic viability, and corporate eagerness to exploit the opportunities lying in wait. The pioneer will be the winner in business terms. At the same time, the legality of such activities and total appropriation of profits by the engaging private enterprise is under the penumbra of doubt in relation to the OST and allied international instruments that seem braided together for synergy and for public order in space. Business motivations should not overpower the law. The majesty of the OST must be upheld, and changes that are required to revise and contemporise the treaty should be taken as per procedure. Some suggestions propose short-term measures and quick expedients so that the law does not appear to be a handicap in the welfare of humanity and betterment of humankind.

I must concede that some of the concepts and possibilities visualised relating to technical creativity and space applications may appear futuristic, but it is no utopian vision. Science of today is the technology of tomorrow that may become operational ground reality sooner than later. The purpose is that law should not be a laggard and not "look an ass" when it confronts a new situation, challenge or reality. *Non liquet* is the least desirable predicament for the judges. The law scholars also need to be proactive in identifying impending challenges, initiate healthy discussions, and create valid opinions on such issues so that the law is not caught napping. In fact, in

[27] G S Sachdeva, *Space Commercialisation: Prospects, Challenges and Way Forward* (Pentagon Press LLP 2019) 203–225.

[28] *Convention on International Liability for Damage Caused by Space Objects* (adopted 29 March 1972, entered into force 1 September 1972) 961 UNTS 187 (Liability Convention).

[29] *Declaration on International Cooperation in the Exploration and Use of Outer Space for the Benefit and in the Interest of All States, Taking into Particular Account the Needs of Developing Countries* UNGA Res 51/122 (13 December 1996).

the absence of direct and relevant provisions, adverse inferences, ingenious constructions, and devious interpretations motivated by national policy or devoted to vested interests start doing the rounds. A current example of lacunae in law confronting us is in the exploitation of celestial mineral resources by the US, Luxembourg and Saudi Arabia axis, where business corporations are almost ready with the wherewithal, for private appropriation, even without legal permissibility. This is hardly desirable, and only debated solutions with common acceptance and congruence to the established space law should be accepted for the emerging space economic order.

Therefore, let us herald the new space revolution of commercial utilisation of outer space and mineral harvesting from celestial bodies for the betterment of humanity, and welcome it with open arms for its expected benefits, but with a caveat. State practice by a few states for vested interests would not create a valid law. And the acquiescence to or acceptance of infringement of established space law by these activities will come at a cost that will be hard to pay back in the future. Ergo, caution beckons us to adapt and adjust the incumbent laws in conformity to contemporary challenges and impending benefits rather than disrespect the OST and land ourselves in legal chaos and disorder in space. Nor can the interests of posterity be sacrificed.

Therefore, treaty-member-states, as responsible nations, owe a duty *erga omnes,* to usher a development process in consonance with law. It could be to introduce a desired upgrade of the Treaty law or a simultaneous majoritarian UN resolution/agreement/protocol or any other adaptive legal changes to maintain peace and stability in the public order of outer space without its unwanted fall outs on international relations and global peace. Similarly, space law fraternity needs to be part of the solution to build proper *opinio juris* and cultivate new international jurisprudence rather than accentuate a difficult problem by platforming specious arguments and devious logic. The choice rests with us, and we sure have the wisdom to make the right selection of an option.

Prof. (Dr.) G. S. Sachdeva is one of the pioneer scholars on Space Law in India and has been writing on the subject for the last four decades. He has seven books to his credit and has contributed over seventy articles in edited books and law journals in India and abroad. He has Masters in Economics from Delhi School of Economics, Delhi University, Law graduation from Nagpur University, Nagpur (Gold Medallist) and Ph.D. in International Law from Jawaharlal Nehru University, New Delhi. He is currently Professor Emeritus, Chandigarh University and Adjunct Professor to NALSAR University of Law, Hyderabad. He was formerly Adjunct Professor, School of International Studies, JNU, New Delhi. He was Leader of a Think-tank and National Panellists for Revising of Draft Space Activities Bill (2022). He has also been selected as Member of Study Group 3.26 of International Academy of Astronautics, (IAA) Paris. He is a peer-reviewer for many reputed journals like Astropolitics, Space Policy, Indian Journal of InternationalLaw and others and is on the Editorial/Advisory Boards of several Indian and foreign law journals.

National Approaches to Space Resource Utilisation

United States Space Laws: A Roller Coaster of Space Resource Ambition

Michelle L. D. Hanlon

Abstract In the United States (US), the concept of and debate over the governance of space resource utilisation have endured a roller coaster of progress and setbacks. The 1960s buzzed with ambitious visions of lunar bases built with in situ materials. Buoyed by the troves of material astronauts brought back from the Moon, the discussion of space resource utilisation reached a zenith in 1980, when the international community considered the Agreement Governing the Activities of States on the Moon and Other Celestial Bodies. The US decision not to ratify this Agreement, amid vociferous protest, ushered in a period—more than three decades—of research that steadily built to a new height: the signing by President Barack Obama of the United States Commercial Space Launch Competitiveness Act of 2015. This milestone law "entitles" US citizens to possess, own, transport, use, and sell resources obtained from celestial bodies like asteroids and the Moon. The peak was short-lived. By 2019, no space mining missions had materialised, and funding shortages grounded asteroid mining pioneers Planetary Resources and Deep Space Industries. NASA's Commercial Lunar Payload Services (CLPS) programme put space resource utilisation back on an upward trajectory as evidenced by the successful lunar landings of Intuitive Machines' IM-1 in February 2024, and Firefly's Blue Ghost in March 2025—the first commercial spacecraft to achieve soft landings on the Moon. These successes have revived and lent credence to the concept of space resource utilisation and the cadence of missions to the Moon and asteroids to explore—and utilise—their resources promise to increase. This chapter examines the US legal framework steering and evolving these turbulent highs and lows. The ride to space resource utilisation is aiming for new heights, but despite favourable laws, regulatory uncertainty persists, and significant work remains to steer space resource utilisation in the United States off its roller coaster ride and onto a stable, upward trajectory.

Keywords Space resource utilisation · United States · Commercial space launch competitiveness act · Lunar missions · Regulatory uncertainty · Space governance

M. L. D. Hanlon (✉)
Center for Air and Space Law, The University of Mississippi School of Law, Oxford, MS, USA
e-mail: mlhanlon@olemiss.edu

S. Bhat B and A. Variath (eds.), *Legal Responses to Space Resource Utilisation*, Space Law and Policy, https://doi.org/10.1007/978-981-95-8864-0_9

1 Introduction: Historical Context

The United States (US) officially began exploring *in situ* utilisation of space resources in 1959 when President Dwight Eisenhower tasked the US Army with developing concepts for a lunar military base.[1] Early visions included constructing "an underground base to take advantage of the lunar soil as shielding against meteoroids and radiation"[2] and extracting water for life support and propellant. In 1963, a Working Group on Extraterrestrial Resources, "an informal organisation of researchers and engineers," was formed at the National Aeronautics and Space Administration's (NASA) Jet Propulsion Laboratory.[3] The group met seven times between 1963 and 1969. The US Bureau of Mines also hosted a NASA programme on Multidisciplinary Research Leading to Utilisation of Extraterrestrial Resources from 1965 to 1972.[4] Focused on science and engineering aspects, these efforts did not consider the policy or legal implications of monopolising resources on other celestial bodies. Ultimately, momentum waned, likely due to insufficient funding and "general scepticism from the space community."[5]

However, Meurisse and Carpenter suggest that international law may have also dampened enthusiasm for space resource utilisation. As scientists and engineers explored structures on the Moon and other celestial bodies, policymakers at the newly formed United Nations (UN) Committee on the Peaceful Uses of Outer Space (COPUOS) grappled with governing a previously inaccessible domain. The resulting Treaty on Principles Governing the Activities of States in the Exploration and Use of Outer Space, Including the Moon and Other Celestial Bodies[6] (Outer Space Treaty), ratified in 1967, "abstains from regulating specific activities [like space resource utilization], supplying instead the basic legal building blocks for addressing new

---

[1] *See* Alexandre Meurisse and James Carpenter, *Past, Present and Future Rationale for Space Resource Utilisation, 182* Planetary and Space Science (March 2020), https://www-sciencedirect-com.umiss.idm.oclc.org/science/article/pii/S0032063319302077#bib208.

[2] *Id.*

[3] The Working Group was formed "To evaluate the feasibility and usefulness of the employment of extraterrestrial resources with the objective of reducing dependence of lunar and planetary exploration on terrestrial supplies; to advise cognizant agencies on requirements pertinent to these objectives, and to point out the implications affecting these goals." Proceedings on the Working Group on Extraterrestrial Resources, Fourth Annual Meeting, 29 November 1965, https://ntrs.nasa.gov/api/citations/19660026216/downloads/19660026216.pdf.

[4] "The objective of the program is to help provide basic scientific and engineering knowledge needed to use extraterrestrial mineral resources in support of future space missions. U.S. Bureau of Mines NASA Program of Multidisciplinary Research Leading to Utilization of Extraterrestrial Resources, Progress Report, April 1965,

in support of future space missions." https://ntrs.nasa.gov/api/citations/19710026674/downloads/19710026674.pdf.

[5] *Id.*

[6] Treaty on Principles Governing the Activities of States in the Exploration and Use of Outer Space, Including the Moon and Other Celestial Bodies, Jan. 27, 1967, 610 UNTS 206 [hereinafter Outer Space Treaty].

activities and capabilities."[7] Nevertheless, Article II of the Treaty states in full that "[o]uter space, including the moon and other celestial bodies, is not subject to national appropriation by claim of sovereignty, by means of use or occupation, or by any other means."[8] As this language precludes appropriation "by means of use," it can be interpreted to prohibit resource utilisation. As Senator and Apollo 17 astronaut Harrison Schmitt noted, "[t]here is this basic conflict between not being able to lay an internationally valid legal claim and conducting an operation that is going to exploit resources."[9]

The international community largely agrees that no sovereign State may make claim to extraterrestrial territory, a principle so embedded in the bedrock of space exploration that many consider it customary international law.[10] However, a colourable argument exists that Article II does not apply to non-State entities. First, the article plainly indicates that space shall not be subject to *national* appropriation. Second, the primary restriction is against appropriation by claiming *sovereignty*. Thus, a non-State entity asserting proprietary rights arguably does not engage in "national appropriation," or "claim of sovereignty." This interpretation is supported by Article 17 of the United Nations Universal Declaration of Human Rights, which states that "[e]veryone has the right to own property alone as well as in association with others."[11] Article III of the Outer Space Treaty requires States to "carry on activities in ... space in accordance with international law"[12] which would include supporting these property rights in space. As Stephen Gorove interpreted it: "an individual acting on his [or her] own behalf or on behalf of another individual or a private association or an international organisation could lawfully appropriate any part of outer space, including the moon and other celestial bodies."[13]

The Apollo missions brought realism to this academic discussion.

> NASA returned nearly 50 pounds of material to Earth, including samples of lunar regolith and rocks consisting mainly of basalts and breccias. Once scientists found resources (silicon, calcium, aluminum, oxygen, magnesium, iron, and significant quantities of titanium) in these samples, pundits, politicians, and many scientists reawakened old fantasies about exploring mineral frontiers on the Moon.[14]

---

[7] Brian Israel, *Space Resources in the Evolutionary Course of Space Lawmaking,* Symposium on the New Space Race at 1 (2019).

[8] Outer Space Treaty, *supra* note 6, art. II.

[9] Senator Harrison Schmitt, Hearings before the Subcommittee on Science, Technology, and Space, of the Committee on Commerce, Science, and Transportation, United States Senate, 96th Congress, (29–31 July, 1980) 96–115, at 55.

[10] Fabio Tronchetti, The Non-Appropriation Principle Under Attack: Using Article II of the Outer Space Treaty in Its Defence, 50 Proc. L. Outer Space 526, 530 (2007).

[11] U.N. Declaration of Human Rights, art. 17, https://www.un.org/en/about-us/universal-declaration-of-human-rights.

[12] Outer Space Treaty, *supra* note 6, art. III.

[13] Stephen Gorove, *Interpreting Article II of the Outer Space Treaty,* 37 Fordham L. Rev. 349, 352 (1969).

[14] Stephen S. Buono, Governing the Moon, A History 18–19, Monographs in Aerospace History No. 59 (2025)(citations omitted).

Having samples in hand added credence to the possibility even as funding and public enthusiasm for lunar exploration dissipated. The lack of international objection to the Apollo samples claimed by the US government suggests that Article II does not impose a blanket ban on space resource appropriation. As Senator Schmitt observed in 1980, "[t]he momentum seemed to be for the United States to move forward with a continuous lunar exploration program and implicit in that is the potential exploitation of resources on the Moon in the foreseeable future ..."[15] This perception shifted when Apollos 18 and 19 were cancelled in 1970, and the Nixon administration redirected focus to low Earth orbit and the shuttle programme in 1972.[16]

Meanwhile, COPUOS continued to refine outer space governance, particularly regarding resource—a term absent from the Outer Space Treaty—characterisation and use. The resulting Agreement Governing the Activities of States on the Moon and Other Celestial Bodies[17] (Moon Agreement) renewed US interest in space resource activities and, for the first time, opened debate about a legal framework for them.

This chapter examines the US legal framework for space resource utilisation by starting with the US response to the Moon Agreement and its ultimate rejection, which has steered US law. It then describes commercialisation attempts, current US legal treatment, the rebirth of the commercial space resource industry as supported by the NASA Commercial Lunar Payload Services (CLPS) programme, and the development of the Artemis Accords. Finally, it assesses the state of US law today and suggests future directions.

2 The Moon Agreement and US Reaction: Roller Coaster Rising

2.1 The Moon Agreement

The genesis and development of the Moon Agreement are expertly described by Stephen Buono in NASA Monographs in Aerospace History No. 59, Governing the Moon, A History.[18] Article 11 is critical to understanding the US approach to space resource utilisation. Article 11(1) characterises "the moon and its natural resources"

---

[15] Senator Harrison Schmitt, Hearings before the Subcommittee on Science, Technology, and Space, of the Committee on Commerce, Science, and Transportation, United States Senate, 96th Congress, (29–31 July, 1980) 96–115, at 54 [hereinafter Moon Agreement Hearings].

[16] John M. Logsdon, *Lessons from the Past, Why the U.S. Decided Not to Explore After Going to the Moon,* IAC-14-E3.2.4, https://bpb-us-e1.wpmucdn.com/blogs.gwu.edu/dist/7/314/files/2018/11/Logsdon-Lessons-from-the-Past-Why-the-U.S.-Decided-Not-to-Explore-After-Going-to-the-Moon-1vcit1f.pdf.

[17] Agreement Governing the Activities of States on the Moon and Other Celestial Bodies, Dec. 18, 1979, 1363 U.N.T.S. 3 [hereinafter Moon Agreement].

[18] This book, which is highly recommended by this author, is available free of charge at: https://www.nasa.gov/history/governing-the-moon/.

as "the common heritage of [hu]mankind."[19] Lacking a formal legal definition, Article 11(1) goes on to note that the concept of "common heritage" will "find its expression" in Article 11(5), which delays addressing the issue by mandating the development of an international regime when the exploitation of resources "is about to become feasible."[20] The purpose of the regime, Article 11(7) clarifies, is to assure: "the orderly and safe development of the natural resources of the Moon; ... the rational management of those resources; ... the expansion of opportunities in the use of those resources; ... and an *equitable sharing by all States Parties in the benefits derived from those resources.*"[21] Article 11(3) fills any gaps left by Article II of the Outer Space Treaty and makes it very clear that

> Neither the surface nor the subsurface of the moon, nor any part thereof or natural resources in place, shall become property of any State, international intergovernmental or non-governmental organization, national organization or non-governmental entity or of any natural person. The placement of personnel, space vehicles, equipment, facilities, stations, and installations on or below the surface of the moon, including structures connected with its surface or subsurface, shall not create a right of ownership over the surface or the subsurface of the moon or any areas thereof.

2.2 2.2. The Hearings

In the United States, proponents of space exploration and resources utilisation, no matter how speculative, were outraged by the Moon Agreement, particularly the characterisation of natural resources in space as the common heritage of mankind, and the requirement for "equitable sharing of any benefits" which they viewed as antithetical to free-market principles.[22] When the US Senate held hearings on the Moon Agreement on July 29 and 31, 1980, those detractors testified and lobbied aggressively.[23]

Although at the time of the hearing, the industry "had no real plans for exploitation of nonterrestrial minerals,"[24] a number of the participants anticipated mining and utilising space resources "within ten years after the STS (space shuttle) [became] available."[25] Indeed, Senator Schmitt argued that "the scenario of mining the Moon is not as far-fetched as it is painted to be by some who would like us to just agree to

---

[19] Moon Agreement, *supra* note 17, art. 11(1). It is important to note that Article I of the Moon Agreement is clear that its provisions are intended to cover more than just the Earth's moon by stating that the provisions of the Agreement relating to the Moon "shall also apply to other celestial bodies within the solar system, other than the Earth." *Ibid.* at art.1.

[20] *Ibid.* at art. 11(5).

[21] *Ibid.* art. 11(7) (emphasis added).

[22] Buono, *supra* note 14, at 67–69. The Moon Agreement was often compared disfavorably to the United Nations Convention on the Law of the Sea. *Ibid.*

[23] *Ibid.*

[24] Louis Friedman, Moon Agreement Hearings, *supra* note 15 at 96, 161.

[25] James Arnold, Moon Agreement Hearings, *supra* note 15, at 154.

[the Moon Agreement] for other political purposes not having to do with the realities of actually getting resources in space."[26]

2.2.1 Common Heritage and Economic Barriers: Seeking to Create a New Economic Order

The Moon Agreement's characterisation of the Moon and all its resources as the "common heritage of mankind"[27] and the mandate for a new international regime or "UN agency," created under Article 11(7) that would "rationally manage" any resources and assure an "equitable sharing … in the benefits,"[28] sparked fierce opposition as critics saw these provisions as both a geopolitical manoeuvre and an economic deterrent.

Negotiated in parallel with the United Nations Convention on the Law of the Sea[29] (UNCLOS), which similarly labelled deep seabed resources as the "common heritage of mankind," critics saw both treaties as being "tilted in favor of the Third World as against the developed nations, including the United States,"[30] and promoting a "political model of control by the votes of the Third World" and an "economic model [of] socialism.[31] They suspected a broader agenda, supported by the Soviet Union and developing nations, to establish a "New International Economic Order (NIEO), the goal of which [was] a fundamental redistribution of global wealth."[32]

The American Mining Congress warned that the Moon Agreement too closely mirrored UNCLOS's restrictive regime, which they believed had hindered seabed mining."[33] Companies, it was asserted, "cannot make the investments necessary for seabed mining because they could not operate profitably under the regime envisaged by the Law of the Sea Treaty."[34] Alexander Haig, Jr., speaking for United Technologies, stated that the aerospace industry could not "undertake the development of space resources under a similar system applied by the Moon Treaty."[35] Senator Schmitt summed it up:

---

[26] Schmitt, Moon Agreement Hearings, *supra* note 15 at 138.

[27] Moon Agreement, *supra* note 17, art. 11(1).

[28] Moon Agreement, *supra* note 17, art. 11(7).

[29] United Nations Convention on the Law of the Sea, art. 136, Dec. 10, 1982, 1833 U.N.T.S. 397.

[30] Roberts B. Owen, Moon Agreement Hearings, *supra* note 15, at 5.

[31] American Bar Association Report, With Recommendation, of the Section of Natural Resources Law to the House of Delegates, Moon Agreement Hearings, *supra* note 15, at 83.

[32] American Bar Association Report, With Recommendation, of the Section of Natural Resources Law to the House of Delegates, Moon Agreement Hearings, *supra* note 15, at 83.

[33] Marne A. Dubs, Moon Agreement Hearings, *supra* note 15, at 133.

[34] Letter from Alexander M. Haig to Robert Owens, Moon Agreement Hearings, *supra* note 15, at 220.

[35] Letter from Alexander M. Haig, Jr. to Roberts B. Owen, June 4, 1980, Moon Agreement Hearings, *supra* note 15, at 220.

> [t]he Law of the Sea negotiations clearly have retarded the pace at which we exploited technologies developed by this country for the harvesting of the resources of the ocean floor. Are we, with [the Moon Agreement], going to similarly retard development of U.S. technology to the harvesting, if you will, of the resources of ... the Moon?[36]

Critics noted that, through the negotiations of the UNCLOS, it was apparent "119 developing countries and six East European Socialist States" interpreted "common heritage," to mean "property which is owned by the U.N. agency created by treaty, and subject to private acquisition only with the consent of a new supranational regime, pursuant to a global agreement."[37] In fact, one critic pointed out, "in Spanish, in French, in Russian and in Chinese, the term 'heritage' does not appear at the United Nations."[38] The term used instead is "property."[39] Thus, the "common heritage of mankind is the common property of mankind,"[40] further reinforcing concern about the viability of financial returns on mining investments.

The "equitable sharing" requirement compounded these concerns and was derided as akin to taxation without representation[41]:

> It is like a windfall tax; isn't it? You have not discovered the oil yet, but you know you will have to pay or have an added cost in the form of a tax. You have not developed the resource. You have not started manufacturing ball bearings in space or mining the titanium and hydrogen of the Moon, but you already know you will have to pay something out that your stockholders will not get.[42]

The financial ambiguity produced by the indeterminate amount the equitable sharing might require was also criticised: "How much of gross or net sales must go to the other signers of the Moon Treaty? No one can begin to answer that."[43] Finally, Article 11(5)'s mandate for the establishment of an international regime raised fears of a moratorium on space resource utilisation until such a regime is in place. While it was

---

[36] Schmitt, Moon Agreement Hearings, *supra* note 15, at 35, 43.

[37] American Bar Association Report, With Recommendation, of the Section of Natural Resources Law to the House of Delegates, Moon Agreement Hearings, *supra* note 15, at 84.

[38] Leigh Ratiner, Moon Agreement Hearings, *supra* note 15, at 114.

[39] Leigh Ratiner, Moon Agreement Hearings, *supra* note 15, at 114. *See also* Charles Scheffield, Moon Agreement Hearings, *supra* note 15 at 89–90 ("the common heritage of all mankind [is] a phrase which is believed by the vast majority of nations to mean common property. Indeed, in several of the official languages of the UN the term used to translate heritage is 'patrimony' the legal equivalent of property in most of the world....").

[40] Scheffield, Moon Agreement Hearings, *supra* note 15, at 89–90 (citations omitted).

[41] Schmitt, Moon Agreement Hearings, *supra* note 15, at 43 (as the sharing would not be something "over which elected representatives of the country have any control").

[42] Schmitt, Moon Agreement Hearings, *supra* note 15, at 42. *See also* Hearing page 107: ("The 'equitable sharing of benefits' principle stands for two additional subsidiary principles of a more concrete character—taxation, and international taxation code, and an international taxation machinery which is for enforcement of taxes.").

[43] Louis Friedman, Moon Agreement Hearings, *supra* note 15, at 90. *See also* Sheffield, Moon Agreement Hearings, *supra* note 15, at 87 ("you could go to the Moon and even perhaps mine what is there, but you would have to pay something, an amount still completely undefined, for the privilege of mining those resources.").

conceded that the language itself does not necessarily mandate a moratorium, such a claim was "readily foreseeable"[44] and yet another[45] manner in which the Agreement would discourage investment. Even without a moratorium, "what company will put $1 billion or $2 billion into a mining venture on the Moon, or to harness the resources of an asteroid, until it knows exactly what those new rules and regulations will be?".[46]

2.2.2 Property Ownership: No Pre-Mining Legal Claim?

Article 11(3)[47] "also caused consternation as it greatly expanded the nonappropriation principle found in Article II of the Outer Space Treaty." While the Moon Agreement seems to permit entities to "actually take material and use it,"[48] how, it was wondered, would that party lay "claim to a particular area while … not taking material and using it"?[49] Senator Schmitt called the Agreement "clearly inconsistent on its face."[50]

> You can't have a legal claim but you can mine and exploit the resource. You don't mine and exploit the resource instantaneously. It happens over decades of time. If in that decade of time you don't have a valid claim for the area you need to make that resource economically viable and then you have a problem.[51]

This was also seen as a deterrent making long-term investment in space mining operations economically unviable as companies could not protect their claims during development phases.[52]

2.2.3 Premature Negotiations

Opponents also deemed the Moon Agreement premature for two main reasons. First, even though few questioned the potential resources on the Moon and elsewhere in space, many pointed out that limited knowledge of space resources rendered the Agreement a "theoretical exercise" that could prove ultimately inapplicable if when

---

[44] American Bar Association Report, With Recommendation, of the Section of Natural Resources Law to the House of Delegates, Moon Agreement Hearings, *supra* note 15, at 85.

[45] American Bar Association Report, With Recommendation, of the Section of Natural Resources Law to the House of Delegates, Moon Agreement Hearings, *supra* note 15, at 85.

[46] Ratiner, Moon Agreement Hearings, *supra* note 15, at 109–110.

[47] Moon Agreement, *supra* note 17, art. 11(3). ("neither the surface nor the subsurface of the Moon, nor any part thereof of natural resources in place, shall become property of any State, international intergovernmental or nongovernmental organization, national organization or non-governmental entity or of any natural person.").

[48] Frosch, Moon Agreement Hearings, *supra* note 15, at 39.

[49] Frosch, Moon Agreement Hearings, *supra* note 15, at 39.

[50] Schmitt, Moon Agreement Hearings, *supra* note 15, at 55.

[51] Schmitt, Moon Agreement Hearings, *supra* note 15, at 55.

[52] Frosch, Moon Agreement Hearings, *supra* note 15, at 39.

"one arrives at the situation, it is a different situation than was invented beforehand."[53] Moreover, it was pointed out that "[y]ou don't know whether you are negotiating away your interest or not because you do not know what they are."[54] The Agreements scope, applying to the Moon ***and*** "all other celestial bodies within the solar system"[55] was criticised as overly broad given unknown compositions.[56] One critic went so far as to suggest there was no need to rush as it was clear "[t]he real (and intended) significance of the treaty is the precedent which it would establish with respect to the resources of the seabed and Antarctica."[57]

The second concern about timing laid blame at the feet of US President Carter for failing to articulate a national space policy to include space resource utilisation. Senator Schmitt, a Republican, criticised the president, a Democrat, asking "[h]ow can we properly negotiate and then examine and ratify a treaty of this kind when we have no space policy with respect to exploitation of either inner space or outer space?"[58] Without defined objectives, others noted, the US risked conceding vital interests unknowingly.[59]

2.2.4 Lack of Industry Input

The hearings also highlighted the lack of industry involvement in the Agreement's negotiation. Critics noted that the US negotiating team lacked mining expertise, with Schmitt stating, "[n]o one with any experience in mining operations was on the negotiating team"[60] Industry representatives expressed frustration, stating, "[w]e don't feel the administration is too worried at the moment about the health of the space industry, and we feel as if we were left outside when the treaty was developed"[61] This exclusion fuelled perceptions that the treaty was a *fait accompli* without industry blessing,

---

[53] Frosch, Moon Agreement Hearings, *supra* note 15, at 40.

[54] Frosch, Moon Agreement Hearings, *supra* note 15, at 40.

[55] Moon Agreement, *supra* note 17, art. 1(1).

[56] Charles Sheffield, Moon Agreement Hearings, *supra* note 15, at 91.

[57] American Bar Association Report, With Recommendation, of the Section of Natural Resources Law to the House of Delegates. Moon Agreement Hearings, *supra* note 15, at 83.

[58] Schmitt, Moon Agreement Hearings, *supra* note 15, at 32.

[59] Ratiner, Moon Agreement Hearings, *supra* note 15, at 116: I think, that if this nation knew its space policy, particularly with respect to resources in the future, it would make it a whole lot easier to write instructions for our negotiators and submit them to a normal clearance process up to and including the President. Then, the negotiator would be told, "Don't agree to anything that doesn't include free access on a first-come, first-serve basis. You may agree to the sharing of certain benefits up to the following limits," and so on. But if we don't know what our space policy is with respect to the resources, then we don't have the slightest idea what we have given away in this treaty and in the subsequent treaty when we start negotiating.

[60] Schmitt, Moon Agreement Hearings, *supra* note 15, at 54.

[61] Sheffield, Moon Agreement Hearings, *supra* note 15, at 86.

undermining confidence in its provisions.[62] In addition failing to bring experts to the table, this exclusion was deemed yet another deterrent as " so long as industry perceives that it is on the outside" not only will it not commit to funding research and development, but it could also deter investment in the US space programme overall.[63]

2.2.5 Against National Interest and American "Frontier" Ethos

Finally, the hearings exposed a clash between the treaty's international framework and US national interests, amplified by Cold War tensions and the American frontier ethos. Critics like Leigh Ratiner warned that the treaty's "common heritage" principle and potential for international control over space activities were "directly contrary to the national interest of the United States," risking access to "all of the resources of the twenty-first century needed by men on Earth."[64] This concern was heightened by fears of a "celestial bodies" regime that could constrain US space investigations, forcing the nation to share technology and resources funded by American taxpayers.[65]

Cold War dynamics added a sharp edge, with questions about the Soviet Union's role in shaping the Agreement, and there were many questions about how much the Soviet Union conceded during the negotiation process.[66] Many were suspicious of Soviet support for the Agreement, and it was a ploy to align with Third World interests to solely counter US technological dominance, heightening fears of a loss of strategic control in space.

For many, too, the treaty threatened to derail the American frontier spirit, which viewed space as the next great frontier. One participant captured this ethos, stating, "The challenge of the space frontier is something that speaks very directly to the American heritage… The intellectual, moral effect of activating and moving into this thing is so important for the kind of country we are."[67] Ratifying the treaty, it was argued, risked chaining this pioneering spirit to an international regime, stifling US innovation and leadership in a domain seen as critical to national identity and security.

---

[62] The treaty appeared without industry participation, and without industry blessing; and only now, when the treaty is a finished written report and a fait accompli, is industry being heard from and asked to provide inputs. Charles Sheffield, Moon Agreement Hearings, *supra* note 15, at 86.

[63] Sheffield, Moon Agreement Hearings, *supra* note 15, at 86.

[64] Ratiner, Moon Agreement Hearings, *supra* note 15, at 116.

[65] American Bar Association Report, With Recommendation, of the Section of Natural Resources Law to the House of Delegates, Moon Agreement Hearings, *supra* note 15, at 84–85.

[66] S. Neil Hosenball, Moon Agreement Hearings, *supra* note 15, at 63.

[67] Richard G. Darman, Moon Agreement Hearings, *supra* note 15, at 164.

2.3 *Conclusion*

Ultimately, the detractors prevailed and the Carter administration "buried the [Moon Agreement] in legislative review, and the Senate never voted on ratification."[68] The hearings revealed deep tensions between international cooperation and US national interests, particularly regarding free-market principles and the potential for space resource exploitation. Critics, including industry leaders and free-market advocates, saw the treaty as a premature and restrictive framework that threatened innovation and investment. By drawing parallels with UNCLOS, emphasising the lack of industry input, and highlighting conflicts with American values, opponents effectively challenged the treaty's alignment with US priorities and shaped the discourse on the future of space exploration and resource utilisation.

3 The 2015 Space Resource Law: Peak and Peril

The US rejection of the Moon Agreement was followed by a decades-long lull in space resource utilisation discussions. By 2014, the US commercial space industry had grown markedly. Two commercial companies were handling cargo resupply missions to the International Space Station. This progress revived interest in space resource utilisation activities. Two companies, not with lunar but asteroid mining ambitions, emerged: Planetary Resources (founded in 2010) and Deep Space Industries (founded in 2013). Both firms aimed "to split asteroid water into its constituent hydrogen and oxygen, which are the chief components of rocket fuel. Asteroid-derived propellant could be dispensed from off-planet 'gas stations,' allowing satellites and journeying spacecraft to top up their tanks cheaply and efficiently."[69] Ambitiously, Deep Space Industries planned to being "prospecting for asteroid in 2015, begin returning samples to Earth by 2016, and mining asteroids by 2023."[70] Planetary Resources planned to create, by 2020, fuel depots in space that would "convert water extracted from asteroids into liquid oxygen and hydrogen rocket fuel, which could then be shipped to Earth orbit for refuelling commercial satellites or spacecraft."[71]

---

[68] Buono, *supra* note 14 at 3.

[69] Mike Wall, *The Beginning of An Industry: Second Asteroid Mining Company Enters the Picture,* SpaceNews.com (Jan.22, 2013).

[70] Matt Williams, *US House of Representatives Passes Asteroid Mining Bill,* Herox.com, https://www.herox.com/blog/381-us-house-of-representatives-passes-asteroid-mining.

[71] *Ibid.*

3.1 ASTEROIDS Act of 2014

Likely spurred by this growth, the next serious consideration of space resource utilisation from a legal and policy perspective by the US Congress came in 2014 with the introduction of the American Space Technology for Exploring Resource Opportunities in Deep Space Act (ASTEROIDS Act), a precursor to current US law. It was introduced to provide the "legal certainty" for American companies "that if they obtain resources from an asteroid that they can own them."[72] The Act sought to create a legal framework with four foundational principles that addressed critiques raised during the Moon Agreement hearings. These were:

a. property rights in any resource obtained from an asteroid—a clear rebuttal of the common heritage principle.
b. freedom from harmful interference expressed as the right that "any assertion of superior right to execute specific commercial asteroid resource utilisation activities in out space shall prevail if it is found to be first in time, derived upon a reasonable basis,"—addressing the issue raised by Senator Schmitt regarding rights during developmental stages[73]
c. avoidance of harmful interference, and
d. exclusive jurisdiction of US courts over any disputes[74]—rebutting the concept of an international governance regime.

This was the first national legislation in the United States or elsewhere to "address important issues about the relatively new commercial intent to obtain resources from space."[75] In September 2014, the House Subcommittee on Space gathered experts to discuss the ASTEROIDS Act.[76] It was generally argued that "the future of human activity in space depends on our ability to cost-effectively recover and utilize resources from near-Earth asteroids"[77] and that "an agreed-upon policy regarding property rights for resources mined from asteroids will eventually be important for commercial investment."[78] The proposed legislation was criticised because it "could lead to efforts to claim, and hold hostage, large numbers of potential resource targets long before the capability to exploit those targets is developed."[79] It was also pointed out that international "disagreement should be expected as to the meaning

[72] Rep. Posey, 2014 Hearing, *supra* note 73, at 13–14.

[73] *See supra* note 51 and accompanying text.

[74] H.R. 5063, To Promote the Development of a Commercial Asteroid Resources Industry for Outer Space in the United States and to Increase the Exploration and Utilization of Asteroid Resources in Outer Space (10 July 2014) https://www.congress.gov/bill/113th-congress/house-bill/5063/text.

[75] Lamar Smith, Exploring Our Solar System: The ASTEROIDS Act as a Key Step, Hearing before the Subcommittee on Space, Committee on Science, Space, And Technology, House of Representatives, 113–98 (10 Sept. 2014), at 17 [hereinafter 2014 Hearing].

[76] 2014 Hearing, at 9.

[77] Mark Sykes, 2014 Hearing, at. 61.

[78] Jim Bell, 2014 Hearing, at 36.

[79] Mark Sykes, 2014 Hearing, at 64.

of this kind of legislation."[80] Ultimately, this legislation did not advance beyond the Subcommittee.

3.2 *The 2015 Act*

A revised version of the ASTEROIDS Act, the Space Resource Exploration and Utilisation Act (2015 Act), was introduced in March 2015 and this time it was successfully included in the Commercial Space Launch Competitiveness Act,[81] signed into law by President Obama in November 2015. The law begins by defining both the terms "space resource"—as an abiotic resource in situ in outer space, including waters and minerals—and "asteroid resource"—as a space resource found on or within a single asteroid.[82] It goes on to recognise commercial property rights in resources extracted from celestial bodies:

> A United States citizen engaged in commercial recovery of an asteroid resource or a space resource under this chapter shall be entitled to any asteroid resource or space resource obtained, including to possess, own, transport, use, and sell the asteroid resource or space resource obtained in accordance with applicable law, including the international obligations of the United States.[83]

The law also directs the President to "facilitate commercial exploration for and commercial recovery of space resources by United States citizens" and "promote the right of United States citizens to engage in commercial exploration for and commercial recovery of space resources free from harmful interference, in accordance with the international obligations of the United States and subject to authorisation and continuing supervision by the Federal Government."[84] Finally, in a nod to the Outer Space Treaty, the Act included a "disclaimer" that "It is the sense of Congress that by the enactment of this Act, the United States does not thereby assert sovereignty or sovereign or exclusive rights or jurisdiction over, or the ownership of, any celestial body."[85]

3.3 *Conclusion*

The evolution from the discourse on the Moon Agreement to the ASTEROIDS Act to the 2015 Act demonstrates a maturation of the US explication of its space resource utilisation policy and goals. The 2015 Act does not address the developmental aspect

---

[80] Joanne Gabrynowicz, 2014 Hearing, at 74.

[81] U.S. Commercial Space Launch Competitiveness Act, Public Law 114–90 (Nov. 2015).

[82] 51 U.S.C. § 51,301.

[83] 51 U.S.C. § 51,303.

[84] 51 U.S.C. § 51,302.

[85] U.S. Commercial Space Launch Competitiveness Act, Public Law 114–90, § 403 (Nov. 2015).

of resource prospecting but does make clear that the US interprets Article II of the Outer Space Treaty to permit ownership of resources once extracted. It also reaffirms the US commitment to meet its obligations under Article VI of the Outer Space Treaty to authorise and continually supervise the activities of its nationals. However, the 2015 Act does not direct which agency of the federal government will take on that responsibility or offer any specifics on parameters. Notably, the US State Department, in 2016, reaffirmed the US view that "Article II of the [Outer Space Treaty] would preclude the US government from recognising property rights in celestial bodies or 'their resources in place.'".[86]

Despite addressing concerns raised by the critics of the Moon Agreement, resource activities did not flourish immediately. In fact, the opposite was the case. By 2022, both Planetary Resources and Deep Space Industries were "acquired by other companies that have nothing to do with space mining" and as of 2025, humanity "has yet to commercially mine even a single asteroid."[87] The fact remains that space mining is a very long-term investment.

Nevertheless, it is fair to say that since at least 2014, the United States has been instigating—on a bipartisan basis—an effort to address a lacuna in international space law and assure that commercial space mining companies may retain such property rights in the minerals and elements they extract from the Moon and other celestial bodies as to be able to sell the resources to others for their own profit.

4 CLPS to Artemis Accords Regaining Momentum

4.1 Commercial Lunar Payload Services: Reviving Space Resource Activities

This effort was supported by NASA which, in 2018, introduced the Commercial Lunar Payload Services (CLPS) initiative. Authorized under the NASA Reauthorization Act of 2018, CLPS leverages commercial capabilities in support of the Artemis Programme's[88] goal of returning humans to the Moon. It allows NASA to contract with private US companies to deliver NASA scientific and technological payloads to the lunar surface, with a focus on resource-rich regions like the lunar south pole.

[86] Brian R. Israel, *Space Resources in the Evolutionary Course of Space Lawmaking,* Symposium on the New Space Race, Am. Soc. of Int'l L. (2019), at 116 (citing Brian J. Egan, Legald Adviser, *The Next Fifty Years of the Outer Space Treaty,* Galloway Symposium on Critical Issues in Space Lw (Dec. 7, 2016).

[87] Magdalena Petrova, *The First Crop of Space Mining Companies Didn't Work Out, But a New Generation is Trying Again,* CNBC (9 Oct. 2022) https://www.cnbc.com/2022/10/09/space-mining-business-still-highly-speculative.html.

[88] NASA, *Artemis,* https://www.nasa.gov/humans-in-space/artemis/ (last visited 29 June 2025).

Backed by $2.6 billion in funding through 2028[89] at this writing, CLPS supports 14 companies across diverse lunar missions.[90] Successes include:

a. Intuitive Machines' IM-1 (February 22, 2024): first commercial lunar landing.[91]
b. Firefly's Blue Ghost (March 2, 2025): successful lunar mission.[92]

These discrete missions, funded primarily by NASA, serve a dual purpose: first, to generate foundational research on the nature, distribution, and accessibility of lunar resources—particularly water ice and regolith-derived materials; and second, to foster the development of a sustainable lunar economy by enabling commercial actors to engage in the discovery, extraction, and eventual use of space resources. Unlike the more speculative ambitions of companies like Planetary Resources and Deep Space Industries, which focused on asteroid mining with limited infrastructure, the CLPS offers a pragmatic and incremental path towards space resource utilisation, anchored in tangible missions, proven technologies, and public–private collaboration. Through CLPS, NASA is not only advancing its own scientific and exploration goals but also laying the groundwork for a future in which the commercial exploitation of space resources is viable, regulated, and economically sustainable.

4.2 Fully Repudiating the Moon Agreement

The Trump Administration reaffirmed the US position on ownership of extracted space resources with an Executive Order issued in April 2020 which, among other things, recognises that "successful long-term exploration and scientific discovery of the Moon, Mars, and other celestial bodies will require partnership with commercial entities to recover and used resources … in outer space."[93] The Order further states that "American should have the right to engage in commercial exploration, recovery, and use of resources in outer space," and urges executive departments to "take all appropriate actions to encourage international support for the public and private recovery and use of resources in outer space."[94] The Order goes further to say that the United States "does not view [space] as a global commons" and, in one last swipe at the Moon Agreement:

> The United States is not a party to the Moon Agreement. Further, the United States does not consider the Moon Agreement to be an effective or necessary instrument to guide nation states regarding the promotion of commercial participation in the long-term exploration,

[89] NASA, *Commercial Lunar Payload Services*, https://www.nasa.gov/commercial-lunar-payload-services/ (last visited 29 June 2025).

[90] NASA, *Commercial Lunar Payload Services Providers*, https://www.nasa.gov/commercial-lunar-payload-services/clps-providers/ (last visited 29 June 2025).

[91] https://moonregistry.forallmoonkind.org/im-1/.

[92] https://moonregistry.forallmoonkind.org/3558-2/.

[93] Exec. Order No. 13914, 85 Fed. Reg. 20,381 (Apr. 6, 2020).

[94] *Ibid.* at 20,381–20,382.

> scientific discovery, and use of the Moon, Mars, or other celestial bodies. Accordingly, the Secretary of State shall object to any attempt by any other state or international organization to treat the Moon Agreement as reflecting or otherwise expressing customary international law.[95]

It is notable that the Biden Administration did not reverse this Executive Order and it remains US policy.

4.3 Regolith Purchase Request for Proposal

The Trump Administration took it one more step further. On September 10, 2020, NASA Administrator Jim Bridenstine revealed on Twitter (now X) that the NASA "is buying lunar soil from a commercial provider!"[96] Indeed, NASA announced that it will purchase 50 g to 500 g of lunar regolith, the equivalent of "three tablespoons to 2.5 cups," for which it "will pay between $15,000 and $25,000."[97] The proposed purchase amount would hardly cover the cost of implementation. Moreover, it is intended to be an "in-place" transfer, meaning it will take place on the lunar surface.[98]

---

[95] *Ibid.* at 20,381.

[96] Jim Bridenstine (@JimBridenstine), Twitter (Sep. 10, 2020, 8:31 AM, https://twitter.com/JimBridenstine/status/1304049845309669376. *See* Jim Bridenstine, *Space Resources Are the Key to Safe and Sustainable Lunar Exploration,* Blogs.NASA.Gov, (Sep. 10, 2020), https://blogs.nasa.gov/bridenstine/2020/09/10/space-resources-are-the-key-to-safe-and-sustainable-lunar-exploration/.

[97] Cat Hofacker, *NASA Plans to Pay Companies to Extract Tablespoons of Lunar Regolith,"* Aerospace America (Sep. 10, 2020), https://aerospaceamerica.aiaa.org/nasa-plans-to-pay-companies-to-extract-teaspoons-of-lunar-regolith/#:~:text=NASA%20will%20pay%20between%20%2415%2C000,is%20turned%20over%20to%20NASA.

[98] Jim Bridenstine, *supra* note 95. Ultimately, four companies were selected by NASA:

- Lunar Outpost proposed collection for $1 following arrival of a lander to the lunar South Pole in 2023.
- ispace Japan proposed collection for $5,000 following arrival in 2022 of a lander to Lacus Somniorum on the Moon's northeastern near side.
- ispace Europe proposed collection for $5,000 following arrival in 2023 of a lander to the lunar South Pole.
- Masten Space Systems proposed collection for $15,000 following arrival in 2023 of a lander to the lunar South Pole.

Press Release, Nat'l Aeronautics and Space Admin., NASA Selects Companies to Collect Lunar Resources for Artemis Demonstrations (Dec. 3, 2020), https://www.nasa.gov/press-release/nasa-selects-companies-to-collect-lunar-resources-for-artemis-demonstrations. Note the ambitious dates. As of this writing, none of the contracts have been fulfilled. Masten Space Systems was acquired by Astrobotic in September 2022, and the press release regarding the deal does not mention the regolith contract. *See* Astrobotic, *Astrobotic Acquires Masten Space Systems,* (13 Sept. 2022) https://www.astrobotic.com/astrobotic-acquires-masten-space-systems/. ispace has had two hard landing on the Moon in March 2023 and June 2025. Lunar Outpost reached the Moon in March 2025 as part of the Intuitive Machines IM-2 mission. Unfortunately, the Lunar Outpost rover, the Lunar Voyage 1 MAPP, ended up on its side under the IM-2 lander and was unable to deploy. LunarOutpost, *Lunar Voyage 1 Update,* https://www.lunaroutpost.com/post/lunar-voyage-1-update.

This purely symbolic transaction seeks to set a controversial legal norm—namely, allowing the extraction and sale of space resources by private companies for profit. According to Administrator Bridenstine, the purpose of these contracts is to "make sure that there is a norm of behavior that says that resources can be extracted and that we're doing it in a way that is in compliance with the Outer Space Treaty."[99] In 2022, the US reiterated its intent to comply with the Outer Space Treaty describing, on the international stage—at the Meeting of the Legal Subcommittee of the United Nations Committee on the Peaceful Uses of Outer Space—this particular NASA programme as "an opportunity to begin working through practical issues of space resources."[100]

4.4 Artemis Accords

In keeping with US policy to "encourage international support for the public and private recovery and use of resources in outer space, consistent with applicable law,"[101] the US spearheaded an effort to reach multilateral agreement on the ownership of resources extracted from other celestial bodies. On October 13, 2020, the civilian space agencies of eight nations announced that they had agreed to 12 Principles for Cooperation in the Civil Exploration and Use of the Moon, Mars, Comets, and Asteroids, known colloquially as the Artemis Accords.[102] As a starting point, it must be made clear that the Artemis Accords are not binding international law. They are not even a treaty exacting obligations from Signatories. As indicated in Section 1 of the Accords, they represent "a political commitment" by each Signatory to implement the principles contained therein "through their own activities by taking, as appropriate, measures such as mission planning and contractual mechanisms with entities acting on their behalf."[103] That said, even as a political commitment, the Accords advance the discourse regarding many important legal aspects of space exploration, including the commercial use of space resources.

As an initial matter, the Accords reaffirm the most important provisions of the Outer Space Treaty. For example:

---

[99] Rachel Lindbergh, *Space Resource Extraction: An Overview and Issues for Congress,* Congressional Research Service, (29 July 2024), at 9 *quoting* Jim Bridenstine, keynote address at the 2nd Summit for Space Sustainability, 10 Sept. 2020.

[100] U.S. Mission to International Organizations in Vienna, "61st Session of the COPUOS Legal Subcommittee – Agenda Item 15: Potential Legal Models for Activities in Exploration, Exploitation and Utilization of Space Resources," statement to the UN Committee on the Peaceful Uses of Outer Space (COPUOS) Legal Subcommittee, March 28, 2022, https://vienna.usmission.gov/2022-copuos-lsc-u-s-on-space-resources/. *See also,* Lindbergh, *supra* note 98, at 9.

[101] Exec. Order No. 13914, 85 Fed. Reg. 20,381, at 20,381.

[102] *The Artemis Accords,* NASA, https://www.nasa.gov/specials/artemis-accords/img/Artemis-Accords-signed-13Oct2020.pdf (last visited Apr. 17, 2022) [hereinafter *The Artemis Accords*]. The initial eight signatories were: Australia, Canada, Italy, Japan, Luxembourg, the United Arab Emirates, the United Kingdom and the United States.

[103] *Id.* §1.

a. Section 3 of the Accords mirrors the Treaty's Article IV, indicating that the Moon and other celestial bodies shall be used "exclusively for peaceful purposes."
b. Section 4 of the Accords gives meat to Article XI of the Outer Space Treaty by committing the Signatories to transparency and sharing of scientific information.
c. Section 6 of the Accords supports Article V of the Outer Space Treaty and recommit the Signatories to the Return and Rescue Agreement.
d. Section 7 of the Accords recognises the importance of the registration of space objects and commits the Signatories to meeting the obligations and responsibilities set forth in the Registration Agreement.
e. Section 8 of the Accords reiterates the Signatories' commitment to share scientific data, embodying both Article XI and the requirement, captured in Article I of the Outer Space Treaty, that States "shall facilitate and encourage international co-operation" in scientific investigation.[104]

But the Artemis Accords go further than confirming decades-old treaties. They seek to bridge some of the gaps in those treaties that—while always present—have widened considerably, with the advancement of technology and the new reality that the resources of space, once only accessible by sovereign States, are now attracting non-State and commercial actors. Specifically, the Artemis Accords affirm "that the extraction of space resources does not inherently constitute a national appropriation under Article II of the Outer Space Treaty …"[105]

The Artemis Accords also aim to establish guidelines that address the concerns raised by Senator Schmitt during the Moon Agreement hearings.[106] In a section entitled "Deconfliction of Space Activities," the Accords require signatories to "respect the principle of due regard" and "refrain from any intentional actions that may create harmful interference."[107] The signatories to the Accords further commit to "provide notification of their activities and commit to coordinating with any relevant actor to avoid harmful interference. The area wherein this notification and coordination will be implemented to avoid harmful interference is referred to as a 'safety zone'…"[108] Finally, the Signatories agree to "respect reasonable safety zones to avoid harmful interference with operations under [the] Accords, including by providing prior notification to and coordinating with each other before conducting operations" in the safety zone of another Signatory.

Of course, the Artemis Accords are non-binding principles, but they nevertheless advance the discourse regarding Article II, the nonappropriation principle, and the concept of due regard in that they signal a multilateral consensus that using space resources even for commercial purposes is permitted by Article II and offer initial proposals as to what due regard will look like in practice. As of this writing, 55

[104] *The Artemis Accords*, *supra* note 102.

[105] *The Artemis Accords*, *supra* note 102, § 10.

[106] *See supra* note 51 and accompanying text.

[107] *Ibid.*, §§ 11(3) and 11(4). Note here that the Artemis Accords actually strengthen the harmful interference provision found in the Outer Space Treaty. The Treaty requires only a consultation in advance of interference, whereas the Accords require that such interference be avoided altogether.

[108] *Id.* § 11(7).

nations that have signed the Artemis Accords, a number well above the 17 nations that are still bound by the Moon Agreement.

5 Future Horizons: Bumps into Breakthroughs?

The landscape of space resource utilisation is poised for transformative growth, yet significant challenges remain in establishing a stable, predictable regulatory framework to support this emerging industry. As a matter of first instance, the US Congress has yet to assign responsibility for the authorisation and continuing supervision of space mining activities as mandated by Article VI of the Outer Space Treaty. This causes confusion and may perhaps inadvertently set a dangerous precedent.

5.1 Astroforge

Undeterred by the failures of Planetary Resources and Deep Space Industries, and perhaps buoyed by the Artemis Accords and the CLPS programme, several intrepid commercial companies have formed with the mission of mining asteroids.[109] In a notable milestone, in November 2024, a company called Astroforge, became the first-ever commercial entity to receive a license for a deep space mission.[110] Granted by the US Federal Communications Commission (FCC), for its mission to prospect and extract resources from a near-Earth asteroid, this license, only governs the use of radio frequencies for spacecraft communication. The FCC's approval underscores the US government's willingness to facilitate private sector innovation in space, aligning with the principles of the 2015 Act. However, the license addresses only communication aspects, leaving broader questions of mission authorisation and resource extraction governance unresolved. This development highlights both the progress

---

[109] *See* Sam Farnham, *The Next Goldrush: 3 Top Startups Leading the Asteroid Mining Revolution,* Nasdaq.com (14 Aug, 2024), https://www.nasdaq.com/articles/next-gold-rush-3-top-startups-leading-asteroid-mining-revolutionYasmin Tamanna, *10 Top Space Mining Companies and Startups to Watch 2025,* StartUs.com (2 May 2023) https://www.startus-insights.com/innovators-guide/space-mining-companies/

[110] Keith Cooper, *Asteroid-Mining Company Astroforge Gets 1st-Ever FCC License for Commercial Deep-Sapce Mission,* Space.com (5 Nov. 2024) https://www.space.com/space-exploration/tech/asteroid-mining-company-astroforge-gets-1st-ever-fcc-license-for-commercial-deep-space-mission (Deep space is defined by the International Communications Union as being farther than 2 million kilometers from Earth). The licensed mission, a spacecraft named Odin, launched on February 26, 2025, but the Astroforge team was unable to sustain communications with the spacecraft thereafter. Mike Wall, *Hope Is All But Lost for Private Asteroid Probe in Deep Space—'The Chance of Talking With Odin is Minimal,* Space.com (6 Mar. 2025) https://www.space.com/space-exploration/private-spaceflight/hope-is-all-but-lost-for-private-asteroid-probe-in-deep-space-the-chance-of-talking-with-odin-is-minimal.

made and the gaps that persist in the US regulatory framework for space resource activities.

5.2 *Competing Approaches for Mission Authorisation*

The absence of a clear, centralised regulatory process for authorising space resource missions has sparked ongoing debate over how the United States should oversee these activities while complying with the mandate of Article VI of the Outer Space Treaty for States to authorise and continuously supervise their national non-governmental entities. Meeting this obligation requires a careful balancing act: clear regulations can encourage innovation and attract investment by providing legal certainty, while overly restrictive or ambiguous rules risk stifling progress or pushing companies towards jurisdictions with more permissive frameworks.

Two competing approaches have emerged to address this gap. The first approach proposes consolidating all mission authorisation for "space objects" within the Office of Space Commerce (OSC) in the Department of Commerce (DOC), which currently oversees the licensing of remote sensing satellites and operations.[111] This approach, captured in legislation proposed in 2023, emphasises a streamlined, business-friendly framework that leverages the OSC's existing role in promoting commercial space activities. It would designate the OSC as the "single authority responsible for authorisation and supervision certification process,"[112] which would include space resource activities. The proposal envisions a "light-touch" regulatory model, requiring companies to submit mission plans for review to ensure compliance with international obligations. While this legislation was not reported out of Committee, it is interesting to note that it included a provision that "[n]otwithstanding any other provision of law, outer space may not be considered a global commons"[113] effectively codifying current US space policy.

The second approach would split the assignment of these responsibilities to the Department of Transportation's (DOT) Federal Aviation Administration (FAA) Office of Commercial Space Transportation, which currently oversees commercial launch and re-entry operations and the OSC.[114] This model, set forth in a proposal prepared by President Biden's National Space Council, would have the FAA license human activities in space and the transportation of goods between destinations in space. Marcia Smith parses it out as such:

---

[111] Commercial Space Act of 2023, H.R. 6131 (1 Nov. 2023). The press release for the bill.

[112] Executive Summary, Commercial Space Act of 2023, https://republicans-science.house.gov/_cache/files/1/1/11aa5e55-bdef-4cd5-8ab9-955a1eb434db/2D19014E3D1714EB09AA39D022DEC69B.executive-summary---commercial-space-act-of-2023.pdf.

[113] H.R. 6131 § 80,208.

[114] The White House, *FACT SHEET: U.S. Novel Space Activities Authorization and Supervision Framework,* December 20, 2023.

a. DOT currently regulates the safety of humans during launch and re-entry. Now it would also regulate their safety while in space—in Earth orbit, on a commercial space station, or beyond Earth orbit.
b. DOT currently licenses space launch and Earth re-entry. Now it would also regulate in-space transportation for delivering goods from one point to another in space, including delivering fuel from a space fuel depot to a space station and to/from the lunar surface.
c. DOC currently licenses commercial remote sensing satellites. Now it would license all novel space activities not human-rated or otherwise assigned to DOT, including in-space assembly and manufacturing and active debris removal.
d. DOC would have the authority to provide space traffic coordination and warnings to avoid collisions in space.[115]

It can be argued that this approach ensures a comprehensive review, leveraging the FAA's experience in space transportation. However, splitting authority between agencies could introduce regulatory redundancy, create interagency conflicts, and ultimately undermine the efficiency and predictability that commercial actors need—especially in a field as capital-intensive and time-sensitive as space resource utilisation.

Both approaches aim to balance innovation with compliance but differ in their emphasis on speed versus thoroughness. The lack of consensus on which agency should lead—or whether a new entity is needed—underscores the regulatory uncertainty that continues to challenge the industry. As Astroforge and others prepare for missions, resolving this debate will be critical to providing the legal clarity investors and companies need. Nevertheless, it is important to point out that neither of these proposal addresses the myriad other issues related to space resource utilisation, two of the most important of which are the obligations imposed by Article IX of the Outer Space Treaty.[116]

5.3 *Harmful Contamination and Harmful Interference*

In meeting their obligations under Article VI of the Outer Space Treaty, States must ensure that national activities conducted even by non-governmental entities must be done in conformity with the provisions of the Treaty.[117] The foundational premise

---

[115] Marcia Smith, *White House Reveals Its Mission Authorization Plan for Novel Space Activities*, Space Policy Online (15 Nov. 2023) https://spacepolicyonline.com/news/white-house-reveals-its-mission-authorization-plan-for-novel-space-activities/.

[116] Outer Space Treaty, *supra* note 6, art. IX.

[117] There remains not insignificant consideration of how to interpret "national activities," but that is well beyond the scope of this chapter. *See e.g.*, Steven E. Bartz, *Reframe: A New Interpretive Framework for Non-Governmental Entity Responsibility in Outer Space*, 90 J. of Air Law and Commerce 135 (2025).

of the Outer Space Treaty is the freedom of exploration and use.[118] This freedom is limited only by Article IX, which, in addition to seeking to prevent harmful interference, requires that space exploration be conducted with due regard for the activities of others and done in such a way "as to avoid their harmful contamination."[119]

5.3.1 Harmful Contamination

As commercial space activities expand, the rise of vanity payloads—objects sent to space for symbolic, commemorative, or personal purposes—poses new challenges for ensuring responsible stewardship of celestial environments, particularly the Moon. The increasing accessibility of space has led to a proliferation of vanity payloads. For example, in 2024, commercial missions attempted to send capsules containing human ashes to the lunar surface, sparking debate about the environmental and cultural impacts of such actions[120] and raising the question of who should be involved in the decisions regarding what is appropriate to send to the Moon or other celestial bodies.

The Moon, in particular, risks becoming a repository for terrestrial waste if unregulated payloads accumulate, potentially interfering with scientific research or future resource utilisation. For instance, mining equipment and processes, if not carefully managed, could release dust, chemicals, or debris that degrade the lunar surface or disrupt other missions. The question of who determines the propriety of objects sent to space remains unanswered, as no US agency is explicitly tasked with evaluating the environmental impact of payloads beyond launch safety.[121] The Committee on Space Research (COSPAR) provides planetary protection guidelines, but these are non-binding and primarily focus on biological contamination, leaving gaps in addressing environmental or cultural impacts.[122]

Having taken the lead in cementing the ability of commercial entities to extract and own extraterrestrial resources, the United States can also lead in developing, nationally or with the Artemis Accords partners, a framework to assess the environmental implications of all activities on other celestial bodies, including vanity

[118] Outer Space Treaty, *supra* note 6, art. I.

[119] Outer Space Treaty, *supra* note 6, art. IX. There are other restrictions, including Article III which makes it clear that international law applies in outer space and Article IV which prohibits the placement of nuclear weapons or weapons of mass destruction anywhere in space.

[120] *See* Meghan Bartels, *Human Remains Are Headed to the Moon Despite Objections,* Scientific American (5 Jan. 2024) https://www.scientificamerican.com/article/human-remains-are-headed-to-the-moon-despite-objections/.

[121] The Federal Aviation Administration reviews payload that is not covered by the Federal Communications Commission (telecommunications) or the Department of Commerce (remote sensing satellites) and, by regulation, is required only to evaluate potentially hazardous or explosive payloads. 14 C.F.R. § 450.43.

[122] *See* COSPAR Policy on Planetary Protection, https://cosparhq.cnes.fr/assets/uploads/2024/07/PP-Policy_SRT_220-July-2024.pdf.

payloads and mining activities. Such a framework would need to balance commercial freedom with the need to preserve celestial environments for future exploration and utilisation. After all, without clear oversight, the Moon risks becoming a cluttered frontier, undermining the principles of responsible exploration enshrined in the Outer Space Treaty.

5.3.2 Harmful Interference

The 2015 Act fails to address Senator Schmitt's concerns regarding the rights of the miner/operator during the developmental stages of a mining operation.[123] What characteristic do the minerals take before they are extracted? The Artemis Accords attempt to address this concern by developing, not property rights, but so-called "safety zones" to be determined in a "reasonable manner,"[124] a delicate balance between Article II of the Outer Space Treaty and the 2015 Act. It remains to be seen how this concept will be implemented and how it will evolve.

The One Small Step to Protect Human Heritage in Space Act, signed by President Trump in 2020, adds another dimension to the interference debate by requiring US-licensed entities to avoid harmfully interfering with historic lunar sites, such as Apollo landing sites.[125] The Act mandates that federal agencies ensure compliance with heritage preservation guidelines in licensing agreements, recognising the cultural significance of these sites.[126] However, it does not clarify how to balance heritage protection with commercial activities like mining, which could inadvertently disturb these sites, nor does it designate an agency to enforce compliance, adding complexity to the regulatory landscape.

Notably, though perhaps not surprisingly, the United States may be one of the only States in the world that has actually had an extraterrestrial property claim brought in a national court. In 2003, Gregory W. Nemitz, through his Orbital Development company, filed a complaint for declaratory judgement in the United States District Court for the District of Nevada, asserting ownership of Asteroid 433, Eros, based on a registration claim he made on an online registry, the Archimedes Institute, in March 2000.[127] Nemitz claimed that NASA's landing of the NEAR Shoemaker spacecraft on Eros in February 2001 infringed his property rights, and he sought compensation for "parking and storage fees" of $20 for a century of rent, along with

---

123 *See supra* note 51 and accompanying text.

124 *Artemis Accords, supra* note 101, § 11.

125 P.L. 116–275 (31 Dec. 2020) § 3. This Act is the first national legislation in the world to recognise the existence of human heritage beyond Earth.

126 *Ibid.*

127 Robert Kelly, *Case Note: Nemitz v. United Sates, A Case of First Impression: Appropriation, Private Property Rights and Space Law Before the Federal Courts of the United States*, 30 J. Space L. 297, 297–98 (Fall 2004).

special damages.[128] The court dismissed Nemitz's complaint, finding that he did not present a claim for which relief could be granted.[129]

As humanity's presence in space becomes more routine and economically driven, disputes like Nemitz's—once considered fringe—are likely to grow in frequency and complexity. While this court rejected this private claim to a celestial body, emerging commercial activities, such as in situ resource utilisation and long-duration habitation, will test the boundaries of existing legal frameworks. These developments raise fundamental questions about the nature and recognition of property rights beyond Earth, especially as private actors seek assurances for investments in space infrastructure and extraction. As legal systems begin to grapple with property claims in outer space, the human right to own property cannot be ignored and may become increasingly relevant, prompting a re-evaluation of how property, ownership, and sovereignty are conceptualised in the extraterrestrial context. The Nemitz case, while unsuccessful, may thus prove to be an early signal of a coming era of contested legal terrain in the cosmos.

6 Conclusion

The journey of space resource utilisation in the United States has been marked by a roller coaster of progress and setbacks. From the ambitious lunar base concepts of the 1960s to the international debates surrounding the Moon Agreement in the 1980s, and the passage of the 2015 Commercial Space Launch Competitiveness Act, the United States has made significant strides in establishing a legal framework that supports commercial resource extraction. The recent successes of NASA's CLPS programme, exemplified by Intuitive Machines' IM-1 landing in February 2024 and Firefly's Blue Ghost mission in March 2025, have revitalised interest and demonstrated the feasibility of lunar resource utilisation. Astroforge's FCC license further signals momentum for asteroid mining ventures. However, significant challenges persist. The absence of a clear mission authorisation process remains a critical hurdle. The failure of early pioneers like Planetary Resources and Deep Space Industries due to funding shortages highlights the high financial risks and long-term investment horizons of space resource ventures. Moreover, international disagreement over resource ownership, particularly from nations adhering to the Moon Agreement's "common heritage" principle, could complicate US efforts to establish a stable legal norm. To move beyond this roller coaster and achieve a stable, upward trajectory, the United States must address these regulatory gaps, clarify mission authorisation processes, address environmental implications, and foster international cooperation through frameworks like the Artemis Accords.

[128] *Ibid.* at 298.

[129] *Nemitz v. United States*, No. CV-N030599-HDM (RAM), 2004 WL 3,167,042, at *1 (D. Nev. Apr. 26, 2004).

Professor Michelle L. D. Hanlon is Executive Director of the Air and Space Law Program at the University of Mississippi and its Center for Air and Space Law. She is the Editor-in-Chief of the Journal of Space Law, the world's oldest law journal dedicated to the legal problems arising out of human activities in outer space. She is also the Editor-in-Chief of the Journal of Drone Law and Policy, the first legal journal focused on law and regulations governing the operation and use of uncrewed aircraft in both civilian and military capacities. Her research and advocacy centers upon the concept of "due regard" in space law and evolving the framework necessary to assure that human exploration of space is responsible, successful and sustainable. In this regard she has done considerable work and written and presented extensively on topics related to orbital debris remediation, space solar power, small satellite constellations, environmental considerations and the protection of human heritage in space. Michelle is also Co-Founder and President of For All Moonkind, Inc., a nonprofit corporation that is the only organization in the world focused on obtaining international legal recognition for and protection of human cultural heritage in outer space. In this capacity, she was instrumental in the development of the One Small Step Act in the United States, the first national legislation to acknowledge the existence of human heritage in outer space. For All Moonkind has been recognised by the United Nations as a Permanent Observer to the United Nations Committee on the Peaceful Uses of Outer Space.

Chinese Legal Response to Space Resource Utilisation

Fabio Tronchetti

Abstract This chapter examines China's evolving approach to the governance of space resource activities, with particular attention to its lunar programme and recent submissions to the United Nations Committee on the Peaceful Uses of Outer Space (COPUOS). It begins by situating China within the contemporary "space mining gold rush," noting its technical achievements under the Chang'e missions and its ambitious plans for the International Lunar Research Station (ILRS). Unlike the United States-led Artemis Programme, which has been supported through national legislation and the Artemis Accords, China has thus far advanced its legal position through multilateral engagement rather than unilateral or minilateral instruments. The chapter contextualises China's perspective within broader debates on international and national space law, including the ambiguities of the 1967 Outer Space Treaty, the limited acceptance of the 1979 Moon Agreement, and the proliferation of domestic space resource utilisation laws. Against this backdrop, China's submissions to COPUOS in 2024 and 2025 are analysed in detail. These emphasise the centrality of the Outer Space Treaty, prioritisation of scientific research, reaffirmation of the non-appropriation principle, sustainability, enhanced supervision of private actors, and benefit-sharing. Notably, China distinguishes between scientific and commercial uses of resources, appearing to reject the latter as potentially constituting unlawful appropriation. In contrast to China's multilateral, law-preserving approach with the United States' unilateral initiatives, the chapter argues that China positions itself as a defender of the international space law framework, while the U.S. acts as a disruptor through national legislation and "safety zones." Finally, the chapter underscores China's strategy of leveraging international law and COPUOS to shape the future governance of lunar and space resource activities.

Keywords China · Space resource governance · International Lunar Research Station (ILRS) · Outer Space Treaty · Non-appropriation principle · Multilateral approach

F. Tronchetti (✉)
Northumbria Law School, Northumbria University, Newcastle upon Tyne, UK
e-mail: f.tronchetti@northumbria.ac.uk

S. Bhat B and A. Variath (eds.), *Legal Responses to Space Resource Utilisation*, Space Law and Policy, https://doi.org/10.1007/978-981-95-8864-0_10

1 Introduction

The space mining gold rush is on. From companies developing technology to mine resources from a celestial body[1] to international discussions on legal principles to govern space resources activities,[2] the day in which mankind will be able to collect resources from a celestial body and use them to sustain and extend presence therein no longer seems a far-fetched dream. While Western media[3] and legal analysts[4] have largely focused their attention on the position of the United States and its plan to return to the Moon via the Artemis Programme, no analysis of the topic of space resources would be complete without an assessment of the Chinese position. Not only is China at the forefront of mankind's plans to establish a (semi) permanent presence on the Moon, through its International Lunar Research Station (ILRS) initiative, but, at the time of the writing of this chapter, it seems to be better placed to realise these plans earlier than its American counterpart. Indeed, while delays and budget uncertainties have negatively affected the implementation of the original schedule of the Artemis missions,[5] China appears to be on track to launch its upcoming missions to the Moon intended to lay down the foundations for the placement of facilities on the Moon and the landing of the first Chinese astronaut therein. As a matter of fact, it is possible that China will be the first country to set up a non-temporary presence on the lunar surface, rendering, thus, even more relevant to understand what its approach towards the utilisation of lunar (space) resources and, lunar activities in general, is. The present chapter will attempt to achieve this goal by reviewing the official submissions delivered in 2024 and 2025 by China to the United Nations Committee on the Peaceful Uses of Outer Space (COPUOS)'s Working Group on the Legal Aspects of Space Resources Activities.[6] The analysis will aim at highlighting the core aspects of

[1] Among these companies, the following can be listed: the Asteroid Mining Corporation (https://www.asteroidminingcorporation.co.uk/ (accessed 8.8.2025), Ispace https://ispace-inc.com/ (accessed 8.8.2025), Lunar Outpost https://www.lunaroutpost.com/, (accessed 8.8.2025), Astrobotic Technology https://www.astrobotic.com/ (accessed 8.8.2025).

[2] These discussions are taking place within the United Nations Working Group on Legal Aspects of Space Resources Activities, see Report of the Chair and Vice-Chair of the Working Group on the Legal Aspects of Space Resources Activities, UN Doc. A/AC.105/C.2/2025/SRA/L.1 (14.5.2025) https://www.unoosa.org/oosa/oosadoc/data/documents/2025/aac.105c.2sra/aac.105c.22025sral.1_0.html (accessed 8.8.2025).

[3] Robert Lea, 'Artemis Accords: What are they & which countries are involved?'(2025) Space.com https://www.space.com/artemis-accords-explained (accessed 8.8.2025).

[4] Frans G. von der Dunk, 'The Artemis Accords as a Tool of Cooperation' (2022) Proc. Int'L Institute of Space L.145; Chris Newman &William Ralston, 'Only a paper moon: The Artemis Accords and future human settlements' in Charles S. Cockell (Ed.), *The Institutions of Extraterrestrial Liberty* (Oxford University Press 2022) 335.

[5] Mike Wall, 'Reshaping our return to the moon: Trump's 2026 budget gives Artemis a major facelift' (2025) Space.com https://www.space.com/space-exploration/reshaping-our-return-to-the-moon-trumps-2026-budget-gives-artemis-a-major-facelift (accessed 8.8.2025).

[6] This working group was established under the Legal Subcommittee agenda item "General exchange of views on potential legal models for activities in the exploration, exploitation and utilization of space resources" at the sixtieth session of the Legal Subcommittee in 2021.

the Chinese position and assessing its consistency with the established principles of international space law. In doing so, this chapter will also compare the Chinese position to that of the United States, as contained in the Artemis Accords[7] and the 2015 US Space Resources Exploration and Utilization Law.[8] Before reviewing the Chinese position, this chapter will provide a brief overview of the Chinese lunar programme and review the international and national laws applicable to space resources activities.

2 China's Lunar Exploration Programme: Achievements and Future Plans

China's lunar exploration programme is a story of success; from placing a rover on the Moon's surface to collecting samples from its far side, China has achieved remarkable milestones that have gained the admiration of observers around the world.[9] What makes Chinese achievements stand out even further is the fact that they have been realised without any noticeable failure and by exclusively relying on domestic space capabilities.

Overall, the Chinese lunar exploration programme consists of six phases, three of which have been completed and the last one envisions the building of permanent facilities on the Moon's surface.

The first phase of the lunar exploration programme entailed the launch of two lunar orbiters, respectively Chag-e 1 in 2007 and Chang-e 2 in 2009. These missions, which mapped the surface of the Moon to provide reference for future landings and to assess the distribution of its chemical elements, were completed successfully. The second phase of the lunar exploration programme, which included the launch of spacecraft capable of soft landing on the Moon and deploying lunar rovers, respectively the Chang-e 3 spacecraft in 2013 and the Chag-e 4 in 2018, has been completed. The Chang-e 4 mission marked the first time that a spacecraft landed on the South Pole of the Moon. With the third phase of the programme, China attempted to achieve its most ambitious goals, namely the automated collection of samples from the near and the far side of the Moon and their transportation to China; these goals were successfully achieved with the Chang-e 5 and Chang-e 6 missions, respectively, in 2020 and 2024.

The fourth, fifth and sixth phases of the Chinese lunar exploration programme have been planned but have yet to be implemented. The fifth mission envisions the launch of the first Chinese manned journey to the Moon; to achieve this goal, China

---

[7] The text of the Artemis Accords, officially signed on 13 October 2020, is available at https://www.nasa.gov/specials/artemis-accords/img/Artemis-Accords-signed-13Oct2020.pdf (accessed 8.8.2025).

[8] Title IV—Space resources exploration and utilization, US Commercial Space Launch Competitiveness Act, H.R. 2262, 114th Cong. (2015).

[9] Katie Holland, 'Chinese lunar exploration program' (2024) National Space Centre https://www.spacecentre.co.uk/news/space-now-blog/chinese-lunar-exploration-program/ (accessed 8.8.2025).

is developing the Mengzhou crewed spacecraft and the Lanyue crewed lunar lander. China intends to land two Chinese astronauts on the Moon by 2030.[10] The fourth and sixth phases of the lunar programme must be analysed together, as the former is instrumental to the latter. The goal of these two missions is the establishment of an extended lunar research facility at the Moon south pole, the so-called International Lunar Research Station, sometime in the 2040's. To achieve this goal, China has decided to take a gradual approach that will include a preparatory/testing, stage and a final stage. The preparatory stage will be undertaken during the upcoming fourth phase of the Chinese lunar exploration programme, where the Chang-e 7 and Chang-e 8 missions will be launched, respectively in 2026 and 2028. The Chang-e 7 mission will study the amount and quality of the lunar south pole's resources, particularly those contained in the lunar craters. The Chang-e 8 mission will verify in-situ resources development and utilisation technologies; accordingly, it will include a lander, a rover, a flying detector, and a 3D printing experiment using in-situ utilisation. This equipment will be used to test the technology necessary to build a basic/preliminary lunar research base, which, pursuant to the official announcement, should be completed by 2035. This base will act as the model to build an extended lunar research facility, the ILRS, in the 2040s.

The ILRS is intended to operate as a scientific experimental facility where multidisciplinary, multiple-objective scientific and technological activities will be conducted.[11] The ILRS will possess capabilities such as Earth-Moon transportation, energy supply, centralised control, communication, navigation, lunar surface scientific exploration and ground support capabilities.[12] To implement the ILRS project, China has opened it to international partners, thus initiating an initiative to build a station on the lunar south pole parallel to the US-led Artemis Programme. Among the differences between the two, there is the fact that the IRLS initiative is not only open to States but also to subnational entities and firm; indeed, at the time of this

[10] Leonard David, 'China's lunar lander aces touchdown and takeoff tests ahead of planned 2030 crewed moon mission' (2025) Space.com https://www.space.com/astronomy/moon/chinas-lunar-lander-aces-touchdown-and-takeoff-tests-ahead-of-planned-2030-crewed-moon-mission-video (accessed 8.8.2025); Zhao Lei, 'China advances manned lunar program for 2030 moon landing' (2025) China Daily https://global.chinadaily.com.cn/a/202503/03/WS67c55e07a310c240449d8414.html (accessed 8.8.2025).

[11] Statement by Wu Weiren, the chief designer of China's lunar exploration program on April 23, 2024, available at The State Council of the people's Republic of China'International Lunar Research Station attracts more partners: CNSA' https://english.www.gov.cn/news/202504/24/content_WS680976d7c6d0868f4e8f2045.html.

[12] Ibid.

writing, 17 countries and international organisations,[13] as well as more than 50 international research institutions, have joined the ILR project.[14] Another difference lies in the fact that China has not published a document providing the legal foundation to ILRS initiative in a manner similar to what the US have done with the release of the Artemis Accords, a document intended to guide the activities of the States that have joined the Artemis Programme.

3 Space Law and Space Resources Activities: International and National Perspectives

The legality of the extraction and utilisation of space resources is a controversial matter in international space law.[15] Indeed, while the removal and use of resources in the context of scientific investigation seems to be an uncontroversial practice, the acceptability of the exploitation of such resources for the purpose of gaining a profit remains highly debatable. The problems arise from the fact that international space law, at least in its most accepted form, does not clarify the legal status of such resources nor the purpose for which they might be used. In this respect, two international treaties governing human activities in outer space are relevant, the 1967 Outer Space Treaty[16] and the 1979 Moon Agreement.[17] These treaties differ from each other in two substantial ways. First, while the former has been largely accepted by States, the latter has been ratified by only 17 countries, not including the main space players. Second, while the Outer Space Treaty does not specifically address space resources, the Moon Agreement contains detailed rules governing their recovery and use.

---

[13] These include Russia, Venezuela, Belarus, Pakistan, Azerbaijan, South Africa, Egypt, Nicaragua, Thailand, Serbia, Kazakhstan and Senegal as national-level ILRS participants, international institutions such as the Belt and Road Alliance for Science and Technology, the Foundation for Space Development Africa, Africa Business Alliance and Institutions from the United Arab Emirates (Orbital Space), Serbia (Astronomical Observatory Belgrade), Switzerland (Spacetalk SA), Indonesia (University of Bandar Lampung), Pakistan (National University of Sciences & Technology), Panama (Space Science Innovation Center of Panama) and South Africa (South African Radio Astronomy Observatory (SARAO).

[14] Global times, 'China to include more African members in its lunar research program in latest effort to boost South-South cooperation' available at https://www.globaltimes.cn/page/202409/1319316.shtml.

[15] Fabio Tronchetti, 'Legal aspects of space resources utilization', in Frans G. von der Dunk & Fabio Tronchetti (eds), *Handbook of Space Law*, (Edward Elgar, 2015) 769; Stephan Hobe & Kuan-Wei Chen, 'Legal status of outer space and celestial bodies', in Ram S. Jakhu& Stephen Dempsey (eds), *Routledge Handbook of Space Law*, (Routledge, 2017) 25–41.

[16] Treaty on Principles Governing the Activities of States in the Exploration and Use of Outer Space, including the Moon and other Celestial Bodies, 18 U.S.T. 2410; TIAS 6347; UNTS 205 (the Outer Space Treaty).

[17] Agreement Governing the Activities of States on the Moon and Other Celestial Bodies, 18 December 1979, 18 ILM 1434; 1363 UNTS 3 (the Moon Agreement).

3.1 Space Law Treaties

The Outer Space Treaty clarifies that celestial bodies cannot be appropriated by means of use, occupation, or by any other means;[18] additionally, it confers States the right to freely access, explore and use outer space, upon the condition that such an exploration and use are for the benefits and in the interests of all countries and the province of all mankind.[19] Furthermore, it recognises the right to conduct scientific investigation of outer space. The Treaty does not clarify: (a) whether the non–appropriative nature of celestial bodies also extends to the resources contained therein; (b) whether the right to explore and use outer space also covers the exploration and use of space resources; (c) how the province of all mankind clause affects space resources' utilisation activities.[20] Scholars have given diverging views on how to answer these questions, ranging from those claiming that celestial bodies' resources cannot be appropriated to those arguing that the right to use outer space extends to space resources.[21] Regardless of this debate, the fact remains that the Outer Space Treaty does not give a clear answer to the legal status of celestial bodies' resources nor to the circumstances and rules according to which these resources might be used.

The Moon Agreement was intended to fill the gaps left open by the Outer Space Treaty with respect to the status and utilisation of lunar resources as well as the conduct of human activities on a celestial body.[22] Accordingly, the Agreement distinguishes between two forms of use, respectively, scientific and commercial (for the purpose of profit). In the context of scientific operations, States can: (a) build facilities;[23] (b) collect and use lunar samples to sustain their scientific missions; (c) use the portion of the lunar surface necessary to carry out their operation.[24] On the contrary, the use of resources for commercial purposes must be governed by an international regime that States Parties undertake to develop once the utilisation of such resources becomes feasible.[25] Importantly, while not setting up such a regime, the Agreement recommends some of its principles, including the equitable sharing of the benefits derived from lunar activities.[26] The requirement for an international regime derives from Article 11 (1) of the Agreement that declares the Moon and its resources to be

---

[18] Art. II, Outer Space Treaty.

[19] Art. I(1), Outer Space Treaty.

[20] Fabio Tronchetti, 'The non-appropriation principle as a structural norm of international law' (2008) 33(3) Air& Space L. 277; Philip de Man, *Exclusive use in an inclusive environment* (Springer, 2016) 73–136.

[21] For an analysis of these diverging views see Fabio Tronchetti, 'Title IV – Space Resource Exploration and Utilization of the US Commercial Space Launch Competitiveness Act: A Legal and Political Assessment', Air & Space Law, Vol. 41, No. 2, 2016, pp. 143–156. At 145–6.

[22] Fabio Tronchetti, *The exploitation of natural resources of the Moon and other celestial bodies* (Brill, 2009) 38–60; Fabio Tronchetti, 'Moon Agreement' in Sandeep Bhat & Dilip Ukey) *International Space Law in the New Space Era* (Oxford University Press 2024) 165.

[23] Art. 9, Moon Agreement.

[24] Art. 9(1), Moon Agreement.

[25] Art. 11 (5), Moon Agreement.

[26] Art. 11(7), Moon Agreement.

the common heritage of mankind; traditionally, the concept of the common heritage of mankind demands activities undertaken within an international area, such as outer space, to be governed by internationally agreed rules clarifying rights and obligations of States.[27]

Ultimately, due to the lack of acceptance of the Moon Agreement, one is left with the Outer Space Treaty which, as said above, does not clarify the status of space resources nor the purpose of their utilisation; this uncertainty has opened the door to different interpretation of the Treaty's provisions by States and diverging initiatives based on such interpretations in the form of national space resources utilisation law and quasi-multilateral initiatives such as the Artemis Accords.

3.2 National Space Resources Utilisation Laws

The current race to the Moon is characterised by the presence of private companies having as their primary target the recovery and commercial utilisation of the Moon's (and other celestial bodies) resources. Considering the uncertainty surrounding the legality of these projects under the Outer Space Treaty, these companies have pushed their respective governments to pass domestic laws supporting their plans.[28] Their lobbying has been successful and has led to the enactment of four domestic space resources utilisation laws in the United States (2015[29]), Luxembourg (2017[30]), the United Arab Emirates (2019[31]) and Japan (2021[32]). The essence of these laws is pretty similar, in the sense that they recognise under domestic law the right of private entities to collect, own, use and sell celestial bodies' resources, and they establish (or at least set the foundation for the establishment of) the procedure for the authorization of private space mining activities.[33] Such a procedure is meant to meet the requirements of Article VI of the Outer Space Treaty, which requires States to authorise and continuously supervise national space activities, including private ones. Importantly, the issuing of such laws has faced the opposition of several countries,

[27] On the concept of the 'common heritage of mankind' see Rudiger Wolfrum, 'The principle of the common heritage of mankind' (1983) 43 HJIL 312; Christopher C. Joyner, 'Legal Implications of the Concept of the Common Heritage of Mankind' (1986) 35 International Comp. L Q. 376.

[28] For a legal analysis of the various national space resources utilization laws see Frans G. von der Dunk, 'Property Rights Over the Moon or On the Moon? The Legality of Space Resource Exploitation on Celestial Bodies' (2023) 6 Journal of L. & Innovation 95.

[29] See *supra* (n 21).

[30] Luxembourg Law 674 on the Exploration and use of space resources of July 20, 2017.

[31] Arts. 14, 18, Federal Law No. (12) of 2019 on the Regulation of the Space Sector.

[32] Law No. 83, Space Resources Act of 15 June 2021.

[33] Tronchetti (n 16); Philp De Man, 'Luxembourg Law on Space Resources Rests on Contentious Relationship with International Framework'(2017) KU Leuven Working Paper No. 189; P.J. Blount & Christian J. Robinson, J., 'One small step: the impact of the US Commercial Space Launch Competitiveness Act of 2015 on the exploitation of resources in outer space' (2016) 18(1) North Carolina J.L. & Tech 160.

such as Germany, Russia, and Brazil. Their criticism is centred on the idea that the private use of space resources for the purpose of profit is inconsistent with the non-appropriative nature of outer space and the notion of the province of all mankind. Furthermore, in their opinion, rules governing the recovery and utilisation of celestial bodies' resources could only be developed by the international community as a whole, ideally in COPUOS, and not by States acting unilaterally.

3.3 Principles for Cooperation in the Civil Exploration and Use of the Moon, Mars, Comets and Asteroids for Peaceful Purposes: The Artemis Accords

One of the most interesting recent regulatory developments has been the release of the Artemis Accords, a US-led regulatory initiative aimed at supporting the realisation of NASA's Artemis Programme.[34] Despite being an American initiative, the Artemis Programme has been opened to other willing countries; so far, 56 countries have accepted the Accords. A detailed analysis of the Accords goes beyond the purpose of this chapter; however, it is important to point out that, even though they reaffirm numerous clauses of the Outer Space Treaty and declare that signatories will act consistently with international (space) law, the consistency of some of the Accords' provisions with the Outer Space Treaty is questionable. This is true for the section dealing with space resources utilisation, namely Section 10, which declares that: "The Signatories affirm that the extraction of space resources does not inherently constitute national appropriation under Article II of the Outer Space Treaty, and that contracts and other legal instruments relating to space resources should be consistent with that Treaty." Evidently, thus, the Artemis Accords fully embrace the American view on the utilisation of space resources as elaborated in the 2015 US Space Resources Exploration and Utilisation Act which, as explained above, attributes to US companies the right to collect, use, and sell space resources. A second controversial point concerns the possibility for Signatories to establish 'safety zones' on a celestial body's surface as a means to avoid interference with nominal operations.[35] Even though these 'zones' should be limited in size, temporary in nature, guarantee 'free access' to all areas[36] and Signatories are committed to contribute to multilateral efforts to further develop this concept,[37] it is questionable whether the establishment of such zones would be consistent with the Outer Space Treaty;[38] indeed, not only the Treaty never foresee such a possibility but also,

---

[34] The Artemis Program aims at returning men to the Moon by 2024. For more information on the Program see at https://www.nasa.gov/artemisprogram/ (accessed 8.8.2025).

[35] Ibid Section 11.

[36] Ibid Section 11 (7).

[37] Ibid Section 11 (6).

[38] Melissa de Zwart, Stacey Henderson &Michelle Neumann, 'Space resource activities and the evolution of international space law' (2023) 211 Acta Astronautica 155;

depending on the Signatories' behaviours, such a practice could result in the appropriation of a portion of a celestial body's surface and subsurface by means of use or occupation, a possibility specifically prohibited by Article II of the Treaty.

3.4 *Multilateral Efforts Within the United Nations Committee on the Peaceful Uses of Outer Space*

Alongside the above-mentioned unilateral and quasi-multilateral initiatives, there is an ongoing multilateral effort within COPUOS aimed at drafting internationally agreed principles to govern space resource activities.[39] Accordingly, since 2021, a Working Group on Legal Aspects of Space Resource Activities has been established within COPUOS's Legal Subcommittee; among the Working Group's tasks is the finalisation of a set of initial recommended principles for space resource activities.[40] The first draft version of these principles has been presented by the Working Group's Chair during the 2025 session of the Legal Subcommittee;[41] notably, among the listed principles, there is Principle 6:'Prioritization of Scientific Research and Investigation' that, after reaffirming that there shall be freedom of scientific investigation in outer space, provides that: "In developing, planning and conducting space resource activities, *States shall give priority to: (a) scientific research and investigation missions that [aim to give rise to benefits to and] promote the interests of all humankind* (emphasis added); and (b) space resource activities that support such missions". The idea that States, when planning their space resources activities, shall give priority to scientific research missions, especially those that benefit all humankind, is particularly important for the present work and will play a significant role when distinguishing the American and Chinese approaches to space resources utilisation.

Lucas Mallowan, Lucien Rapp & Maria Topka, 'Reinventing treaty compliant "safety zones" in the context of space sustainability (2021) '8 Journal of Space Safety Engineering 155; Xiaodao Li & Jie Long, 'Developing safety-zone rules: Based on an institutional choice framework' (2025) 71 Space Policy 101; Ben McKeown, Andrew G. Dempster & Serkan Saydam, 'Artemis Accords: Are Safety Zones Practical for Long Term Commercial Lunar Resource Utilisation?' (2022) 22 Space Policy 504.

[39] UNCOPOUS, 'Report of the Legal Subcommittee on its fifty-sixth session' (Vienna 27 March–7 April 2017) UN Doc. A/AC.105/1122.

[40] Five-year workplan and methods of work for the Working Group on Legal Aspects of Space Resource Activities, Appendix II, Report of the Legal Subcommittee on its sixty-first session, UN Doc. A/AC.105/1260 19 April 2022.

[41] Initial draft set of recommended principles for space resource activities, note by the Vice-Chair of the Working Group on Legal Aspects of Space Resource Activities, A/AC.105/C.2/L.339, 27 March 2025.

4 Chinese Approach to Space Resources Utilisation

Until recently, China's approach towards space resources utilisation remained difficult to ascertain. This difficulty was largely due to three factors. First, China had not gone beyond delivering somewhat general statements on this subject within COPUOS's discussions. Second, China had not elaborated its position through the enactment of a national space resources utilisation law. Third, China did not provide the ILRS initiative with a regulatory document that could mirror what the Artemis Accords represent in the context of the Artemis Programme.

In the last two years, this situation has begun to change thanks to two written submissions that China made to the UNCOPUOS's Working Group on the Legal Aspects of Space Resources Activities, respectively in 2024 and in 2025. While not providing a comprehensive overview of China's approach towards space resources utilisation, they provide the first public insight into how China interprets the Outer Space Treaty and how this interpretation is going to guide China's ILRS activities. The following sections analyse these two submissions and reflect upon their legal and regulatory implications.

4.1 Submission by the Delegation of China to the Working Group on Legal Aspects of Space Resource Activities of the Legal Subcommittee of the Committee on the Peaceful Uses of Outer Space (6th March 2024)

On March 6th, 2024, China submitted to the Working Group on Legal Aspects of Space Resource Activities of the Legal Subcommittee of the Committee on the Peaceful Uses of Outer Space (COPUOS) a document highlighting the country's position towards the utilisation of space resources and indicating issues that the Working Group should address as a matter of priory importance.[42] China's submission addresses three main areas: (1) The mandate of the Working Group; (2) The existing legal framework for space resources activities; (3) Elements for the Development of Initial Recommended Principles.

Mandate: China encourages the Working Group to tie discussions to the space exploration projects of member States; in China's view, pursuant to this approach, the Working Group could provide a useful contribution to the challenges that States might face when undertaking lunar activities rather than engaging in theoretical discussion with little practical implications.

---

[42] The text of the submission is available at https://www.unoosa.org/documents/pdf/copuos/lsc/space-resources/LSC2024/English_Chinas_submission_to_the_working_group_on_space_resources.pdf (accessed 8.8.2025); Andrew Jones, 'China outlines position on space resources' (2024) Space News https://spacenews.com/china-outlines-position-on-use-of-space-resources/ (accessed 8.8.2025).

Existing legal framework for space resources activities: China views the Outer Space Treaty as the cornerstone of the legal framework applicable to space activities; accordingly, the Treaty shall maintain its central role also in the era of space resources utilisation. Importantly, China recommends that the Working Group seek a uniform interpretation and application of the principles of the OST. This suggestion is significant for two reasons: (1) China is mindful of the issues that diverging interpretations of the Treaty may lead to, including the promotion of controversial regulatory initiatives, such as the adoption of space resources utilisation laws, and the risk of increased tension among space-faring countries. (2) China promotes a multilateral, rather than unilateral, approach to advancement in international space law. Indeed, by demanding that the Working Group take the lead in seeking a uniform interpretation of the Outer Space Treaty, China emphasises the central role that the United Nations should play in the evolution of the law applicable to space activities

Elements for the Development of Initial Recommended Principles: China argues that, due to the novelty of space resource activities, recommended principles should be structured in a practical and measured way. Accordingly, China gives input on five issues that it deems to have priory importance, respectively:

4.1.1 Reaffirming and Operationalising the Principle of Non-appropriation

China believes that any form of exploration, exploitation and utilisation of space resources shall be consistent with the non-appropriation principle set forth in Art. II, OST, and that the Working Group shall make efforts to operationalise its meaning in the context of space resources activities. While there does not seem to be anything remarkably innovative in such a statement, China's remark is relevant for two reasons. First, once again, it restates the central role to be played by COPUOS in the space law-making process. Second, it indicates that China is in principle not opposed to the use of space resources for different purposes, including, possibly, also for commercial ones. What China opposes is the method used by certain countries, particularly the United States, to affirm the legality of the use of space resources for profit (and the possibility of licencing private companies to do so) through national and quasi-multilateral initiatives rather than seeking recognition and acceptance of such interpretation within COPUOS first. This is why China is calling upon the Working Group to make efforts to reach a uniform understanding of the meaning of the non-appropriation principle to space resources activities.

4.1.2 Encouraging Space Resource Activities for Scientific Investigation

Space resources might be used for two purposes: scientific and commercial (for the purpose of profit). China argues that while the legality of the latter under the Outer Space Treaty remains controversial, that of the former is unquestionable because it falls within the scope of the freedom of scientific investigation of outer space provided by Article I(3) of the Treaty. This principle has also been confirmed in practice, as both the United States and China have collected samples of lunar rocks in the course of their scientific investigation of the Moon and shared some of these samples with international partners. Based on this approach, China recommends that the Working Group develop principles to mitigate the potential negative impact that commercial space resources activities might have on activities undertaken for the purpose of scientific investigation. It is evident from this that China is encouraging the Working Group to take steps to promote space resources activities for scientific purposes and to ensure coordination among the different lunar plans proposed by States.

4.1.3 Enhancing the Coordination of Space Resource Activities

China recognises the existence of parallel lunar projects and the possibility that these projects might be undertaken simultaneously in certain areas of the Moon. Mindful of this scenario, China invites the Working Group to formulate initial recommended principles to give effect to the terms of Article IX, OST, which provides that space activities shall be guided by the principle of cooperation and mutual assistance and be conducted with due regard to the interests of other States. The reference that China makes to Art. IX, OST, is particularly relevant because it indicates China's belief that the two major existing lunar projects, namely, the Artemis Programme and the ILRS, might peacefully co-exist and non-necessarily be the cause of tensions between the United States and China. For these two projects to co-exist, however, it is necessary for the Working Group to prepare principles favouring cooperation between the two sides, for example, in the areas of technical assistance, rescuing of astronauts and sharing of scientific data. Significantly, in its submission, China does not comment on the permissibility of 'safety'/non-access zones around lunar facilities; therefore, it remains to be seen how China will approach the question of the prevention of harmful interference in the upcoming Working Group's sessions.

4.1.4 Sufficient Supervision of Space Resource Activities of Non-governmental Entities

China's submission repeatedly the need for space resources activities to be consistent with the OST. This is true not only in relation to their compliance with Art. II of the Treaty but also with the requirements set forth in Article VI, a key provision of the Treaty that establishes the principle of States' international responsibility for national activities in outer space, including those undertaken by non-governmental entities, and the duty of States to continuously supervise the private space activities that they have authorised. In China's view, the means to implement Article VI's requirements to space resources activities remain to be determined; indeed, because these activities will occur on a celestial body, it is uncertain how States could effectively and continuously supervise them. For this reason, China recommends that the Working

Group to investigate methods to implement Art. VI to space resources projects, for example, through the formulation of standards and procedures for authorisation and supervision.

4.1.5 Protecting Sustainability of Space Resource Activities

Finally, China recommends the Working Group to formulate principles to render space resources activities sustainable and accessible to future generations. While China does not elaborate upon the meaning of the concept of sustainability, it indicates that the principles should aim at preventing the disruption of lunar resources (and, indirectly, of the environment that contains them). Arguably, the implementation of this idea would require effective means to control space resources activities, a point that links this recommendation to what China had already expressed in connection to the supervision of private activities (see supra IV.d).

4.2 Submission by the Delegation of China to the Working Group on Legal Aspects of Space Resource Activities of the Legal Subcommittee of the Committee on the Peaceful Uses of Outer Space (May 2025)

The submission made by the Chinese Delegation to the Working Group on the Legal Aspects of Space Resources Activities in 2025 significantly differs from the one made in 2024.[43] First, the submission was intended to provide input for the drafting of the initial recommended principles for space resource activities, hence its content is more specific than that one of the 2024 version. Secondly, the submission addresses areas of the Outer Space Treaty that had not been covered in 2024, thus providing a broader and deeper window to understand the Chinese interpretation of the Treaty.

The submission is divided in two parts, the first providing 'definitions' and the second indicating the recommended principles to be inserted in the initial draft of the Working Group.

(a) Definitions: Two definitions are provided: (a) 'space resources' and 'space resources activities'. 'Space Resources' are defined as "*abiotic material that does not originate from and would not naturally enter the Earth.*" This definition comes from the willingness of China to facilitate the timely completion of the five-year workplan of the Working Group; hence, in China's view, the Group should exclusively focus its attention on abiotic material resources and exclude non-material resources, such as solar energy, radio frequency and orbits being defined as space sources under other circumstances. Furthermore, China defines 'space resources activities' as "*activities*

[43] The text of the 2025 submission is available at https://www.unoosa.org/documents/pdf/copuos/lsc/space-resources/LSC2024/English_Chinas_submission_to_the_working_group_on_space_resources.pdf (accessed 8.8.2025).

taking place in outer space, including the Moon and other celestial bodies, with the purpose of exploring, exploiting and utilising space resources. Such activities may include but are not limited to, exploration, extraction, process and transportation". Interestingly, these definitions are reflected nearly verbatim in the initial set of draft principles prepared by the Working Group's Chair which, in its preamble, states: "*Noting that [, whilst there are currently no internationally agreed definitions of 'space resources' or 'space resource activities',] the following clarifications apply to the application of these Principles: a) for the purposes of these Principles, space resources include extractable and/or recoverable natural, abiotic resources [material] [such as minerals, water, ice, other liquids, and gases] located [in situ] in outer space, including [on the surface or in the subsurface of] the Moon and other celestial bodies [and do not include satellite orbits, radio spectrum, solar energy or other resources not of a physical nature]; and (b) for the purposes of these Principles, space resource activities include activities conducted in outer space, including the Moon and other celestial bodies, for the [primary] purpose of searching for space resources, the extraction of space resources, and the exploitation and utilization of space resources [as well as related activities such as processing and transportation].*

(b) Principles to govern space resources activities: While reaffirming the concept stated in its submission 2024, namely that space resources activities must be consistent with international law and, particularly, the Outer Space Treaty, China expands its view and recommends a series of principles to be inserted in the initial draft of the Working Group.

First, China recommends space resources activities to be carried out for peaceful purposes only; in doing so, it does not elaborate upon the notion of 'peaceful purposes' in the context of space resources activities, beside a reference to Article IV of the OST, thus one should assume that its interpretation should be consistent with the one given by China in the Conference of Disarmament and in documents such as the PPWT.[44]

Second, China emphasizes the importance of the principles of freedom of access and benefit and interests of all countries. Accordingly, it stresses the fact that, while conducting space resources activities, States should not compromise the freedom to access all areas of the Moon on a basis of equality and in accordance with international law. While not specifically mentioning it, this statement may be viewed as an indirect reference to (and criticism of) the Artemis Accords 'safety zones' and the challenges that their establishment would pose to the right of States to freely access all areas of the Moon. China also refers to Article I of the Outer Space Treaty to point out that space resources activities shall be conducted for the benefit and in the interest of all countries. In doing so, China is, however, not merely repeating the words of the Outer Space Treaty, but it is providing an innovative contribution to the matter. Indeed, it emphasises that the right of scientific investigation is the prerequisite for the sharing of scientific and technological results and for implementing the idea "benefit and in the interests of all countries". For this reason, the freedom of scientific

[44] Fabio Tronchetti & Hao Liu, 'The 2014 updated Draft PPWT: Hitting the spot or missing the mark?' 33 Space Policy 38.

investigation, as provided in Article I(3) of the Outer Space Treaty, shall be fully preserved in the era of space resources utilisation. This statement reinforces once again the primary role that China gives to scientific undertakings in the context of activities conducted on a celestial body. This primary role is further reaffirmed in the section of the submission dealing with the non-appropriation principle, where China affirms that space resources activities shall not result in national appropriation of the Moon surface or subsurface by claim of occupation or by any other means in line with Article II of the Outer Space Treaty. Importantly, the Chinese delegation states that space resources activities for purposes not for profits, such as for scientific investigation and public interests, should not be considered as national appropriation; one, thus, may argue that, even if never explicitly mentioned, China believes that activities undertaken to gain a profit may constitute national appropriation, therefore illegal under the Outer Space Treaty.

Third, the prohibition to appropriate the Moon and any of its parts leads us to another valid suggestion made by China, that of the promotion of benefits sharing. China views outer space as a shared domain, a global common, and space exploration as a common cause of humanity. Therefore, the benefits deriving from it must be shared. Notably, China is not calling for forced forms of sharing, but it is simply encouraging States to promote it through enhanced participation of all countries, especially developing countries, in space resource activities, capacity-building and the sharing of information as well as scientific and technological outcomes. This approach is in line with the terms of the 1996 UNGA Declaration on Space Benefits.

Fourth, China calls for States to implement the principles of cooperation and mutual assistance set forth in Article IX of the Outer Space Treaty and to, thus, conduct space resources activities with due regard to the corresponding interests of other States.[45] Accordingly, China believes that, to ensure coordination and safety of space resources operations, States shall undertake consultations in good faith. It is interesting that, while reflecting upon Article IX, China only considers the concept of good faith while not addressing the notion of 'harmful interference'; arguably, this might be the result of a conscious decision dictated by the reluctance to take an open position on this controversial notion, especially in relation to the establishment of safety zone, or it might simply be that China views the implementation of the concept of 'good faith' a sufficient foundation to ensure safety of operations. Notably, when discussing about coordination and safety of activities, China puts forward a suggestion that, to the author's knowledge, had not been made before, namely the establishment of a special register for man-made objects on the Moon and other celestial bodies; this register which would find its legal foundation in the Outer Space Treaty and the 1975 Registration Convention, should be maintained by the United Nations Secretariat to ensure that space resources activities are conducted in a safe and orderly fashion. This suggestion is important not only for its innovative character but also because it shows a different approach to the problems that the United States is

[45] On the concept of 'due regard' see Julia Gaunce, 'On the Interpretation of the General Duty of "Due Regard"' in Aldo Chircop, Scott Coffen-Smout and Moira L McConnell (eds), Ocean Yearbook 32 (Brill Nijhoff, 2018) 27.

attempting to address through the hypothetical formation of 'safety zones'. Indeed, it appears that in China's view, the safety of operation of facilities on the Moon and the coordination of activities between co-existing projects could be managed through the functioning of such a register and the above-mentioned consultations in good faith rather than through the establishment of 'exclusion', 'safety' zones.

China concludes its submission by emphasising the importance of assessing the environmental impact of space resources activities; importantly, China indicates that attention should be paid to both backwards and forward contamination, thus expanding the notion of harmful contamination as opposed to Art. Article IX of the Outer Space Treaty,[46] which is exclusively concerned with backwards contamination. In doing so, China appears to take a broader approach than the Artemis Accords, which, in Section 12,[47] are merely concentrated on the mitigation of space debris generated during the mission without containing any requirement to assess and limit the environmental impact of activities on the lunar environment.

5 Chinese Approach to Space Resources Activities: A Commentary

After having reviewed the Chinese submissions to the Working Group on the Legal Aspects of Space Resources Activities, it is now time to assess China's position both in terms of its content and form. In doing so, the analysis will also compare it, when needed, with the American approach.

First, China's position is firmly rooted in international space law, especially the Outer Space Treaty; indeed, since the Legal Subcommittee began discussing about space resources activities in 2017, and even more so during the Working Group's discussions, China has repeatedly stated that the Outer Space Treaty shall maintain its central role in the era of space resources activities and that such activities shall not contravene it. From this specific perspective, the Chinese and American positions are not significantly different, as the United States, in its statement and in the Artemis Accords, has affirmed the centrality of the Outer Space Treaty. The question, then, becomes whether China and the United States have been true to their statements or if they have, instead, acted in a manner arguably inconsistent with the Treaty.

This consideration leads us to the second point of this analysis, which is, indeed, to assess if the proposed interpretations by China and Russia, as well as the actual initiatives that they have undertaken, are in line with the Outer Space Treaty and international space law in general. As far as China is concerned, there seems to be little doubt that its actions and interpretations are consistent with the Outer Space

[46] On Art. IX, Outer Space Treaty see Michael C Mineiro, 'FY-1C and USA-193 ASAT Intercepts: An Assessment of Legal Obligations under Article IX of the Outer Space Treaty' (2008) 34(2) Journal of Space Law 321; Andrea J. Harrington, 'Due Regard as the Prime Directive for Responsible Behavior ' (2023) 20 Loy. U. Chi. Int'l L. Rev. 57.

[47] Artemis Accords, Section 12—Orbital Debris.

Treaty and aligned with State practice. The importance that China places on the right of scientific investigation and the freedom to access all areas of celestial bodies is firmly rooted in Article I(3) of the Outer Space Treaty. Also, the right to collect and use resources for non-profit purposes (such as scientific research) has found confirmation in practice, is indirectly recognised in the Outer Space Treaty and is specifically provided for in Article 6(2) of the Moon Agreement. Similarly, the approach that China proposes in relation to the safety of space resources activities is based on a literal interpretation of Article IX of the Outer Space Treaty and, particularly, its principles of consultation and good faith, without stretching the meaning and understanding of key terms, such as that of 'harmful interference', or proposing controversial notions, such as 'safety zones', the legality of which is questionable. Even the suggestion that China makes to set up an international registry for objects built on the Moon not only seems to fall within the scope of the Outer Space Treaty and the Registration Convention but also could hardly be seen as containing elements that clash with fundamental principles of space law or as an attempt to force a controversial interpretation of the existing law. The same cannot be said about the United States; in enacting its 2015 space resources utilisation law, not only does it embrace a debatable interpretation of Articles I and II of the Outer Space Treaty but also attributes to itself the right to authorise US companies to undertake activities that are potentially in contradiction with the Treaty. In doing so, the United States weakened the stability of the Outer Space Treaty and, through unilateral action, planted the seed for controversy among space actors. Similar considerations could also be made for the suggestion to set up safety zones on the Moon as per Section 11 of the Artemis Accords, as well as per the recent US plan to establish exclusion zones on the lunar surface to protect nuclear facilities. These initiatives, that attempt to unilaterally stretch the limit of existing law and push for an interpretation favourable to the United States without seeking an international recognition of their legality, compromise the strength of the space law system and may fuel a dangerous 'race to be the first' on the Moon, with all the negative consequences that this entails.[48]

A third significant difference between the Chinese and American positions concerns their approach towards the need to develop international rules to govern space resource activities. Indeed, while China has been supportive of the establishment of a Working Group on the Legal Aspects of Space Resources Activities since this idea was presented in 2020 and in favour of the development of an international legal framework to govern such activities, the United States has taken a different approach. Indeed, by pointing out that there is still so much to understand about space resources, the United States argues that there is a lack of a practical basis for a comprehensive international legal framework governing space resource activities and that the existing core space treaties are sufficient. The United States,

[48] Walker A. Smith, 'Using the Artemis Accords to Build Customary International Law: A Vision for a U.S.- Centric Good Governance Regime in Outer Space' (2022) 86 J. Air L. & Com. 661.

instead, recommends that the Working Group focus its efforts on developing non-binding principles[49] applicable to space resources activities that should be based on the Artemis Accords.[50] Arguably, the US approach has profound implications on the activities conducted by the Working Group; first, it is not conducive to what the Working Group is attempting to achieve; secondly, any outcome that would not match the one that the United States desires, could be blocked from being approved, rendering thus vain the work conducted by the Group.

Another difference between the Chinese and American positions is the primacy that China gives to the use of resources for scientific purposes. Not only does China view such a use as a precondition for the sharing of information and the benefits generated by space resources activities, but it also emphasises the fact that the right of scientific investigation must be preserved in the era of space resources utilisation. This approach differs from the American one, which, since the enactment of the 2015 space resources law, is focused on the use of resources for commercial purposes, namely to make a profit out of them. Interestingly, the Chinese approach is somewhat mirrored in the Draft Principles prepared by the Chair of the Working Group on the Legal Aspects of Space Resources Activities which provides that: "In developing, planning and conducting space resource activities, States shall give priority to: (a) scientific research and investigation missions that [aim to give rise to benefits to and] promote the interests of all humankind; and (b) space resource activities that support such missions."[51]

After having assessed the Chinese approach to space resources activities from the perspective of its legal content, it is now time to discuss the method that China has used to develop and present it. China has embraced an approach that is based on the primacy of international law and that recognises the importance of seeking international agreement on a set of rules to govern space resources activities and to do so through the international body that promotes dialogue and supersedes advancements in international space law, namely COPUOS. In doing so, China has not only promoted the centrality of international system of governance of space activities that has its centre in COPUOS but it has also has counteracted the ongoing trend of minilateralism in space law,[52] that is a process where the space law-making process,

---

[49] On the relation between binding and non-binding rules see Brian Wessel, 'The Rule of Law in Outer Space: The Effects of Treaties and Nonbinding Agreements on International Space Law' (2012) 35 Hastings Int'L & Comp. L. Rev. 289.

[50] Statement by Caitlin Poling, US Representative to the Legal Subcommittee of the UN Committee on the Peaceful Uses of Outer Space, Agenda Item 8, Potential legal models for activities in the exploration, exploitation, and utilization of space resources (May 2025) https://www.unoosa.org/documents/pdf/copuos/lsc/2025/Statements/Wednesday7PM/8_The_United_States.pdf (accessed 8.8.2025).

[51] Principle 6: Prioritization of Scientific Research and Investigation, Initial draft set of recommended principles for space resource activities, UN. Doc. A/AC.105/C.2/L.339 (5–16 May 2025).

[52] Athar ud Din, 'The Artemis Accords: The End of Multilateralism in the Management of Outer Space?' (2022) 20 Astropolitics 135; Fabio Tronchetti, 'The emergence of political commitments related to space: is the international space order under threat?' (2025) 2 Acta Caeli and Astronautica (in press).

instead of occurring in a top-down form takes place from bottom (national) to up.[53] Accordingly, China has not taken any official step outside of COPUOS to present its interpretation or to push for its acceptance on a global scale; in this respect, China has not enacted a space resources utilisation law nor provided the ILRS initiative with a regulatory document that every participant to the initiative is required to accept. Instead, China has engaged in international discussions with other States attempting to seek a common understanding of key principles.

The United States has taken an opposite approach by fully embracing minilateralism and a bottom-up approach to the formation of new law in the area of space resources utilisation, as evidenced by the adoption of the 2015 space resources utilisation law and the Artemis Accords.[54] By doing so, the United States has undoubtedly spurred discussion and interests on this theme, but it has also weakened the role of COPUOS as a centre of governance of space activities and set a dangerous precedent for countries to simply bypass international negotiation and develop their own regime at the national level or with a limited group of like-minded countries. This approach has also some practical implications on the job undertaken by the Working Group on the Legal Aspect of Space Resources Activities; indeed, while it is undeniable that the United States is participating in the Group's discussions, it is also true that is doing so having already developed its regulatory position on this matter and that it is in the position to pressure the signatories to the Artemis Accords to reject any draft principles or report that would not reflect the Accords' form and content.

6 Conclusion

In the last two decades, China has achieved remarkable results in its efforts to explore and study the Moon and is now embarking on the ambitious plan to establish a permanent presence there. Until recently, China focused its attention on the technical side of the matter; however, in the last couple of years, it has begun to elaborate upon its legal approach to space resources activities thanks to two written submissions to the Working Group on the Legal Aspects of Space Resources Utilization of the Legal Subcommittee of the United Nations Committee on the Peaceful Uses of Outer Space.

China's approach is characterised by three elements: (1) the centrality of existing international space law; (2) the fundamental role to be played by COPUOS in

[53] On the space law-making process in the United Nations see Casandra Steer, 'Sources and law-making process relating to space activities', in Ram S. Jakhu & Stephen Dempsey (eds), *Routledge Handbook of Space Law*, (Routledge, 2017) 3–25.

[54] Guoyu Wang, 'NASA's Artemis Accords: the path to a united space law or a divided one?' (2020) The Space Review. https://www.thespacereview.com/article/4009/1#:~:text=by%20Guoyu%20Wang,Monday%2C%20August%2024%2C%202020&text=It%20shows%20that%20the%20US,certainties%20on%20space%20resource%20activities. (accessed 8.8.2025); Carolyn Posey, 'The aftermath of the Artemis Accords: power dynamics past and present in international space law' (2025) 39 Temple Int'L & Compar. L.J. 133.

the formation of principles to govern space resource activities; (3) the absence of behaviours causing disruption to the international system of space law and governance. The present chapter has illustrated how China's actions have significantly differed from those of the United States, as the latter has championed initiatives that have sent shockwaves through the regulatory framework governing space activities. Seen from this perspective, it is possible to make the argument that China has acted as a 'protector' of the established system of space law and governance, while the United States has played more the role of the 'disturber of the order'. There are some that could argue that China's behaviour has merely been the result of a strategic choice intended to gain prestige at the international level and potentially attract to its side countries traditionally aligned with the US, while in reality its intentions are other. Even if this might turn out to be the case at a later stage, it is undeniable that, at the time of the writing of this chapter, China has acted in a manner more consistent with international law than its competitor. Only time will tell if China truly believes in the primacy of international law and is ready to lead a coalition of States and entities in the safe, peaceful, and orderly implementation of space resources activities.

Dr. Fabio Tronchetti is an Associate Professor at Northumbria Law School, Northumbria University in Newcastle, United Kingdom. He worked as the Co-Director of the Institute of Space Law and Strategy and as a Zhuoyue Associate Professor at Beihang University, Beijing (China). He is also an Adjunct Professor of Comparative National Space Law at the School of Law of the University of Mississippi. Previously, he worked as an Associate Professor at the School of Law of the Harbin Institute of Technology (China) and as Lecturer at the International Institute of Air and Space Law, Leiden University (the Netherlands). Dr. Tronchetti's research is focused on the areas of international space law, aerospace law, international security and space economy. Dr. Tronchetti is regularly invited to give lectures and seminars across the world and has participated as a speaker in several international conferences. Dr. Tronchetti has also provided legal advices to governmental and non-governmental entities in various aerospace matters.

Space Resource Utilisation: Legal Developments in Australia

Melissa de Zwart

Abstract Australia has significant expertise in robotics and remote mining operations, making it well placed to participate in space resource activities. Australia is also committed to the peaceful and sustainable uses of outer space, as evidenced by its participation in multiple international fora, including long-term membership of the UN Committee on the Peaceful Uses of Outer Space. In addition to the Outer Space Treaty, the Liability Convention, the Rescue and Return Agreement and the Registration Convention, Australia is a party to the Moon Agreement, as well as being one of the founding signatories to the Artemis Accords. Australia's contribution to NASA's Artemis mission through its Moon to Mars Initiative includes the Trailblazer program—the supply of a lunar rover whose primary task will be the collection of lunar regolith and delivery of that regolith to a NASA ISRU facility for exploration and scientific purposes. A fundamental aspect of this project is also the demonstration of 'Australia's commitment to the development and implementation of rules and norms that seek to support the safety, stability and sustainability of outer space.' This chapter will identify the role that the Trailblazer program will play in clarifying the applicable legal framework for exploration, exploitation and utilisation of space resources from an Australian legal perspective, including key issues of due regard, benefit to humanity, interoperability, jurisdiction and control, non-appropriation, environmental protection and sustainability.

Keywords Australia · Space resource activities · Artemis Accords · Moon Agreement · Trailblazer program · Sustainability

---

M. de Zwart (✉)
Andy Thomas Centre for Space Resources, ARC Centre of Excellence in Plants for Space, Adelaide University, Adelaide, South Australia, Australia
e-mail: Melissa.dezwart@adelaide.edu.au

S. Bhat B and A. Variath (eds.), *Legal Responses to Space Resource Utilisation*, Space Law and Policy, https://doi.org/10.1007/978-981-95-8864-0_11

1 Introduction

In 2020, Australia was one of the original eight signatories to NASA's Artemis Accords.[1] Whilst signing the Artemis Accords does not necessitate or guarantee participation in NASA's Artemis program, to return humans to the Moon and on to Mars, Australia has committed to working with NASA on the development of a lunar rover as part of that program. Australia's contribution to NASA's Artemis mission through its Moon to Mars Initiative includes the Trailblazer program—the supply of a lunar rover whose primary task was identified as the collection of lunar regolith and delivery of that regolith to a NASA ISRU facility for exploration and scientific purposes. Whilst the specific design and mission have evolved and are likely to continue to evolve over ongoing design reviews and testing, a fundamental aspect of this project remains the demonstration of 'Australia's commitment to the development and implementation of rules and norms that seek to support the safety, stability and sustainability of outer space.'[2]

Australia's significant expertise in robotics and remote operations means that it is well placed to participate in space resource activities. Australia is also committed to the peaceful and sustainable uses of outer space, as evidenced by its participation in multiple international fora, including long-term membership of the UN Committee on the Peaceful Uses of Outer Space (UNCOPUOS). In addition to the Outer Space Treaty, the Liability Convention, the Rescue and Return Agreement and the Registration Convention, Australia is a party to the Moon Agreement, as well as being one of the founding signatories to the Artemis Accords.[3]

Whilst Australia does have a specific and active role both in the development of space resource activities, through its role in the NASA Artemis program, as well as other space activities, and also through participation and leadership in UNCOPUOS and the Working Group on Legal Aspects of Space Resource Activities ('Working

[1] NASA, The Artemis Accords, Principles for Cooperation in the Civil Exploration and Use of the Moon, Mars, Comets, And Asteroids for Peaceful Purposes, 13 October 2020, https://www.nasa.gov/wp-content/uploads/2022/11/Artemis-Accords-signed-13Oct2020.pdf?emrc=689536250a8fa.

[2] Statement by Australia, 63rd Session of the COPUOS Legal Subcommittee, 15–26 April 2024, Agenda Item 9: Australia—Input to the Working Group on Legal Aspects of Space Resource Activities, 'Consideration of general themes and obligations, commitments or guidelines to an activity contributing to exploration and scientific investigation of the Moon.'

[3] Treaty on Principles Governing the Activities of States in the Exploration and Use of Outer Space, including the Moon and Other Celestial Bodies, opened for signature 27 January 1967, 18 UST 2410, 610 UNTS 205 (entered into force 10 October 1967) ('Outer Space Treaty'); Convention on International Liability for Damage Caused by Space Objects, opened for signature 29 March1972, 961 UNTS 187 (entered into force 1 September 1972) ('Liability Convention'); Agreement on the Rescue of Astronauts, the Return of Astronauts and the Return of Objects Launched into Outer Space, opened for signature 22 April 1968, 672 UNTS 119 (entered into force 3 December 1968) ('Rescue and Return Agreement'); Convention on Registration of Objects Launched into Outer Space, opened for signature 14 January 1975, 1023 UNTS 15 (entered into force 15 September 1976) ('Registration Convention'); Agreement Governing the Activities of States on the Moon and Other Celestial Bodies, opened for signature 18 December 1979, 1363 UNTS (entered into force 11 July 1984) ('Moon Agreement').

Group') Australia has no domestic law which permits or regulates space resource activities. It therefore provides an interesting case study in the current context of the formation of principles and laws applicable to space resource utilisation.

This chapter will identify the role that the Trailblazer program will play in clarifying the applicable legal framework for exploration, exploitation, and utilisation of space resources from an Australian legal perspective, including key issues of due regard, benefit to humanity, interoperability, jurisdiction and control, non-appropriation, environmental protection, and sustainability. It will also identify Australia's role in the development of international space resource frameworks and the responsible uses of outer space.

Part two of the chapter will outline the Trailblazer Program and the development of the Australian ROO-ver. Part three will outline the current legal framework relevant to space resource activities. Part four addresses the particular obligations of the Moon Agreement upon Australia as one of the few members of that treaty. Part five considers whether Australia has a unique role to play as an advocate for further adoption of the Moon Agreement. The final section concludes that, in the context of the number of planned lunar missions in the coming years, there is an urgent need to address the sustainable and responsible approach to space resource utilisation and to develop an international framework that establishes how lunar resources may be identified and managed in compliance with existing international law.

2 ROO-Ver and the Trailblazer Program

When Australia signed the Artemis Accords in 2020, as one of the founding partners, it was regarded as a major achievement for Australia's recently established Australian Space Agency. The creation of the Australian Space Agency was announced by the Australian Government at the Opening Ceremony of the International Astronautical Congress in Adelaide, Australia, in 2017.[4] At the time, Australia was also undertaking a major review of its national space industry capability and its domestic space law legislation, which resulted in the implementation of the Space Law (Launches and Returns) Act 2018 (Cth) and a number of new pieces of subordinate legislation which revamped Australia's space activities regulatory environment. Creation of an Australian Space Agency was regarded as essential to providing a 'single point of contact for international engagement and partnerships' which had been lacking up to that point.[5]

The Australian Space Agency commenced operation on 1 July 2018, with its headquarters to be located in Lot Fourteen, an entrepreneurship and innovation hub in Adelaide, South Australia. The Agency has responsibility for the regulation of

[4] Melissa de Zwart, 'South Australia's Role in the Space Race: Then and Now' (2019) 40(1) Adelaide Law Review 63, 70–71.

[5] Expert Reference Group, 'Review of Australia's Space Industry Capability – Report from the Expert Reference Group' (Report, March 2018), Appendix 3, 69.

the Australian space industry, as well as Australia's compliance with international obligations. It is also tasked with the growth and transformation of the Australian space industry. As one of the Agency's first major initiatives, Australia signed a letter of intent establishing co-operation between NASA and the Australian Space Agency in 2019.[6] As a consequence of this partnership, Australia created a five-year program- the Moon to Mars initiative- with $150 million in funding to support Australian businesses and researchers to develop space technologies that would support NASA's return to the Moon and on to Mars. That initiative comprised three programs: Supply Chain program, Demonstrator program, and the Trailblazer program, which would comprise the 'flagship mission supporting NASA's Moon to Mars activities that demonstrates and develops Australia's strengths, ingenuity and capabilities, while inspiring the nation.'[7] Pursuant to a further agreement with NASA, the Trailblazer mission called for the development of a small 'foundation services rover' which could demonstrate the collection of regolith from the lunar surface.[8] The attributes of this project were intended to showcase: 'Australian industry's world-leading skills and experience in remote operations and autonomous systems… [and draw] … from our expertise in the resources and mining sector.'[9] The proposed mission involved several stages, including lunar surface operations, with the primary task identified as the delivery of 'lunar regolith to the ISRU facility or equivalent.'[10] In 2025, the ASA announced that the ELO2 Consortium had been successful in the bid to build the Australian ROO-ver.[11] It was also announced that the ROO-ver mission would now include the deployment of a NASA instrument. Whilst it may now appear that the ROO-ver mission may focus more on remote and autonomous operations, rather than physical regolith collection, the engagement of Australia in the Artemis Accords and the particular objects of the Trailblazer program have become the focus for consideration of the particular legal status of regolith collection and utilisation for Australian space operators.

---

[6] NASA, Australian Government Commits to Join NASA in Lunar Exploration and Beyond, 21 September 2019, https://www.nasa.gov/news-release/australian-government-commits-to-join-nasa-in-lunar-exploration-and-beyond/.

[7] Australian Space Agency, Moon to Mars initiative -Trailblazer program, Consultation paper, 8 April 2021.

[8] https://www.space.gov.au/news-and-media/headed-moon-trailblazer-program-and-nasa-space-act-agreement.

[9] Above.

[10] Consideration of general themes and obligations, commitments or guidelines to an activity contributing to exploration and scientific investigation of the Moon, A/AC.105/C.2/2024/CRP.24.

[11] The name of the rover, 'ROO-ver', was selected following a public competition, and is intended to represent the 'iconically Australian' kangaroo, Australian Space Agency, G'day Roo-ver, 5 December 2023, https://www.space.gov.au/news-and-media/gday-roo-ver.

3 Legal Framework for Space Resource Activities and the UNCOPUOS Working Group on Legal Aspects of Space Resource Activities

The current international space law prescribes no specific regulatory regime for space resource activities. Certain provisions of the Outer Space Treaty establish principles that affect the conduct of space resource activities, but only the Moon Agreement foreshadows the development of an international regime that will govern the extraction of natural resources beyond Earth.[12] As will be discussed below, the Moon Agreement has had little uptake, and no such international regime has been created. There have, however, been several efforts to develop guidelines and principles outside of the UN treaty process. These include the Hague International Space Resources Working Group's Building Blocks for the Development of an International Framework on Space Resource Activities (2019), the Outer Space Institute's Vancouver Recommendations on Space Mining (2020), the Moon Village Association's Recommended Framework and Key Elements for Peaceful and Sustainable Lunar Activities (2022) and the Lunar Policy Platform's Lunar Policy Priorities (2023).[13]

Of course, there are also the Artemis Accords, which consist of ten high-level principles 'to enhance the governance of the civil exploration and use of outer space' to be implemented by signatories in their own space activities.[14] The original eight signatories have increased exponentially to fifty-six as of July 2025.[15] It is beyond the scope of this chapter to examine the Artemis Accords in detail, as has been done elsewhere.[16] However, it should be noted that the provisions include several principles that repeat and reinforce the provisions of the UN Space Treaties and other

[12] See discussion below in Part 3.

[13] The Hague International Space Resources Governance Working Group, Building Blocks for the Development of an International Framework on Space Resource Activities, November 2019 https://www.universiteitleiden.nl/binaries/content/assets/rechtsgeleerdheid/instituut-voor-publiekrecht/lucht--en-ruimterecht/space-resources/bb-thissrwg--cover.pdf; Outer Space Institute, Vancouver Recommendations on Space Mining, April 20, 2020 https://outerspaceinstitute.ca/osisite/wp-content/uploads/Vancouver_Recommendations_on_Space_Mining.pdf; Moon Village Association, Recommended Framework and Key Elements for Peaceful and Sustainable Lunar Activities https://moonvillageassociation.org/gegsla/documents/gegsla-recommended-framework/; Lunar Policy Platform, Lunar Policy Priorities, 11 November 2023, https://static1.squarespace.com/static/653ffafa2edf841ff28639a4/t/674f0e4bcdaf8306da13bd38/1733234253483/Lunar+Policy+Priorities+Report+%5BDec+2024+online+version%5D.pdf.

[14] NASA, The Artemis Accords: Principles for Cooperation in the Civil Exploration and Use of the Moon, Mars, Comets, and Asteroids for Peaceful Purposes, 2020 https://www.nasa.gov/wp-content/uploads/2022/11/Artemis-Accords-signed-13Oct2020.pdf?emrc=689821095e5ee.

[15] NASA, The Artemis Accords, https://www.nasa.gov/artemis-accords/.

[16] Melissa de Zwart, To the Moon and Beyond: The Artemis Accords and the Evolution of Space Law, in de Zwart and Henderson (eds) Commercial and Military Uses of Outer Space, Springer, 2021, 65–80; Rosana Deplano, The Artemis Accords: Evolution or Revolution in International Space Law? (2021) 70(3), The International and Comparative Law Quarterly 799–819; Arthar ud Din 'The Artemis Accords: The End of Multilateralism in the Management of Outer Space?' (2022) 2–3 Astropolitics 136; Melissa de Zwart and Joel Lisk 'The Artemis Accords and Subsequent

UN instruments, such as the conduct of activities for peaceful purposes and within the tenets of the Outer Space Treaty, transparency of policies and plans, provision of emergence assistance to astronauts, registration of space objects in accordance with the Registration Convention, release and sharing of scientific data, debris mitigation, and safe disposal of spacecraft as well as a number of new (and more controversial) principles, including the development of standards for the interoperability of space systems, protection of historic sites and artefacts, the extraction and utilisation of space resource and the deconfliction of activities through the use of 'safety zones.'[17]

It should also be noted that the Artemis project is not the only multistate project planning for a sustained human presence on the Moon. China announced its plans for the creation of a permanent base on the Moon in 2021 and has called for partners to join its International Lunar Research Station 'to cooperate in the areas of planning, substantiating, engineering, development, implementation and operation of ILRS, to strengthen scientific research exchanges and promote the peaceful exploration and use of space in the interests of all of humankind.'[18] As of July 2025, thirteen states have joined as participants in the ILRS: China, Russia, Venezuela, Pakistan, Azerbaijan, Belarus, South Africa, Egypt, Thailand, Nicaragua, Serbia, Senegal, and Kazakhstan. A number of non-governmental entities, such as corporations, agencies, universities and astronomical associations have also signed agreements with the ILRS, as well as the Asia–Pacific Space Cooperation Organisation (APSCO). Notably, a number of Artemis Accord signatory states already have some involvement with China's space program, including Bahrain, UAE and Peru, either through membership of APSCO, or involvement in other space projects through commercial, educational or government operators. Notably NASA is prohibited from participating in the ILRS or engaging China in the Artemis project, due to the effect of the Wolf Amendment, which, subject to some exceptions, prohibits both NASA and the White House Office of Science and Technology Policy (OSTP) from spending money on any bilateral policy, program, order, or contract of any kind' or participating or collaborating with Chinese government agencies or Chinese-owned companies.[19] Further, no government funds may be used to host visits from Chinese nationals to any NASA facilities, effectively restricting any collaboration or exchange of information.[20]

Unlike the Artemis Accords, China has stated that participants will be involved in the '[j]oint development of the legal documents regulating relations, including the

---

Developments' in Bhat, Ukey and Variath (eds.) International Space Law in the New Space Era, Oxford University Press, 2024, 231–254.

[17] NASA, The Artemis Accords, above.

[18] Joint Statement Between CNSA and ROSCOSMOS Regarding Cooperation for the Construction of the International Lunar Research Station, 29 April 2021, https://www.cnsa.gov.cn/english/n6465668/n6465670/c6811967/content.html, https://www.cnsa.gov.cn/english/n6465668/n6465670/c6811967/content.html.

[19] Wolf Amendment, within Public Law 112–55, Section 539, https://www.govinfo.gov/content/pkg/PLAW-112publ55/html/PLAW-112publ55.htm.

[20] Dan Hart, Dean Cheng and Brian Weeden, Should the Wolf Amendment Be Repealed? Center for Space Policy and Strategy, The Aerospace Corporation, July 2025, https://csps.aerospace.org/sites/default/files/2025-07/WolfAmendmentDebate_20250728.pdf.

involvement in cooperation of third parties, in the framework of creation of ILRS' as one of the designated co-operation domains.[21] Strategically, the existence of the ILRS and the invitation for other states to join in the program provide an alternative to the US-led Artemis program and ensure that the Artemis Accords will not become the sole rule book for operations on the Moon. However, it is also evident that despite attempts to distinguish the two approaches, it is likely that, for practical and safety reasons alone, the two projects may end up producing a very similar set of principles needed to survive and operate on the surface of the Moon.[22] Both projects will involve resource extraction utilisation on the Moon to support a sustained presence, and both also profess compliance with the Outer Space Treaty. Of course, neither the Artemis Accords nor any ILRS rules may be recognised as international law, although operation pursuant to these principles may, in time, become evidence of state practice.

The key international space law provisions which affect the conduct of space resource activity include Article I of the Outer Space Treaty which provides that the 'exploration and use of outer space…shall be carried out for the benefit and in the interests of all countries… and shall be the common province of all mankind,' further it affirms that exploration and use of outer space 'shall be carried out for the benefit and in the interests of all countries, irrespective of their degree of economic or scientific development' and there 'shall be freedom of scientific investigation in outer space'; Article II which establishes the non-appropriation principle, stating that 'Outer space, including the Moon and other celestial bodies, is not subject to national appropriation by claim of sovereignty, by means of use of occupation, or by any other means'; Article III which requires States to 'carry on activities in the exploration and use of outer space… in accordance with international law'; Article VI which provides that States 'bear international responsibility for national activities in outer space' and that activities of non-governmental entities 'require authorization and continuing supervision' of the relevant State; and Article IX, which requires activities in outer space to be conducted 'with due regard to the corresponding interests' of other States and to avoid harmful contamination and 'potentially harmful interference with activities of other States.'[23] However, these are all high-level principles and do not provide any clear permissions either for or against the extraction and use of space resources. The only UN Space Treaty that provides further specific guidance on this activity and the purposes for which it may be undertaken is the Moon Agreement, which, after many years languishing behind the other UN Space Treaties, has once again become a focus of attention. The Moon Agreement will be considered in the next section.

---

[21] International Lunar Research Station (ILRS) Guide for Partnership, 16 June 2021, https://www.cnsa.gov.cn/english/n6465652/n6465653/c6812150/content.html.

[22] Submission by the Delegation of China to the Working Group on Legal Aspects of Space Resource Activities of the Legal Subcommittee of the Committee on the Peaceful Uses of Outer Space, March 2024, https://www.unoosa.org/documents/pdf/copuos/lsc/space-resources/LSC2024/English_Chinas_submission_to_the_working_group_on_space_resources.pdf.

[23] Melissa de Zwart, Stacey Henderson, Michelle Neumann 'Space resource activities and the evolution of international space law' (2023) 211 *Acta Astronautica*,155–162, 156–157.

In the absence of an international framework, states such as the United States, Japan, the UAE, and Luxembourg have introduced their own domestic laws that support the development of and investment in space resource activities.[24] There remains some disagreement whether any of these initiatives are compliant with international space law; however, as the expansion of commercial and government missions to the Moon indicates, there is some urgency in clarifying the legal framework.[25] For example, in 2024, the Chang'e 6 mission succeeded in landing in the South Pole-Aitken Basin, on the far side of the Moon and collecting and returning to Earth two kilograms of soil and rock.[26] This was the first time this had been achieved. Hence, there is a significant need for these complex and overlapping issues to be addressed.

The Working Group on Space Resource Activities was established at the 60th Session of the UNCOPUOS Legal Subcommittee in 2021, under the agenda item on the general exchange of views on potential legal models for activities in the exploration, exploitation, and utilisation of space resources.[27] Creation of the Working Group was widely supported by a diversity of States, including the US and Russia. In its endorsement of the initiative, Russia stated that any 'activity that is related to the exploration, exploitation and utilisation of space resources must be based on a clear legal framework which is widely recognised on a global level.'[28] The Group of 77 and China asserted the urgency of addressing the growth of domestic laws 'allowing for the exploitation of celestial bodies for economic purposes' in light of international space law 'in order to avoid contradictions or gaps within the legal framework in this area and in order to provide a clear understanding of the legal obligations of States in space exploration.'[29]

[24] Adam Janikowski 'Bold steps forward: The investment impact of enacting space resources legislation' (2025) 72 *Space Policy*,101,675, https://doi.org/10.1016/j.spacepol.2024.101675. United States Commercial Space Launch Competitiveness Act (2015); Japan, Space Resources Act (2021); United Arab Emirates, Federal Law on the Regulation of the Space Sector (2019); Luxembourg Law on the Exploration and Utilisation of Space Resources (2017).

[25] Güneş Ünüvar, Can National Laws on Space Resources Serve as Evidence of Customary International Law? Opinio Juris, 8 November 2022, https://opiniojuris.org/2022/11/08/can-national-laws-on-space-resources-serve-as-evidence-of-customary-international-law/; Yannick Radi, Space Mining in Practice An International Space Law Perspective on Upcoming Challenges, (2024)13.8 ESIL Reflections, 6 May 2024, https://esil-sedi.eu/wp-content/uploads/2024/05/Radi-Vol.13-Issue-8_final.pdf.

[26] NASA, NASA Space Science Data Coordinated Archive, Chang'e 6,10 August 2025 https://nssdc.gsfc.nasa.gov/nmc/spacecraft/display.action?id=CHANG-E-6.

[27] United Nations General Assembly, Report of the Committee on the Peaceful Uses of Outer Space, 64th Sess, UN Doc A/76/20 (25 August—3 September 2021).

[28] Russian Delegation, General Exchange of Views on Potential Legal Models for Activities in Exploration, Exploitation and Utilization of Space, 60th Session of the COPUOS Legal Subcommittee (June 2021).

[29] Alejandro Solano Ortiz, G-77 and China Statement during the Sixtieth Session of the Legal Subcommittee of the United Nations Committee on the Peaceful Uses of Outer Space, 60th Session of the COPUOS Legal Subcommittee (June 2021).

The Working Group created a five-year work plan to finish in 2027, involving an extensive information gathering and consultation process. Views of member States have been exhaustively sought and views exchanged over this period, including a meeting at the International Space Resources Week in Luxembourg in 2024. Submissions have been made to the Working Group over the course of its activities from a range of non-government organisations and other interested parties.[30] These include groups such as the Space Generation Advisory Council, Open Lunar Foundation, For All Moonkind and the Moon Village Association.[31] All submissions made to the Working Group have noted the importance of, and need for, international regulation of space resource activities. With the work continuing until 2027, it is likely that during this time, more states will enact domestic laws and plan more missions dedicated to resource extraction. Certainly, the current cadence of robotic lunar missions suggests that this will be the case.

Chair of the Working Group, Professor Steven Freeland, is an Australian lawyer and academic.[32] The Initial draft set of recommended principles for space resource activities was circulated in 2025.[33] Part 1 of those draft principles identifies concepts on which there 'appears to be, in general terms, a common understanding in the Working Group' and acknowledged issues such as the lack of a clear definition of 'space resources.'[34] Part 1 addresses the following key issues of resource activity:

a. Compliance with international law, including international space law;
b. Freedom of access and non-appropriation;
c. Peaceful purposes;
d. Safety;
e. Sustainability and protection of the environment of Earth and Outer Space;
f. Prioritisation of scientific research and investigation;

---

[30] COPUOS, Updated summary by the Vice-Chair of the Working Group on Legal Aspects of Space Resource Activities of views and contributions received on the mandate and purpose of the Working Group, A/AC105/C.2/L.340 (20 March 2025).

[31] United Nations General Assembly, Committee on the Peaceful Uses of Outer Space, UN Doc A/AC.105/C.2/2021/CRP.13 (27 May 2021).

[32] ABC Listen, Cooperating over space resources, 26 June 2025, https://www.abc.net.au/listen/programs/latenightlive/getting-countries-to-cooperate-space-resources-steven-freeland/105462916.

[33] UNCOPUOS, Initial draft set of recommended principles for space resource activities, A/AC.105/C.2/L.339, 27 March 2025.

[34] The following clarifications are provided:

> (a) for the purposes of these Principles, space resources include extractable and/or recoverable natural, abiotic resources [material] [such as minerals, water, ice, other liquids, and gases] located [in situ] in outer space, including [on the surface or in the subsurface of] the Moon and other celestial bodies [and do not include satellite orbits, radio spectrum, solar energy or other resources not of a physical nature]; and (b) for the purposes of these Principles, space resource activities include activities conducted in outer space, including the Moon and other celestial bodies, for the [primary] purpose of searching for space resources, the extraction of space resources, and the exploitation and utilization of space resources [as well as related activities such as processing and transportation].

g. Sharing of information and data;
h. Coordination, cooperation, and consultation;
i. International responsibility for activities of non-governmental entities;
j. Process for review and revision of the principles.

Part 2 of the Draft includes a set of possible additional principles, which do not currently reflect a common understanding of the Working Group. In particular, it lists a number of matters which may require further elaboration in the future revision and amendment of the principles. This list includes many factors will be vital to the co-operative and peaceful conduct of space resource activities, including, for example:

a. Development of rules and/ or norms for space resource (and related) activities and equitable sharing of benefits derived from such activities;
b. Capacity building, knowledge exchange and technology transfer for developing states;
c. Sharing of space resources with the scientific community and procedures for the sharing of data;
d. Development of interoperability standards and common infrastructure;
e. Rules for notification, nature, size, and operation of ‘safety’ zones.

This work represents a key initiative in the ongoing task of establishing the legal framework for space resource utilisation, but there remain some key points that create rivalry and possible friction across the many planned lunar operations. The next section considers whether there is any residual role for the Moon Agreement in this context, given its application specifically to Australia, and Australia’s efforts to address some of the questions raised above concerning the Trailblazer project.

4 The Moon Agreement

The Moon Agreement (formally Agreement Governing the Activities of States on the Moon and Other Celestial Bodies, (1979) 1363 UNTS 3) entered into force in 1982. As of 2025, there are seventeen States party to the Moon Agreement and four signatories following the withdrawal of Saudi Arabia in 2023. Whilst the Moon Agreement replicates and extends many provisions of the Outer Space Treaty, it has received renewed interest in the last few years, due to the specific provisions which address resource utilisation on the Moon (and other celestial bodies). The Moon Agreement was the last of the major space law treaties negotiated in the UN. Despite its more recent reputation as a failure, it was negotiated over a period of seven years and the final text was adopted unanimously by UNGA Resolution.[35]

[35] IISL, The Role of the Moon Agreement in the governance of lunar activities, including the utilization of lunar resources, Submission of 15 November 2024 by the International Institute of Space Law (IISL) study group on the Moon Agreement to the Working Group on the Legal Aspects

Article 11(1) provides that 'the moon and its natural resources are the common heritage of [hu]mankind.' Significantly, Article 11(3) of this treaty provides that:

> Neither the surface nor the subsurface of the moon, nor any part thereof or natural resources in place, shall become property of any State, international intergovernmental organization, national organization or non-governmental entity or of any natural person.

Most importantly from the perspective of space resource utilisation, Article 11(5) provides that state parties 'undertake to establish an international regime, including appropriate procedures, to govern the exploitation of the natural resources of the moon as such exploitation is about to become feasible.' According to Article 11(7), the main purposes of this international regime include 'the orderly and safe development of the natural resources of the moon, the rational management of those resources, [and] the expansion of opportunities in those resources.' Further, the Moon Agreement commits parties to the principle of 'equitable sharing ... in the benefits derived from those resources' and special consideration of the 'interests and needs of the developing countries, as well as the efforts of those countries which have contributed either directly or indirectly to the exploration of the moon.' Of course, it is not certain what the sharing of benefits may specifically entail, such as the potential sharing of actual resources (whether on the Moon or on Earth is also unclear) or the allocation of royalties.[36]

As a foundation signatory of the Artemis Accords, Australia's position with respect to compatibility with its membership of the Moon Agreement has been called into question by some commentators, even going so far as to say that Australia may be 'forced' to pick a side or withdraw from the Moon Agreement.[37] Critics of Australia's position have claimed that, by signing the Artemis Accords, Australia is in violation of its treaty obligations under the Moon Agreement, specifically with respect to the obligation under Article 11(5) to develop an international framework for resource utilisation and the application of the common heritage of humankind principle. These arguments are based on the assertion that the Artemis Accords, as a product of NASA, attempt to replace the international regime identified by the Moon Agreement. Further, they focus on the fact that the Artemis project is led by the US, which has rejected the common heritage of humankind principle in the context of resource utilisation.[38]

---

of Space Resources Activities of the United Nations Committee on the Peaceful Uses of Outer Space in response to Circular OOSA/2024/43, CU 2024/221 of 22 July 2024.

[36] Ben McKeown, Andrew G. Dempster, Serkan Saydam, Jeff Coulton, The space resources fund: A solution to the space resources benefit sharing dilemma? (2025) 71 Space Policy,101,663, https://doi.org/10.1016/j.spacepol.2024.101663.

[37] Fabio Tronchetti and Hao Liu 'Australia Between the Moon Agreement and the Artemis Accords' *Australian Outlook*, 3 June 2021, https://www.internationalaffairs.org.au/australianoutlook/australia-between-the-moon-agreement-and-the-artemis-accords/; Fabio Tronchetti 'Moon Agreement: Key Provisions' in Bhat, Ukey and Variath (eds.) International Space Law in the New Space Era, Oxford University Press, 2024, 165–182, 179.

[38] In 2020, the US 'Executive Order on Encouraging International Support for the Recovery and Use of Space Resources' clearly stated that: "[T]he United States does not consider the Moon Agreement to be an effective or necessary instrument to guide nation states regarding the promotion

Australia has been explicit that its membership of both the Artemis Accords and the Moon Agreement is consistent and has made many statements to this effect. In its statement to the 60th Session of the UNCOPUOS Legal Subcommittee, Australia notes that it is a founding member of the UNCOPUOS and a State Party to all five of the UN space treaties, including both the Outer Space Treaty and the Moon Agreement.[39] The statement also notes Australia's role as 'a founding signatory to the Artemis Accords, which establish a practical set of principles to guide space exploration and aim to increase the safety of operations, reduce uncertainty, and promote the sustainable and beneficial use of space.' The statement asserts: 'Australia's position is that the Artemis Accords are consistent with Australia's international legal obligations.' That Statement goes on to identify the role and effect of each of the Moon Agreement and the Artemis Accords.

Australia's official position on the Moon Agreement is that it:

> …provides a framework specific to the exploration, exploitation and use of the Moon and its natural resources. It can also be useful in helping to identify and develop principles for activities in the exploration, exploitation, and utilisation of space resources.[40]

Building upon the paper submitted by Australia, Austria, and the Kingdom of the Netherlands at the 60th session of the Legal Subcommittee, Australia has also stated that the general themes that underpin the four instruments that may be applicable to space resource activities (namely the Outer Space Treaty, the Moon Agreement, the Hague Building Blocks and the Artemis Accords) 'can meaningfully inform considerations regarding the development of a set of initial recommended principles for activities in the exploration, exploitation and utilisation of space resources.'[41]

Australia later further elaborated its views with respect to space resources and the Moon Agreement:

> Australia supports the exploration, exploitation and use of space resources in a manner consistent with international law. Australia's view is that the Moon Agreement provides a framework specific to the exploration, exploitation and use of the Moon and its natural resources. Article 6 of the Moon Agreement permits States Parties to use the Moon's minerals and other substances in carrying out scientific investigations, including in quantities appropriate for the support of their missions.

---

of commercial participation in the long-term exploration, scientific discovery, and use of the Moon, Mars, or other celestial bodies. Accordingly, the Secretary of State shall object to any attempt by any other state or international organization to treat the Moon Agreement as reflecting or otherwise expressing customary international law." Whilst the accompanying White House statement affirms the position that this Executive Order addresses "US policy regarding the recovery and use of resources in outer space' it reaffirms the position that the US will object to "any attempt to treat the 1979 Moon Agreement as expressing customary international law", Melissa de Zwart, Stacey Henderson, Michelle Neumann 'Space resource activities and the evolution of international space law' (2023) 211 *Acta Astronautica*,155–162.

[39] Statement—Australia, Item 14, 60th Legal Subcommittee, 11 June 2021.

[40] Statement by Australia, 63rd Session of the COPUOS Legal Subcommittee, 15–26 April 2024, Agenda Item 9: Australia- Input to the Working Group on Legal Aspects of Space Resource Activities, 'Consideration of general themes and obligations, commitments or guidelines to an activity contributing to exploration and scientific investigation of the Moon.'

[41] Above.

> This is supplemented by Article 8, which allows States Parties to land space objects on, and launch them from, the Moon in their pursuit of activities in the exploration and use of the Moon. States Parties may also place personnel, space vehicles, equipment and facilities, as well as establish manned and unmanned stations, on or below the surface of the Moon for the same purpose. We note that certain obligations under the Moon Agreement have prospective (that is, future), rather than immediate, effect. This includes consideration of the potential establishment of an international regime at a time when exploitation of the Moon's natural resources is about to become feasible, as per Article 11 of the Agreement.
>
> For Australia, the Moon Agreement provides a framework that clearly enablesenables scientific investigation on the Moon and other celestial bodies, and may provide for other exploratory or experimental activities, prior to the establishment of an international regime. We look forward to further sharing our views during meetings of the working group on space resources.'[42]

Importantly, this Statement also affirmed the view that the Moon Agreement's framework will support Australia's planned activities on the Moon over the coming years as part of NASA's Artemis program.'[43] Australia provided a further paper to the Working Group on Legal Aspects of Space Resource Activities in April 2024.[44] In that paper, Australia affirms that:

In Australia's view, the Moon Agreement clearly contemplates that States Parties may undertake the following activities on the Moon:

1. Scientific investigation (which is permission prior to the establishment of an international regime); and
2. Exploitation of the Moon's natural resources (subject to the requirements of an international regime, established as such exploitation is about to become feasible.[45]

Additionally, it asserted that Article 6(1) 'clearly enables scientific investigation on the Moon and other celestial bodies, and may provide for other exploratory or experimental activities, prior to the establishment of an international regime.' With respect to one of the main questions raised regarding Australia's obligations under the Moon Agreement, the creation of an international regime for the exploitation of the Moon's natural resources, to be established as such exploitation is about to become feasible (Article 11(5) MA), Australia has made the following points:

Article 11(7) prescribes the four main purposes of the envisaged regime are:

a. The orderly and safe development of the natural resources of the Moon;
b. The rational management of those resources;
c. The expansion of opportunities in the use of those resources; and
d. An equitable sharing by all States Parties in the benefits derived from those resources, whereby the interests and needs of the developing countries, as well

---

[42] Statement by Australia, 61st Session of the COPUOS Legal Subcommittee, Agenda Item 15.

[43] Above.

[44] Australia's response to the invitation to provide information on the mandate and purpose of the Working Group on Legal Aspects of Space Resource Activities under the Legal Subcommittee of the Committee on the Peaceful Uses of Outer Space.'

[45] Above.

> as efforts of those countries which have contributed either directly or indirectly to the exploration of the Moon, shall be given special consideration.[46]

However, beyond those four main purposes, states are required to 'negotiate and develop the precise contours of the international regime.'

Clearly, this statement indicates that Australia does not regard the Artemis Accords, in their current state, as constituting the international regime prescribed by Article 11. The question of whether exploitation is 'about to become feasible' is a question of fact. In Australia's view 'the world has not yet arrived at the point where the exploitation of space resources could be considered "feasible". Notably, Australia is no longer the only Moon Agreement signatory to sign on to the Artemis Accords. The current list of Artemis Accords-Moon Agreement members includes Armenia, Australia, Belgium, Chile, Mexico, The Netherlands, Peru and Uruguay, along with France, Romania and India who are signatories to the Moon Agreement.

Despite the general view of the Moon Agreement as 'the most unsuccessful agreement of the entire outer space legislation,' there has been little discussion by the member states regarding their continued participation in this treaty, nor any advocacy by those states to attract further members.[47] However, it is noted that in 2008 Austria, Belgium, Chile, Mexico, the Netherlands, Pakistan and Philippines presented a Joint Statement on the benefits of adherence to the Agreement Governing the Activities of States on the Moon and Other Celestial Bodies of 1979 by States Parties to that Agreement which stated:

> '[I]t must be noted that the participation in the 1979 UN Moon Agreement offers substantial benefits and guarantees with regard to the participation in the other UN outer space treaties: not only a better understanding of concepts of international space law and a better description of concepts or procedures, but also and most of all, a mutual commitment to seek for a multilateral solution for the exploitation of Celestial Bodies' natural resources in accordance with the general principles of outer space law.
>
> The States parties encourage States having signed but not yet ratified the Agreement, as well as other States, to become parties to it, in particular considering their possible involvement in forthcoming missions or projects for the exploration of the Celestial Bodies.'[48]

Australia provided a further paper to the Working Group on Legal Aspects of Space Resource Activities, which provides a detailed analysis of the application of international space law and obligations (notably the Outer Space Treaty, the Moon Agreement and the Artemis Accords) 'that may be applicable to the planning and

---

[46] Above.

[47] Stephan Hobe 'The Moon Agreement—Let's Use the Chance!/ Der Mondvertrag -Ein Neuer Anfang/ Le Traite de la Lune—Profitons de la Chance' (2010) 59 ZLW 372, 372.

[48] Committee on the Peaceful Uses of Outer Space Legal Subcommittee, Joint Statement on the benefits of adherence to the Agreement Governing the Activities of States on the Moon and Other Celestial Bodies of 1979 by States Parties to that Agreement by Austria, Belgium, Chile, Mexico, The Netherlands, Pakistan, and Philippines, Forty-seventh session, 31 March—11 April 2008 (2 April 2008) A/AC.105/C.2/2008/CRP.11.

activities conducted as part of the Trailblazer program.'[49] Attachment 1 to this paper provides a step by step breakdown of the obligations that may be engaged at each stage of the program. These obligations include principles of due regard, cooperation, freedom of scientific investigation, deconfliction, exploration and use, benefit to mankind, interoperability, information regarding activities, jurisdiction and control, non-appropriation, responsibility, safety zones, extraction and utilisation of space resources, environmental protection, and sustainability. The provided table identifies the points of consistency and overlap noting that the examination of each of these topics in the context of an actual activity engaged in the exploration and scientific investigation of the Moon, 'can meaningfully inform considerations regarding the development of a set of initial recommended principles for activities in the exploration, exploitation and utilisation of space resources.'[50] Hence, the actual activities as planned for ROO-ver on the Moon become the test vehicle not only for regolith collection but also for applicable legal principles.

5 Does Australia Have a Role to Play with Respect to the Promotion of the Moon Agreement?

As noted above, Saudi Arabia formally notified the UN on 5 January 2023 of its withdrawal from the Moon Agreement, with that withdrawal to take effect on 5 January 2024 (in accordance with article 20 of the Moon Agreement). No official reason for the withdrawal was communicated, but it occurred six months after Saudi Arabia had signed the Artemis Accords.

Naturally, the withdrawal attracted significant speculation, including assertions that Saudi Arabia and/or the United States perceived incompatibility between the Artemis Accords and membership of the Moon Agreement; pressure or request from the US to withdraw from the Moon Agreement or recognition that the Moon Agreement has failed and therefore should no longer be kept on the books.[51] Other suggestions include preference for supporting the development of bottom-up models for the exploitation of space resources versus the slow or possibly even stalled, top-down legal model of the Moon Agreement and recognition by Saudi Arabia that membership of the Moon Agreement may restrict its ability to develop space resource capability and to hamper its political and economic power in the Middle East.[52] Given that Saudi Arabia's rival state, the UAE, is one of four states which has introduced

[49] Consideration of general themes and obligations, commitments or guidelines to an activity contributing to exploration and scientific investigation of the Moon, 16 April 2024, A/AC.105/C.2/2024/CRP.24.

[50] Above.

[51] Stefan-Michael Wedenig & Jack Wright Nelson ' The Moon Agreement: Hanging by a Thread?' 26 January 2023, https://www.mcgill.ca/iasl/article/moon-agreement-hanging-thread.

[52] Michael J. Listner 'Four key points regarding Saudi Arabia's withdrawal from the Moon Agreement' 11 December 2023, *The Space Review*, https://www.thespacereview.com/article/4706/1.

domestic laws facilitating space resource activity, speculation also exists that Saudi Arabia perceived the need to withdraw from the Moon Agreement before developing its own domestic space resources law.

However, despite speculation that Saudi Arabia's withdrawal would create a cascade of withdrawal from the Moon Agreement (including states such as Mexico and Australia), no other states have announced their intention to withdraw from the treaty. In addition, several other Moon Agreement states have since signed the Artemis Accords. Of course, a major impediment to any advocacy for the further implementation of the Moon Agreement exists due to the US position on the Moon Agreement. The US has, via Executive Order, rejected the validity of the Moon Agreement and any application of the concept of the commons with respect to space resource utilisation.[53] Hence, the Australian position is unlikely to have any impact on the US. One point that could be considered, however, is whether Australia's membership of the Moon Agreement may have any influence on any of its other partners, such as India, or even potentially on its regional rival, China. Suggestions have been made that, as a signatory that has not yet ratified the Moon Agreement, India could make a bold statement by ratifying the Moon Agreement to reinforce its commitment to peaceful and cooperative space exploration.[54] Suggestions have also been made that accession by a space power such as China would send a signal that the Moon Agreement should no longer be considered a failure. Accession to the Moon Agreement would signal that China was endorsing an international framework for resource utilisation, rather than a domestic one, and provide it with the means to continue to shape international space law via participation in any revision of that treaty under its own provisions.[55]

6 Conclusion

Australia's unique physical and environmental attributes equip it with the skills and know-how to operate in complex and remote environments. With a long history in mining and remote operations, Australia stands ready to contribute to the return to the Moon and the provision of foundation and support services. It also has a long history as a provider of space services, from launch and testing services at Woomera to the transmission of the Apollo Moon landing images via Honeysuckle Creek. Importantly, Australia maintains a strong belief in the sustainable and responsible

[53] Executive Order on Encouraging International Support for the Recovery and Use of Space Resources, 6 April 2020.

[54] Mohit Khubchandani, As the First Country to Land on the Moon's South Pole, Should India also be the First Space Power to Ratify the Moon Agreement? EJIL Talk! 13 September 2023, https://www.ejiltalk.org/as-the-first-country-to-land-on-the-moons-south-pole-should-india-also-be-the-first-space-power-to-ratify-the-moon-agreement/.

[55] Huan Yu and Mingyan Nie, Acceding to the Moon Agreement to acquire legal certainty: An optional solution for China in the new era of lunar exploration and exploitation (2023) 212 Acta Astronautica 665–671.

uses of outer space, and as its submissions to UNCOPUOS demonstrate, that standard is applied across the range of treaties and agreements of which it is a member. When the ROO-ver reaches the Moon, it will undertake its designated tasks in accordance with all applicable international and domestic law and provide a platform for further development of specific rules for peaceful operation on the Moon. In the meantime, it seems that rumours of the end of the Moon Agreement are premature.

Prof. Melissa de Zwart is Professor of Space Law and Governance at the Andy Thomas Centre for Space Resources, Adelaide University and the Deputy Director of the ARC Centre of Excellence in Plants for Space. Professor de Zwart is an expert in civil, military and commercial uses of outer space and has published extensively on the domestic and international regulation of the space domain. She has worked closely with the Australian space industry and served on the Board of the Space Industry Association of Australia. She was Dean of the Adelaide Law School, University of Adelaide (2017–2021). She is a Fellow of the Australian Academy of Law and a Member of the International Institute of Space Law.

Industrial Revolution to Economic Evolution: The UK Approach to Outer Space Resources

Christopher J. Newman and J. Lauren Napier

Abstract The space ambitions of the United Kingdom have moved from a standing start in the early part of the twenty-first century to a point where the UK harbours genuine aspirations to become a key player in the emerging space economy. At the present time, the UK has not chosen to enact specific legislation in respect of resource exploitation, focusing instead on regulation for launching and satellite operations. Nonetheless, the UK is an active participant in Project Artemis and is one of the founder signatories of the Artemis Accords, illustrating clear ambitions to be at the forefront of the developing lunar economy. This chapter will consider the UK approach to outer space resources from two crucial perspectives. First will be a comparative evaluation of the UK approach to space mining with the exploitation of resources from the Antarctic and the Deep Seabed. Such a comparison will provide valuable insight as to both the UK position and ambitions from a strategic sense. Given the volatility of global geopolitics, this comparative critique will enable the discussion to move to an exploration of the possible contribution that the UK could make to the mining of resources in outer space. This second perspective will draw on an analysis of the heritage and expertise that the UK has in respect of both coal mining and drilling for oil in the North Sea. Alongside this will be consideration of the embedded environmental and regenerative regulatory processes that form a key part of the resource extraction industry in the UK. Taking these two perspectives together, this chapter will not only evaluate the current position of the UK in respect of celestial mining but also posit a possible strategy for deeper involvement in this area. It will be argued that such engagement could be empowered by a rich regulatory expertise and an understanding of the key infrastructure elements involved in the mining industry. Space resources offer the tantalising possibility of transforming the global economy. This chapter will highlight how the UK could become a cornerstone of this revolution.

C. J. Newman (✉) · J. L. Napier
Northumbria University, Newcastle, UK
e-mail: christopher.newman@northumbria.ac.uk

J. L. Napier
e-mail: napier@northumbria.ac.uk

S. Bhat B and A. Variath (eds.), *Legal Responses to Space Resource Utilisation*, Space Law and Policy, https://doi.org/10.1007/978-981-95-8864-0_12

Keywords United Kingdom · Space economy · Artemis Accords · Space mining · Comparative regulation · Resource exploitation

1 Introduction

The United Kingdom (U.K.) has, in the twenty-first century, sought to significantly enhance the role it plays in the global space economy. Indeed, the UK space sector is expanding to a point where the country can make credible claims to being 'one of the most innovative and attractive space economies in the world'.[1] This chapter will examine the approach of the U.K. to the question of space resources[2] by looking at both contemporary trends in space policy and at the legacy of Britain's industrial past.[3] This discussion will consider the challenge of putting in place governance mechanisms for the extraction and utilisation of space resources. It will be argued that, while the U.K.'s current policy and legislative priorities do not focus on space resources, the nation's rich regulatory heritage, rooted in its experience with coal mining in the nineteenth Century, provides a unique and valuable perspective. By leveraging these lessons, the U.K. could position itself as a powerful global voice as a leader in the regulation of resource extraction and shaping discussions on the future governance of space mining.[4]

At first blush, there appears to be little in common between the coalfields of nineteenth Century Britain as against the technologically demanding attempts to mine the Moon and other celestial bodies. Yet, the 1800s saw Britain's coalfields transformed from largely unregulated hazardous environments into an industry defined by gradually improving standards, legal protection for workers and independent safety

[1] Maureen Haverty, "Op-ed: Britain has all the attributes to create a world-leading space economy" Space News, July 19 2023, available online at < https://spacenews.com/op-ed-britain-has-all-the-attributes-to-create-a-world-leading-space-economy/ > [Accessed 9 August 2025].

[2] For this discussion, 'Space resources' is to be defined as 'an extractable and/or recoverable abiotic resource in situ in outer space'. This definition is taken from Hague International Space Resources Governance Working Group,2019 Building Blocks for the Development of an International Framework on Space Resources Activities available online at: < https://www.universiteitleiden.nl/binaries/content/assets/rechtsgeleerdheid/instituut-voor-publiekrecht/lucht–en-ruimterecht/space-resources/bb-thissrwg–cover.pdf > [Accessed 9 August 2025].

[3] This chapter does not intend to provide a historical analysis of coal mining in Britain. There are numerous texts to consult on this for further details. See, for example, Brian Lewis, *Coal Mining in the Eighteenth and Nineteenth Centuries*, (Prentice Hall Press, 1971), Charles-Francois Mathis, *Coal in Modern Britain: A social and cultural history from 1830 to the Second World War*, (Bloomsbury Academic, 2025).

[4] Department for Business, Energy, & Industrial Strategy (BEIS), *National space strategy* (2021), 6, available online at [https://www.gov.U.K./government/publications/national-space-strategy], 19. [Accessed 9 August 2025] 7 and the aspiration within the Strategy for the UK to be "leading the world in modern space regulation and building new space trading partnerships." https://assets.publishing.service.gov.uk/media/6196205ce90e07043d677cca/national-space-strategy.pdf

and accident investigation.[5] This resonates as humans begin to negotiate the unfamiliar, extreme environment of outer space and attempt to utilise and exploit the vast resources of the solar system.[6] Britain can bring experience of a robust licencing environment, prudent but empowering regulation and diplomatic experience to bear on these discussions. The U.K. is uniquely positioned to navigate the delicate balancing of commercial interests that drove coal mining and will increasingly shape space resource extraction.

This chapter employs multiple methods, utilising policy analysis infused with a legal-historical approach. Considering that space resource utilisation and exploitation will be a highly expensive and risky endeavour, lessons learnt from other relevant industries must be considered. Risk events in space can be mission and life-threatening. Where else is this level of risk to the environment and human life more prevalent than in the mining industry? Looking back to the earlier days of coal mining, especially during the transition to a safer regulatory framework in the U.K., is in line with the challenges faced now in the space sector. This is a critical area of strength that the U.K. could draw upon and contribute to the global conversation on resource utilisation.

Using a legal historical perspective is important because law does not sit in a specific point in time, nor is law in a vacuum. Legal history insights, '*... are important not just for the light they can shed on the manner and extent to which the development of the law influences and is influenced by social, political, and intellectual forces within the broader community, but also for the light shed on the structure and functioning of the modern legal system*'[7] This statement echoes discussions on safety measures in coal mining regulations in the 1850s in Great Britian[8] through to what is being considered at the United Nations Committee on the Peaceful Uses of Outer Space (UN COPUOS) regarding space resource utilisation and exploitation. Societal concerns, political will, as well as economic and environmental concerns, are part of the legal challenges to mining in general.

By analysing the development of law and regulatory frameworks through the lens of coalmining, and given the volatility of global geopolitics, this comparative critique will enable the discussion to move to an exploration of the possible contribution that the UK could make to the mining of resources in outer space. As will be shown, risk and safety-based legal frameworks have inherent similarities, whether the mining is occurring in the North East of England or the Lunar South Pole. As such, lessons learnt from the past, when terrestrial mining technology was developing, can shed

---

[5] Bill Lancaster, *The Making of the Modern World: The History of the North of England Institute of Mining and Mechanical Engineers Volume 1: 1852–1914* (the Common Room, 2021).

[6] For one of the foundational texts advocating space resource exploitation, see: John S. Lewis, Mining the sky, (Addison-Wesley, 1996).

[7] T.T. Arvind, 'Historical Approaches' in Simon Halliday (ed) *An Intro to the Study of Law* (W Green, 2012) 64.

[8] Amongst some fascinating primary historical sources, see Select Committee on the Best Means of Preventing the Occurrence of Dangerous Accidents in Coal Mines, *Report* (HL 1849) (evidence of Matthias Dunn 17 July 1849).

light and potentially give insight into how to address similar challenges and risk events in the domain of outer space.[9]

The chapter will start with a critical examination of the evolution of U.K. space policy to ascertain whether there is or has been any appetite to engage the space sector in the hunt for space resources. The discussion will move on to consider the lessons that can be learned from the incremental development of Britain's industrial journey and what lessons can be drawn from the development of mining in industrial Britain for the broader space sector. Taking these two perspectives together, this chapter will not only evaluate the current position of the UK with respect to space mining but also posit a possible strategy for deeper involvement in this area. It will be argued that such engagement could be empowered by a rich regulatory expertise and an understanding of the key infrastructure elements involved in the mining industry. Space resources offer the tantalising possibility of transforming the global economy. This chapter is a unique and significant contribution, highlighting how the UK could become a cornerstone of this revolution.

2 The Development of U.K. Space Policy

Before considering the historical place of coal mining in Britain and how it can influence contemporary space resource discourse, it is instructive to consider the current strategic direction of the U.K. space ambitions. Such an analysis will establish exactly where space resources align with current policy. The U.K. has gone through a significant shift in its approach to space activity since the start of the twenty first Century. Despite being a founding member of the United Nations Committee on the Peaceful Uses of Outer Space (UN COPUOS), for much of the twenty first Century, the U.K. regressed from a position where it had initially invested in independent launch capability and was the third country to launch a satellite using its own rocket. The economic crisis of the 1970s saw financial retrenchment, which effectively neutered the U.K.'s growth as a space power[10] through to a position, by the year 2000, where national space interests were moribund.[11]

This position had changed dramatically by the start of the 2010s when the *Innovation and Growth Strategy* (IGS) was published by the U.K. Government. Released at a time when small satellite manufacturing was expanding, the IGS coincided with the creation of the U.K. Space Agency[12] (UKSA) and heralded a dramatic step-change in the pace of activity and investment within the sector. The creation of the Space

[9] Arvind (n 7) 64.

[10] Ray Millard, *An overview of United Kingdom space activity, 1957–1987* (2005, ESA Publications Division).

[11] The U.K. had maintained a small, vibrant small satellite community and contributed as a member of the European Space Agency.

[12] https://web.archive.org/web/20100327100337/http://www.bnsc.gov.U.K./17628.aspx [Accessed 9 August 2025].

Agency was matched by a ramping up of the total space budget,[13] and an increase in the contributions that the U.K. paid to the European Space Agency (ESA).

The increasing space budget, at a time of austerity, coupled with the promulgation of the first *National Space Policy* (NSP) in 2015 and the subsequent U.K. National Space Strategy (UK NSS), all signalled that the U.K. was committed to expanding its involvement in space activity. The main driver for this increased focus was, of course, financial. The 2021 UK NSS has as its first goal to 'grow and level up the U.K. space economy'[14] with a related pillar of 'unlocking growth in the U.K. space sector'.[15] The initial focus areas[16] in the UK NSS point towards commercial and defence operations.[17] Similarly, the key priorities of the Strategy, with regard to civil and defence space activity, relate to global satellite communications, Earth Observation (EO) and Intelligence, Surveillance and Reconnaissance (ISR), Space command and control infrastructure, Position, Navigation and Timing (PNT) capabilities, orbital launch capability, in-orbit servicing and manufacturing and expanding the existing sensor networks to enhance the provision of Space Domain Awareness (SDA) data.

These goals, pillars and priorities represent attempts to adopt an integrated approach across the academic, commercial and defence interests within the wider space economy[18] and were all underpinned by a desire to augment the basic infrastructure surrounding U.K. space activity. In 2017, the U.K. Space Agency announced the development of LaunchUK, an ambitious industry-driven programme to provide the U.K. with the ability to launch space objects from spaceports within the country.[19] LaunchUK is committed to developing safe, competitive and sustainable launch capability focused on small satellites and sub-orbital flights and set the priority of the UKSA and the Government to achieve an independent launch from U.K. soil.

The UK NSS also gives prominence to the sustainable, responsible development of outer space. One of the stated goals of the Strategy is to work at the UN to deliver

---

[13] Alok Jah, "UK government to invest extra £300 m in space science research" The Guardian, Fri 9 Nov 2012 available online at: https://www.theguardian.com/science/2012/nov/09/uk-government-invest-space-science-research [Accessed 9 August 2025].

[14] Department for Business, Energy, & Industrial Strategy, BEIS, *National space strategy* (2021) available online at, https://assets.publishing.service.gov.uk/media/6196205ce90e07043d677cca/national-space-strategy.pdf [Accessed 9 August 2025].

[15] Ibid 24.

[16] The ten focus areas are; (1) Capture the European market in commercial small satellite launch, (2) Fight climate change with space technology, (3) Unleash innovation across the space sector, (4) Expand our horizons with space science and exploration, (5) Develop our world class space clusters, (6) Lead the global effort to make space more sustainable, (7) Improve public services with space technology, (8) Deliver the U.K. Defence space portfolio, (9) Upskill and inspire our future workforce, (10) Use space to modernise and transform our transport system. (U.K. NSS, 41–43).

[17] U.K. NSS, 41.

[18] Christopher Newman, "Space Law & Policy: The U.K. Approach to the Regulation of Space Activity" (2022) Oxford Research Encyclopedia of Planetary Science.

[19] For details of LaunchUK see this brochure, which accompanied a series of roadshows across the U.K. https://assets.publishing.service.gov.uk/media/6196205ce90e07043d677cca/national-space-strategy.pdf [Accessed 9 August 2025].

leadership on a safe, sustainable and secure space environment and 'to deliver a new resolution on space threats'. Whilst the attempts to develop dialogue on responsible behaviours in space centred around discussions within the Committee on Disarmament,[20] not UN COPUOS, the UK has still committed to 'taking bold steps towards a cleaner, more sustainable future in space'.[21] As well as engaging with the UN Office of Outer Space Affairs and participating in UN COPUOS, the UK has launched feasibility studies into active debris removal.[22]

Whilst covering a range of activities, there was no mention either in the NSS or the NSP about the utilisation of space resources. The U.K. has, like many other nations, looked to cast the net as widely as possible when trying to capture the range of space activities that were emerging within the U.K. Yet, despite the breadth of activities mentioned in the various policy and strategy documents, there was no discussion of the technical development nor governance infrastructure in respect of the exploitation of space resources. Even the 2018 'Prosperity from Space' report, a follow-up to LaunchUK compiled by the body representing the space industry, UKSpace, focused on global connectivity, Earth information systems, in-space robotics and low-cost access to space.[23] Despite the apparent promise of a trillion-dollar economy in space minerals,[24] there was no meaningful discussion of any aspect of space resources in the numerous strategy or policy documents.

The lack of any mention of space mining and the exploitation of space resources in either the NSP or UK NSS is understandable when considering the broader context of national space activity outlined above. The extant expertise and heritage for U.K. space activity have been in the realm of defence and maritime support, instrumentation and the manufacture of small satellites. The stated target of developing independent U.K. access to space through achieving launch capability meant that resources had to be focused on this. It also meant that when it came to addressing the legislative framework, this too would be where attention was focused. This discussion will now move on to considering those changes and how space resource utilisation was largely overlooked when U.K. space governance was revivified.

[20] UNGA resolution (A/RES/75/36) has since been subsumed into the Open-Ended Working Group (OEWG) on PAROS established under UNGA A/RES/79/512.

[21] For details see: https://www.gov.uk/government/case-studies/space-sustainability.

[22] https://astroscale.com/revolutionising-space-sustainability-astroscale-leads-expert-consortium-to-secure-uk-space-agency-contract-for-adr-refuelling-mission/

[23] See https://www.U.K.space.org/wp-content/uploads/2019/05/Prosperity-from-Space-strategy_2May2018.pdf [Accessed 9 August 2025].

[24] See, for example Shriya Yarlagadda, 'Economics of the Stars: The Future of Asteroid Mining and the Global Economy' in *Harvard International Review* (April 8, 2022) available at https://hir.harvard.edu/economics-of-the-stars/ [Accessed 9 August 2025].

3 Developing U.K. Domestic Legislation

As investment and economic interest coalesced around the space sector within the U.K., it became increasingly obvious to all stakeholders that the current domestic legal framework for regulating and authorising domestic space activity was inadequate and would need to be updated. The extant regime, centred around the Outer Space Act 1986, was drafted at a time when the U.K. had started to develop a 'critical mass of commercial activity centred around the defence industry'.[25] The 1986 Act was, at its core, a licensing regime designed to include within its scope any space activity which, or actor who might, trigger U.K. responsibilities under the international space law treaties to which the U.K. was a signatory.[26]

The 1986 Act and the licensing regime that it begat imposed a 'considerable' regulatory burden on companies and was criticised as leaving U.K. industry at a significant disadvantage.[27] The desire of the Government to stimulate growth in the sector and, crucially, the promulgation of the UKLaunch project with the central aim of creating an independent launch meant that there was significant momentum towards reform of the U.K.'s national space legislation.[28] The government introduced the *Draft Spaceflight Bill* in 2017 and progressed through Parliament at speed, becoming law in 2018. This speed meant that the final, *Space Industry Act 2018*, was skeletal in nature, leaving much of the detail to be established in secondary legislation. In 2021, the *Space Industry Regulations* provided details on the licensing process, the creation and operation of spaceports, together with the regulation of range control.[29]

The UK legislation contains a broad, inclusive definition of permitted activity, which was intended to 'futureproof' the legislation. Space Activity, including launch and return of a space object[30] was augmented with sub-orbital activity to allow for a rocket or any other craft that operates above the stratosphere. Space tourism was not directly addressed, but there was provision to allow for a balloon flight capable of reaching the stratosphere with crew or passengers.[31]

---

[25] Ray Millard, *An overview of United Kingdom space activity, 1957–1987* (2005, ESA Publications Division).

[26] Francis Lyall and Paul B. Larsen, *Space Law: A treatise* (Routledge, 2nd Ed. 2018), 433.

[27] Christopher J. Newman and Michael Listner, "A very British coup: Lessons from the draft UK regulations for CubeSats", The Space Review, (31 August 2015) Available online at http://www.thespacereview.com/article/2816/1 [Accessed 9 August 2025].

[28] Lesley-Jane Smith and Ruairidh Leishman, "Up, up and away: An update on the UK's latest plans for space activities" 44(1) *Air & Space Law* (2019) 1–26, 1–3.

[29] For a more detailed discussion on the provisions of both the Space Industry Act 2018 and the Space Industry Regulations 2021 see Christopher Newman, "Space Law & Policy: The U.K. Approach to the Regulation of Space Activity" (2022) Oxford Research Encyclopaedia of Planetary Science.

[30] Space Industry Act 2018, s1(4).

[31] Space Industry Act 2018, s1(5).

Nonetheless, despite industry[32] lobbying,[33] neither the 2018 Act nor the 2021 Regulations contained any dedicated provisions related to the utilisation of space resources, or even any nod towards future activity in this area. Initially, this may seem surprising. The United States had led the way with Title IV of the 2015 Commercial Space Launch Competitiveness Act (CSLCA).[34] This allowed for a US citizen engaged in commercial space mining to be entitled to possess, own, transport, use and sell any resource obtained.[35] In 2017, while the Draft Spaceflight Bill was being developed and discussed in the UK, Luxembourg passed the Law on the Exploration and Use of Space Resources,[36] stating explicitly that resources can be appropriated and includes a detailed procedure to licence private mining ventures.[37]

The decision to focus the national domestic space legislation on launch and enhancing the existing skills within the space sector is wholly understandable. The need to develop a framework to enable independent launching and licensing of spaceflight activities from the British Isles, and ensuring all of this was done safely, securely and with the appropriate environmental protection, was of paramount concern to the Government. The U.K. had built its ambitions for growth, and its future strategies, upon extant strengths in manufacturing small satellites, space science, instrumentation and satellite applications. The careful alignment of these considerations does not preclude future involvement in lunar exploration,[38] but does indicate that involvement is likely to focus on the key pillars of the U.K. NSS.

[32] The Asteroid Mining Corporation was founded in 2016 and aims to be the first, profitable space resource business. For more details see https://www.asteroidminingcorporation.co.uk/ [Accessed 9 August 2025].

[33] Deganit Paikowsky and Roey Tzezana, "The Politics of space mining – An account of a simulation game" 142 Acta Astronautica (2018) 10–17, 11.

[34] Title IV Space Resource Exploration and Utilization §402, US Commercial Space Launch Competitiveness Act 2015, 129 STAT. 704, Public Law 114–90—November 25, 2015 available online at https://www.congress.gov/114/plaws/publ90/PLAW-114publ90.pdf [Accessed 9 August 2025].

[35] This was further expanded by an Executive Order https://www.whitehouse.gov/presidential-actions/executive-order-encouraging-international-support-recovery-use-space-resources/ [Accessed 9 August 2025] For further discussion on this, see Fabio Tronchetti and Hao Liu, "The White House Executive Order on the recovery and use of space resources: Pushing the boundaries of international space law?" 57 Space Policy (2021) 101,448.

[36] The full text of the law is available online at: https://space-agency.public.lu/en/agency/legal-framework/law_space_resources_english_translation.html [Accessed 9 August 2025].

[37] Adam Janikowski, "Bold steps forward: The investment impact of enacting space resources legislation" 72 Space Policy (2025) 101,675, Section 2.2.

[38] An example of this can be seen in the Lunar Pathfinder mission, a spacecraft designed to showcase commercial communications services to lunar orbiters and surface craft such as rovers and other instrumentation is due to launch in 2026. For further details see: https://www.gov.uk/government/case-studies/lunar-pathfinder [Accessed 9 August 2025].

4 International Approaches: U.K. Leveraging Existing Regulatory Expertise

If the decision not to direct the legislation and policy track of the U.K. space industry towards space resources is understandable, especially given the need to focus on existing expertise within the sector, there nonetheless exists an opportunity for involvement in the global conversation on the issue. The U.K. has a considerable heritage in the exploitation of terrestrial resources, having built much of its industrial strength on mining coal from across the country, as well as gas and oil from the North Sea.[39] Looking at the governance of space resource utilisation and exploitation through the lens of the legal history of coalmining regulation provides (somewhat ironically) a low-tech, yet novel and instructive series of learnings from the rise of safety regulations in the mining industry.

Of the many lessons that can be learnt from nineteenth Century coal mining governance with safety considerations are uppermost when contemplating any upcoming space resource activity. As space mining has yet to begin as an activity, either on the Moon or other celestial bodies, the key participants could look at the safety, sustainability, and security, through the so-called 3S Approach,[40] of these future undertakings at the national and international levels. As has already been considered, the UK NSS specifically refers to safety, sustainability and security.[41] The U.K. also has a space governance framework underpinned by the Outer Space Act 1986 and the Space Industry Act 2018, as well as various strategies, standards, and guidelines that encourage space safety, sustainability, and security. Furthermore, as will be discussed, the UK has joined the Artemis Accords, which provides '… a common set of principles to enhance the governance of the civil exploration and use of outer space' specifically around lunar missions and operations.[42]

As regards the legal dimensions of the approach to space activity, and as stated above, the U.K. was a founding member of UN COPUOS and has signed and ratified four of the international space treaties: the Treaty on Principles Governing the Activities of States in the Exploration and Use of Outer Space, including the Moon

[39] For a detailed discussion on the regulation of Oil and Gas see Greg Gordon, John Paterson and Emre Üşenmez (eds) UK Oil and Gas Law: Current Practice and Emerging Trends. Vol I: Resource Management and Regulatory Law (3rd Edition, Edinburgh University Press, 2018).

[40] For more on the 3S Approach see: Peter Martinez, Peter Jankowitsch, Kai-Uwe Schrogl, Simonetta Di Pippo, and Yukiko Okumura, 'Reflections on the 50th Anniversary of the Outer Space Treaty, UNISPACE + 50, and Prospects for the Future of Global Space Governance' (2019) 47 Space Policy, 28–33; J. Lauren Napier, 'Outer Space Regime Evolution: The Sustainable Governance of Low Earth Orbit for Future Generations' (PhD thesis, Northumbria University, 2023), Chaps. 4 and 7.

[41] See above (n 19.

[42] NASA,'The Artemis Accords' (*NASA*, 2025) < https://www.nasa.gov/artemis-accords/ > [Accessed 9 August 2025].

and Other Celestial Bodies The Outer Space Treaty,[43] Rescue Agreement,[44] Liability Convention,[45] and the Registration Convention.[46] In common with other major space powers, the U.K. has not signed or ratified the Moon Agreement,[47] nor has the U.K. government shown any interest in acceding to it.[48]

Despite showing no interest in signing the Moon Agreement, the U.K. has shown some interest in international discussions in respect of developing lunar governance. There were several UK participants in the Hague Space Resources Governance Working Group,[49] although the U.K. did not make any draft submission to the Legal Subcommittee of UNCOPUOS General exchange of views on potential legal models for activities in exploration, exploitation and utilisation of space resources.[50] Nonetheless, the U.K. has become more active in UNCOPUOS from the start of the 2020s, and there is evidence of increased engagement in the question of space resources at UNCOPUOS by the U.K.

Already at the international level at the UN there is a Working Group on Legal Aspects of Space Resource Activities. Established in 2021 at the United Nations Committee on the Peaceful Uses of Outer Space (UN COPUOS) 60th session of the Legal Subcommittee (LSC) this working group falls under the LSC agenda item 'General Exchange of Views on Potential Legal Models for Activities in the Exploration, Exploitation, and Utilization of Space Resources'.[51] While the work is still in the development stages, there is already an updated draft (as of July 2025) of recommended principles for space resource activities.[52] The UK does make statements under the agenda item on space resources at the LSC and does participate

[43] Treaty on Principles Governing the Activities of States in the Exploration and Use of Outer Space, Including the Moon and Other Celestial Bodies (adopted 27 January 1967, entered into force 10 October 1967) 610 UNTS 205.

[44] Agreement on the Rescue of Astronauts and the Return of Objects Launched into Outer Space (adopted 22 April 1968, entered into force 3 December 1968) 672 UTS 119.

[45] Convention on International Liability for Damage Caused by Space Objects (adopted 29 March 1972, entered into force 1 September 1972) 961 UTS 187.

[46] Convention on Registration of Objects Launched into Outer Space (adopted 14 January 1975, entered into force 15 September 1976) 1023 UTS 15.

[47] Agreement Governing the Activities of States on the Moon and Other Celestial Bodies (adopted 18 December 1979, entered into force 11 July 1984) 1363 UNTS 3.

[48] Christopher Newman, "Space Law & Policy: The U.K. Approach to the Regulation of Space Activity" (2022) Oxford Research Encyclopaedia of Planetary Science.

[49] UN Committee on the Peaceful uses of Outer Space, Legal Subcommittee, "The Hague Spaces Resources Governance Working Group" A/AC105/C.2/2018/CRP.18 (2018).

[50] UN Committee on the Peaceful uses of Outer Space, Legal Subcommittee, "General exchange of views on potential legal models for activities in exploration, exploitation and utilization of space resources." A/AC105/C.2/2021/CRP.8 (2021). Replies were received from Australia, Austria, Bahrain, Belgium, China, Ethiopia, Finland, Germany, Luxembourg, Mauritius, Mexico, Netherlands, Saudi Arabia and Switzerland.

[51] See: United Nations Office for Outer Space Affairs,'Working Group on Legal Aspects of Space Resource Activities' (*UNOOSA*, 2025) < https://www.unoosa.org/oosa/en/ourwork/copuos/lsc/space-resources/index.html > [Accessed 9 August 2025].

[52] UN Committee on the Peaceful uses of Outer Space, Legal Subcommittee, General exchange of views on potential legal models for activities in exploration, exploitation and utilization of space

in the working group as any Member State of UN COPUOS may deliberate on the topic.

In respect of international discussions on space resources, the role of the Co-Chairs of the Working Group on space resources cannot be understated. They ensure that all voices are heard and keep the group on task and within mandate and workplan to find a path towards a 3S Approach to space resource activities governance. The UK has made representations to UN COPUOS in respect of making space activities safe, sustainable, and secure. The UK statement on space resource activities at the 64th session of UN COPUOS LSC in 2025 articulates this:

> We welcome principles on important areas such as, ensuring compliance with existing international law, definition of space resources, ensuring appropriate authorisation and supervision of non-governmental actors, safety, sustainability, information sharing, review and development of principles.[53]

This remark is related to the working group, as it is currently deliberating on an updated draft set of recommended principles for space resource activities during UN COPUOS sessions and via intersessional meetings. Note that the UK has mentioned safety as well as governance, including legal compliance and supervision of non-State actors. This is very much in line with how the UK used coal mining legislation[54] and the inspectorate from the 1850s. Echoes of this can be seen in contemporary space legislation and the way in which space governance is dealt with by the UK.

The updated draft set of recommended principles for space resource activities (as of 2025) includes two parts. Part 1 is the updated draft, and Part 2 sets out possible additional draft provisions. As the work on the document is a work in progress, the document sets out original text, track changes made during working group meetings, and proposed text submitted by working group members.[55] As the document is still in draft mode with many more meetings to be held for deliberation on this topic, there is only a general understanding that can be made now. What is germane to this discussion is that compliance with existing law, as well as safety concerns, is

resources. "Initial draft set of recommended principles for space resource activities" A/AC.105/C.2/L.339 available online at: https://docs.un.org/en/A/AC.105/C.2/L.339 [Accessed 9 August 2025].

[53] United Kingdon,'United Kingdom, Item 8- Legal Models for Activities in the Exploration, Exploitation and Utilization of Space Resources' (*UNOOSA*, 2025, United Nations COPUOS 64th session of the Legal Subcommittee under Agenda Item 8 on Wednesday 7 May 2025 PM session, 1079th meeting) < https://www.unoosa.org/documents/pdf/copuos/lsc/2025/Statements/Wednesday7PM/8_United_Kingdom.pdf > [Accessed 9 August 2025].

[54] Lancaster (n 5) 78–79 and also Report from the Select Committee of the House of Lords, The Best Means of Preventing the Occurrence of Dangerous Accidents in Coal Mines (31 July 1849).

[55] UNOOSA,'Updated Draft Set of Recommended Principles for Space Resource Activities' (*UNOOSA*, 7 July 2025) < https://www.unoosa.org/oosa/en/ourwork/copuos/lsc/space-resources/index.html > [Accessed 9 August 2025]. Note: as of August 2025, this document is still in draft form under the working group deliberations.

addressed.[56] Further safety concerns are under consideration in Part 2 as possible additional provisions; however, the outcome of Part 2 remains to be seen as of 2025.[57]

It would be disingenuous to say that there has been no engagement by the U.K. on the question of space resources. Indeed, the U.K. was one of the founding state signatories[58] to the U.S.-led Artemis Accords,[59] an international framework for cooperation in the exploration and exploitation of the Moon and other celestial bodies. Rather than replace existing treaty arrangements, the Accords are intended to expand upon and operationalise the provisions of the Outer Space Treaty. Classified as falling within the realm of 'adaptive governance',[60] they state that extraction and utilisation of space resources (including any recovery from the surface or subsurface of the Moon) and such activity do not inherently constitute national appropriation under Article II of the 1967 Treaty.[61]

By signing the Accords, the U.K. has indicated an interest and willingness to be a part of the Artemis program. More fundamentally, it has also clearly decided that resource extraction is permitted by the Outer Space Treaty, based on 'usage' under Art I, rather than being prohibited as under Art II. It is not within the purview of this discussion to revisit and rehearse the different positions.[62] Suffice to say that signing the Accords was possibly more a reflection of the economic reality of being involved in a lucrative U.S.-led programme rather than of any fundamental realignment of the U.K.'s priorities in respect of space activity.

---

[56] UNOOSA,'Updated Draft Set of Recommended Principles for Space Resource Activities' (*UNOOSA*, 7 July 2025) < https://www.unoosa.org/oosa/en/ourwork/copuos/lsc/space-resources/index.html > [Accessed 9 August 2025], Part 1, Principle 1 and Principle 4.

[57] Ibid Part 2, Principle 4.

[58] National Aeronautics and Space Administration, "International Partners Advance Cooperation with First Signings of Artemis Accords" (NASA.gov, Media release 20–097, 13 October 2020). Available at: https://www.nasa.gov/press-release/nasa-international-partners-advance-cooperation-with-first-signings-of-artemis-accords [Accessed 9 August 2025].

[59] National Aeronautics and Space Administration, "The Artemis Accords: Principles for Cooperation in the civil exploration and use of the Moon, Mars, Comets and Asteroids for Peaceful Purposes." (NASA, The Artemis Accords, 13 October 2020). Available at: < https://www.nasa.gov/specials/artemis-accords/index.html > [Accessed 9 August 2025].

[60] Rossana Deplano, "The Artemis Accords: Evolution or Revolution in International Space Law?" 70 ICLQ (2021) 799–819, 819.

[61] Artemis Accords, Section 10(2).

[62] For a fuller discussion, see Fabio Tronchetti, "Legal Aspects of space resource utilization" in Frans von der Dunk with Fabio Tronchetti (eds), *Handbook of Space Law*, (Edward Elgar, 2017), 769–813.

5 From Coal and Oil to Space Resource Activities: Regulation from Tragedy

The engagement of the U.K. within UNCOPUOS and the international community on safety and sustainability while not directly linked, clearly has echoes in the Coal mining safety practices established in Britain in the 1850s. In the nineteenth century, industrial risks and mitigation measures were initially poorly understood and primarily unregulated. The legislative work regulating the coal sector pioneered the use of hazard reporting mechanisms and the maintenance of technical standards.[63] The introduction of the coal mining inspectorate through legislation[64] marked a significant shift towards governance with safety at its core. This incremental, iterative approach offers legacy insights towards emerging space resource governance.

Another legacy industry that the UK has experience in regulating is the North Sea oil and gas offshore industry. This industry operates with safety standards, which are also shaped by lessons learnt from disasters. The Piper Alpha disaster of 1988[65] saw an explosion onboard an oil rig in the North Sea result in the death of 165 crew. This ultimately led to a public inquiry[66] and increased safety measures within the industry, showing another example of high-risk activity where safety regulations are critical to a 3S Approach to activities.

Like the British coal mining industry, the UK offshore oil and gas industry follows a safety and risk approach through government oversight, regulations, and technical standards. Currently, within the U.K., the Health and Safety Executive (HSE)[67] oversees offshore oil and gas safety through regulations such as the Offshore Safety Act 1992 and the Health and Safety at Work etc. Act 1974 (Application outside Great Britain) Order 2013. The HSE offers robust guidance on offshore health and safety law. The HSE acts as an authority that can ensure offshore oil and gas activities and risks are properly controlled through inspection, investigation of accidents and incidents, assessment of safety cases, and enforcement of regulations. The HSE can also advise, guide, and inform on offshore activities. Further, the HSE conducts research and development and influences technical standards and policies in the sector.

The guidance, legislation and investigatory function undertaken by the HSE is not the only regulatory voice in the U.K. The UK Office of Communications (Ofcom) provides detailed guidance on licensing and regulations for RF spectrum utilisation

---

[63] Eur Ing B Job, 'The Formation and Early Development of the Mines Inspectorate' (1994) 153 (390) *The Mining Engineer,* 249–255.

[64] Inspection of Coal Mines in Great Britian Act 1850 (13 + 14 Vic), Section VII.

[65] See: Fiona Macleod and Stephen Richardson, 'Piper Alpha: The Disaster in Detail' (*The Chemical Engineer,* 06 July 2018) < https://www.thechemicalengineer.com/features/piper-alpha-the-disaster-in-detail/ > accessed 08 August 2025.

[66] Department of Energy, *The Public Inquiry into the Piper Alpha disaster,* by Lord Cullen, (HMSO, 1990) available online at https://www.hse.gov.uk/offshore/assets/docs/piper-alpha-public-inquiry-volume1.pdf.

[67] Health and Safety Executive, 'Offshore Oil and Gas' (HSE, 2025) < https://www.hse.gov.uk/offshore/index.htm > accessed 08 August 2025.

for satellites and space activities. In the UK, there are numerous authorities (like aviation[68] and maritime[69] as well) where guidance, tech and safety standards, policies, and regulations are overseen by a competent UK government authority.[70] Suggesting that a strong legacy of licensing and regulation regarding high-risk scientific and technical industries is governed in the UK.

Giving the opportunity for many sectors to knowledge share, discuss best practices, and offer lessons learnt with each other, which can then be scaled up to the international level by the UK, offering the same to the relevant committees like UN COPUOS for space activities. At a time when space resource activities have yet to become reality, the U.K. is in a strong position to offer leading lessons learnt and knowledge sharing about high-risk sectors and the importance of safety concerns being addressed through governance frameworks, which could mitigate risk in future space resource activities. The U.K. is trying to position itself as a global leader in space, cyber, and AI.[71] It could facilitate the discussion on lessons learnt from the sectors with strong safety compliance built into the governance frameworks, such as offshore oil and gas or coal mining, as has been discussed here.

The importance of early regulatory intervention in high-risk industries is clear, but it is also undeniable that robust governance tends to emerge *only* following a disaster. This is not a phenomenon that is unique to mining. The 2011 Deepwater Horizon oil spill,[72] the 1988 Piper Alpha fire[73] and, in the space sector, following the US Apollo 1 and 13 disasters as well as the Challenger and Columbia space shuttle disasters. The lessons from these events are a crucial element of establishing the governance framework for resource extraction. There will be considerable pressure to not only re-establish human presence quickly but to ensure that any mining is profitable. One of the key lessons taken from all the above is the danger of prioritising short-term

---

[68] The regulator for both aviation and spaceflight activities is the Civil Aviation Authority for details see: https://www.caa.co.uk/

[69] UK regulation of British and international maritime law and safety is the Maritime and Coastguard Agency (MCA) for details see https://www.gov.uk/government/organisations/maritime-and-coastguard-agency.

[70] The Office for Nuclear Regulation is the regulatory body for the nuclear industry in the U.K. It slightly differs from the others in that it is an independent statutory corporation funded by fees charged to the nuclear industry. For details see: https://www.onr.org.uk/

[71] See above (n13).

[72] The Deepwater Horizon oil rig explosion in the Gulf of Mexico in April 2010 led to the deaths of 11 workers, a significant oil leak that required a $40bn clean up. It resulted in a six-month moratorium on deepwater drilling in the Gulf of Mexico. For further details on this see Nathan Richardson, Molly Macauley, Mark Cohen, Robert Anderson and Adam Stern, "Managing Risk through Liability, Regulation, and Innovation: Organizational Design for Spill Containment in Deepwater Drilling Operations." 2 (2011) Risk, Hazards & Crisis in Public Policy, 2, 1–22.

[73] Department of Energy, *The Public Inquiry into the Piper Alpha disaster*, by Lord Cullen, (HMSO, 1990) available online at https://www.hse.gov.uk/offshore/assets/docs/piper-alpha-public-inquiry-volume1.pdf.

economic gains over long-term environmental and safety.[74] Given the depth of the legacy of coal mining and drilling for oil, it behoves the U.K. to engage with the space resources discussions, to promote safe and responsible regulation.

When translating these lessons to space, the U.K. is well-positioned. Advocating the embedding of professional oversight and regulatory monitoring into international space resource utilisation frameworks would greatly enhance the discourse in this area. Whilst there will undoubtedly be a significant amount of automation in any space mining venture, there must be protection for any human workers who are involved. These protections should be integrated into the design process rather than retrofitted following a tragedy. Evaluating the approach of industries, such as mining, to the question of safety and the subsequent legislative responses will help inform mitigating efforts for the future of space activities—especially when they will be conducted on the Moon or other celestial bodies, which are further away with possible lags in real-time communication with mission control centres back on Earth.

One of the key lessons that can be drawn from the evolution of the British coal sector is the pioneering developments regarding the need for a robust system of safety inspection.[75] A crucial aspect of this was the independent nature of the inspectorate to give safety oversight. According to records from the Northeast of England, before the independent inspectorate was established, mines and collieries had viewers. Viewers were part of senior management and were responsible for supervising mine operations, managing workers, and advising the owner.[76] One of the many criticisms of Viewers was their impartiality and judgement. They often had shares in the mine, and this had the potential to adversely influence their views on safety matters, especially when it might undercut profits. The Inspection of Coal Mines Act 1850[77] introduced an inspection and reporting regime which placed a legal duty on mine owners to facilitate independent inspection.[78]

When conducting the review of the proposed regulations to accompany the Space Industry Act 2018, the recognition of the need for independence of accident investigations was considered. The provision for the creation of the Space Accident Investigation Authority (SAIA)[79] that operates independently[80] and the appointment of

[74] Howard C. Kunreuther and Erwann O. Michel-Kerjan, "Learning from the BP Oil Spill and Other Catastrophes." The Milken Institute Review, (2010) Fourth Quarter 48–57 available online at: https://impact.wharton.upenn.edu/wp-content/uploads/2019/07/J2010Milken_OvercomingMyopia.pdf [Accessed 9 August 2025].

[75] Eur Ing B Job, 'The Formation and Early Development of the Mines Inspectorate' (1994) 153 (390) *The Mining Engineer,* 249–255, 251.

[76] Many thanks to Jennifer Hillyard at the North of England Institute of Mining and Mechanical Engineers for her expertise and explanation about 1800s coal mining operations and safety matters in the North East of England and Great Britain. The contemporary equivalent of Viewers would be a senior mining engineer and site manager.

[77] Inspection of Coal Mines in Great Britian Act 1850 (13 + 14 Vic), Section II.

[78] In line with the ICM Act 1850, government created a select committee within the UK Parliament in 1852 for reporting on the Mines Inspectorate.

[79] Space Industry (Investigations of Spaceflight Accidents) Regulations 2021, s5.

[80] Space Industry (Investigations of Spaceflight Accidents) Regulations 2021, s5(2).

Inspectors of spaceflight accidents[81] can be found in Part 2 of the Space Industry (Investigations of Spaceflight Accidents) Regulations 2021. While having an independent inspectorate body will not eliminate the risk of accidents in space resources activities, it would at least offer impartial input into operations with law, standards, and best practices underpinning inspection recommendations. By advocating independent safety inspection and quality assurance measures as part of a space resources framework, the UK would demonstrate leadership in the regulatory field and support legal accountability and international transparency.

Technical and scientific advancements can also support safety standards in high-risk activities. In the 1850s, the introduction of the safety lamp, designed to be tamper-proof and create better safety measures against gas explosions, was tied to the welfare and safety of miners.[82] Today, support for safety standards and regulations in space resource activities can be found in the inclusion of robotics and artificial intelligence (AI). In addition to having legislation around inspection and general safety, there can also be technical standards and regulations around technology, such as can be seen today with the use of the International Standardisation Organisation (ISO) technical standards, which are also promulgated through British standards in the UK. Having technical advancements and standards can support safety measures and build resilience against error and misuse.

One of the crucial social and economic elements that informed the development of nineteenth century coal mining safety was the enhancement of miners' rights.[83] The support for safe working conditions, miners' welfare, health, and fair compensation was a key element in the push to modernise coal mining. There was a steady expansion of legislation which put in place safeguards on matters such as rules governing working hours, the requirements for ventilation, compensation for accidents and provision for the formation of trade unions.[84] As stated above, there will be a considerable amount of automation embedded within space mining, but should any humans be involved, they will face very similar conditions, specifically physical isolation, exposure to hazardous conditions, and a dependence on safe and reliable machinery.

The updated draft set of Recommended Principles for Space Resource Activities currently suggests:

---

[81] Space Industry (Investigations of Spaceflight Accidents) Regulations 2021, s6.

[82] For more information about safety lamps see: Matthias Dunn, *A Treatise on Winning and Working of Collieries Including Numerous Statistics and Remarks on Ventilation* (Dunn, 1848) from page 177.

[83] The Coal Mining Act 1860 introduced Sections I and II which prohibited boys under 12 years of age from working in mines and collieries and very tight exceptions for boys between 10 and 12 years of age. While today, there are much more stringent employment laws around workers under the 18 years of age, this does add to the general safety principles in Section X of the 1860 Act (and were also in the 1855 Act).

[84] Under Section XIII of the 1860 Act, the Secretary of State may propose alterations or additions to the special rules that owners could adopt on safety on top of the general rules stated in the regulations if the Secretary "… be of the opinion that such rules, or any of them, do not sufficiently provide for the safety of the persons employed in or about such mine or colliery…"An Act for the Regulation and Inspection of Mines 1860 (23 + 24 Vic), Section XIII.

> (A) Space resource activities shall be conducted in a safe manner and with priority for human life and health. (B) States [conducting space resource activities] shall take all possible steps to rescue and render all necessary assistance to any person in the case of accident, distress or emergency landing on the Moon and other celestial bodies, or in any other circumstance, or due to any phenomena, which could constitute a danger to the life or health of that person.[85]

This regard for human safety and health is to be lauded as it is directly reflected in the Guidelines for the Long-Term Sustainability of Outer Space.[86] Although it is outside the scope of the governance framework for resources, there are clearly technical and scientific lessons that can be learnt from the development of coal mining safety standards. Aside from incorporating these into governance discussions, engineers and scientists will also benefit from discussions on improving the quality of life for those living and working on the Moon and other celestial bodies. An example might be the mitigation measures put in place for coal mine roof and shaft collapses. These could bring insight into habitat structure mitigation measures against failures. The literature on ventilation issues in mines and collieries[87] could support critical understanding around live support systems for astronauts and space resource workers.

Labour and workers' welfare could also be discussed around the ethical and legal aspects of including robotics and AI in space resource activities.[88] Finally, independent oversight on safety procedures can lead to national and international laws and licensing procedures, which are already being seen in the space governance framework around space debris, satellite activities, and other space activities. As can be seen, looking to coal mining in 1850s Great Britain offers an abundance of lessons learnt which can support the 3S Approach to space resource activities in future.

---

[85] UNOOSA,'Updated Draft Set of Recommended Principles for Space Resource Activities' (*UNOOSA*, 7 July 2025) < https://www.unoosa.org/oosa/en/ourwork/copuos/lsc/space-resources/index.html > accessed 08 August 2025, Part 1, Principle 4.

[86] See: Treaty on Principles Governing the Activities of States in the Exploration and Use of Outer Space, including the Moon and Other Celestial Bodies (open for signature 27 January 1967, entered into force 10 October 1967) 610 UNTS 205 (also known as the Outer Space Treaty or OST); Agreement on the Rescue of Astronauts, the Return of Astronauts and the Return of Objects Launched into Outer Space (open for signature 22 April 1968, entered into force 03 December 1968) 672 UNTS 9574 (also known as the Rescue Agreement); United Nations General Assembly Report of the Committee on the Peaceful Uses of Outer Space Sixty-second Session (12–21 June 2019) UN Doc A/74/20 Annex II Guidelines for the Long-Term Sustainability of Outer Space Activities of the Committee on the Peaceful Uses of Outer Space 50–69 (also known as the Long-Term Sustainability Guidelines or LTSG).

[87] For a comprehensive review of the available literature, see Yan Xue, Jinmiao Wang, Jun Xiao, "Bibliometric analysis and review of mine ventilation literature published between 2010 and 2023" 10 (2024) Heliyon 10 e26133.

[88] Anne Sophie Martin and Steven Freeland, "Artificial Intelligence – A Challenging Realm for Regulating Space Activities" 45 (2020) Annals of Air and Space Law 275–306.

6 Conclusion

The U.K. has been on a transformative journey with respect to its ambitions in space. From a standing start in the early 2000s, through to a credible claim to be a key part of the global space economy in the middle of the 2020s. The prospect of independent access to space through launch from U.K. soil, a focus on in-orbit servicing and manufacturing (ISAM), as well as active debris removal,[89] all point to a vibrant British space sector. Yet despite this, the U.K. has not passed any specific legislation directly addressing the utilisation of space resources. The various space strategies and policies have not mentioned space mining, and there is a clear focus on other areas of activity.

The future does not augur well for advocates of U.K. involvement in space mining. The 2025 Industrial Strategy, while looking at 'scaling up' the U.K. footprint in the global space economy, is clear that funding will be targeted on technology that will enable ISAM and debris removal and move space technology from research and development into market-ready solutions.[90] The Department for Science, Innovation and Technology (DSIT) is conducting a wide-ranging review of the space regulatory landscape, seeking novel, innovative and agile regulatory solutions for space.[91] This review and the regulatory sandboxes to test out these new solutions tend to focus on ISAM and debris removal.

Nonetheless, this does not mean that the U.K. has no role to play in respect of the emerging law and policy framework regarding the utilisation of space resources. The U.K. is already providing a voice for 'responsible' space behaviour,[92] and this can extend into the realm of space mining. While the national space sector is focusing on its strengths and gaining access to space, diplomatic efforts can leverage existing regulatory expertise as well as the industrial heritage of previous terrestrial resource exploitation. The regulatory journey of coal mining, although harking back to a time long since passed, can be used to inform the global discourse.

The lessons to be learned are not easy ones for an industrial base eager to start operations beyond Earth and not wanting to be tied down by regulatory obstacles. But the experience of the U.K. in respect of coal mining and oil and gas extraction illustrates the value of considering the issues of human safety and dignity, and careful environmental stewardship while the infrastructure is being developed. Drawing the lessons from nineteenth century Britain will allow the global discussion on resources to learn from the past. By ensuring that regulation is pragmatic but rigorous to prevent tragedy, to create a safer, and more sustainable resource industry that will truly be for the benefit of all humans.

---

[89] See above (n 19).

[90] https://assets.publishing.service.gov.uk/media/68595e56db8e139f95652dc6/industrial_strategy_policy_paper.pdf.

[91] https://assets.publishing.service.gov.uk/media/6644d26fb7249a4c6e9d3597/space_regulatory_review_2024.pdf.

[92] See above (n 22).

Prof. (Dr.) Christopher J. Newman is a Professor of Space Law and Policy at Northumbria University at Newcastle in the UK. He is active in the teaching and research of space law and has published extensively on the legal and ethical underpinnings of space governance. Prof. Newman has been Principal Investigator on several grant-funded research projects and is a member of the International Institute of Space Law. He has presented his research to numerous bodies including COSPAR, UNIDIR and the UK Space Agency. He has also been an observer member of the UK delegation to the UN COPUOS. He was appointed Visiting Professor of Space Law at the Open University working on the legal and ethical dimensions of planetary protection. He is an academic consultant to 3SNorthumbria, a leading Space Situational Awareness (SSA) consultancy company. Prof. Newman is a member of the Open Lunar Foundation Lunar Ledger Advisory Board and has made numerous TV and radio appearances in the UK speaking as an expert on space law and policy issues.

Dr. J. Lauren Napier is an Assistant Professor in the Northumbria Law School at Northumbria University at Newcastle in the UK. Dr Napier completed her PhD on the intersection of international law and international relations in respect of the governance of low Earth Orbit. She teaches on the Space Law Masters programme at Northumbria University and is currently programme leader for the Masters framework within the Law School. Dr. Napier has also contributed to the review of the UK space spectrum strategy. In addition, she has recently been part of the UK delegation to UN COPUOS, speaking as part of the Women In Space (WISPER) project.

Japan's Space Resources Act: A Legal Approach to Responsible Space Resource Utilisation

Yumiko Takatori

Abstract This chapter provides an overview of the background, content, and implementation of Japan's Space Resources Act (Act on the Promotion of Business Activities for Exploring and Developing Space Resources). It also examines the significance and challenges of Japan's legal framework in the development of international space law. As various government agencies and private entities continue to plan missions for space resource exploration and sample return, establishing international rules for space resource development has become an urgent issue. The Japanese government is actively promoting space resource exploration policies through technological innovation, international cooperation, and national legislation. This chapter first reviews the history of Japan's space development and international cooperation, followed by an overview of the evolution of Japan's space law and policy. It then examines the rationale behind the enactment of the Space Resources Act. In 2021, Japan enacted the Space Resources Act, which establishes a legal framework for private-sector activities in space resource exploration and development while ensuring compliance with international space law, including the Outer Space Treaty. Thus, this chapter analyses the key provisions and challenges of this eight-article law and examines the current implementation status, including the two licences granted under this Act as of February 2026. Finally, this chapter explores the future of space resource exploration and considers the potential role of Japan's legal approach in the development of international regulatory frameworks for responsible space resource utilisation.

Keywords Japan · Space resources act · International cooperation · International space law · Private sector participation · Space resource regulation

---

Y. Takatori (✉)
Takatori Law Office, Tokyo, Japan

S. Bhat B and A. Variath (eds.), *Legal Responses to Space Resource Utilisation*, Space Law and Policy, https://doi.org/10.1007/978-981-95-8864-0_13

1 Introduction

This chapter aims to clarify the background,[1] objectives, content, implementation status and challenges of the Act on the Promotion of Business Activities for Exploring and Developing Space Resources (Space Resources Act: hereinafter referred to as the SRA) of Japan,[2] while also examining the institutional and normative roles that Japan's legal approach may play in the development of an international regulatory framework for the responsible utilisation of space resources. Exploring and developing space resources has recently emerged as a field of significant interest for governmental agencies, private enterprises, and international cooperation projects.

The development of space resources is a task that cannot be achieved by a single nation alone, and international cooperation is therefore of vital importance. The Government of Japan stated that its 'participation in international space exploration is driven by a commitment to expand human frontiers, to advance the well-being of humanity and to foster international cooperation'.[3] Moreover, space resource development has garnered attention as a new economic sphere, and many private enterprises, including the Japanese startup ispace Inc. (ispace),[4] have continued to enter the field with the aim of engaging in commercial space resource development.

In Japan, there is a growing movement towards private-sector-led industrial creation, with a variety of private enterprises engaging in lunar surface activities across sectors such as transportation, information and communications, media and content, resource and energy, construction and infrastructure, food and biotechnology, and finance and insurance. To support this trend, the Lunar Industry Vision Council (LIVC), established in 2021 and comprising political, academic, and industry representatives in Japan, aims to foster a cislunar ecosystem centred on industrialisation. According to LIVC, more than 100 companies have expressed interest in

[1] Saadia M Pekkanen, Setsuko Aoki and Yumiko Takatori, 'Japan in the New Lunar Space Race' (2023), *Space Policy* <https://www.sciencedirect.com/science/article/abs/pii/S0265964623000449> 2–3 accessed 21 March 2026; Saadia M Pekkanen, Setsuko Aoki and Yumiko Takatori, 'Japan's Strategy in the New Lunar Space Race' in Thomas Hoerber, Mariel Borowitz, Antonella Forganni and Bruno Reynaud de Sousa (eds), *Routledge Handbook of Space Policy* (Routledge 2024) 561–564.

[2] Act on the Promotion of Business Activities for the Exploring and Developing Space Resources (Act No 83 of December 23, 2021) <https://www.japaneselawtranslation.go.jp/ja/laws/view/4816/en> accessed 21 March 2026.

[3] Japan, 'Statement at the 64rd Session of the Legal Subcommittee of COPUOS, Agenda Item 8: General exchange of views on potential legal models for activities in the exploration, exploitation and utilization of space resources' (8 May 2025) <https://www.unoosa.org/documents/pdf/copuos/lsc/2025/Statements/Thursday8AM/8_Japan.pdf> accessed 21 March 2026.

[4] Ispace Inc., 'About' <https://ispace-inc.com/aboutus> (accessed 21 March 2026); ispace, 'ispace Updates Timing for Expected Touchdown on the Lunar Surface to 4:17 a.m. JST' <https://ssl4.eir-parts.net/doc/9348/tdnet/2633951/00.pdf> accessed 21 March 2026.

launching lunar-related businesses and services.[5] These activities are likely to evolve into major investment projects in the future.

Given the finite nature of terrestrial natural resources, the utilisation of water and mineral resources found on asteroids, the Moon, and Martian moons holds the potential to reduce the cost of space exploration and facilitate the construction of space bases, representing a pivotal turning point in humanity's advancement into space. However, as of 1 January 2025, under the Treaty on Principles Governing the Activities of States in the Exploration and Use of Outer Space, including the Moon and Other Celestial Bodies (OST), which has been ratified or acceded to by 116 States, the rules of international law governing the acquisition and use of space resources remain unclear.

The international debate was triggered in November 2015, when the United States of America (US) enacted the Commercial Space Launch Competitiveness Act, whose Title IV, the Space Resource Exploration and Utilization Act of 2015, became the first national legislation to recognise private ownership of space resources. Since then, discussions concerning the ownership of space resources and their extraction for commercial purposes have been actively ongoing. In March 2017, the Legal Subcommittee (LSC) of the United Nations Committee on the Peaceful Uses of Outer Space (UN COPUOS) formally included the item 'General exchange of views on potential legal models for activities in exploration, exploitation and utilization of space resources' in its agenda. In 2021, the Working Group on Legal Aspects of Space Resource Activities was established within the LSC, with the aim of formulating a set of initial principles by 2027 for consideration by UN COPUOS. In March 2025, the 'Initial draft set of recommended principles' was circulated to Member States and subsequently taken up as working document within the LSC, prompting discussions among them and further increasing international attention to the legal framework, prompting discussions among them and further increasing international attention to the legal framework, reflecting the gradual evolution of initial normative approaches.

In June 2021, Japan enacted the SRA, which aims to ensure the accurate and smooth implementation of conventions concerning the development and use of outer space, and to promote business activities for exploring and developing space resources by private business operators. The Act establishes special provisions for licensing under the provisions of the Act on Launching Spacecraft and Launch Vehicle and Control of Spacecraft (Space Activities Act: hereinafter referred to as the SAA).[6] In addition, under certain conditions, licensees shall acquire ownership of the space resources they have mined or extracted, in accordance with their licensed business activity plan. This Act aims to provide legal stability to entities engaged in space resource activities, on the premise of compliance with international space law.

[5] Lunar Industry Vision Council (LIVC), 月面産業ビジョン2024[*Lunar Industry Vision 2024*] (July 2023) 16–21, <https://www.livc.jp/_files/ugd/388a80_6a9b5493178d406ab730c9c4f6524445.pdf> accessed 21 March 2026.

[6] Act on Launching Spacecraft and Launch Vehicle and Control of Spacecraft (Act No. 76 of November 16, 2016) <https://www.japaneselawtranslation.go.jp/ja/laws/view/4813/en> accessed 21 March 2026.

Japan became the fourth country in the world to enact national legislation on space resources, following the US, Luxembourg and the United Arab Emirates (UAE).

In Japan, the prevailing view is that the SRA is national legislation that implements and complements international space law.[7] This chapter is structured into four sections. First, it provides an overview of the history of Japan's space development, including its international cooperation, and explains the historical context in which Japan's space law and policy were formed. Second, it analyses the legislative background and policy objectives of the SRA, with attention to the growing policy focus on the exploration and use of space resources. Particular attention is given to the influence of the US' space law and policy and the business activities of ispace. Third, it explains the structure of the SRA, which consists of eight articles, and examines the implementation status of the two licences granted to ispace as of February 2026. It also analyses institutional challenges from a practical perspective. Finally, it examines the institutional and normative significance of Japan's legal framework and initiatives in the development of international rules for the responsible utilisation of space resources, as well as the challenges to be addressed in realising that contribution.

2 Formation of Japan's Space Law and Policy

This section provides an overview of how Japan's space development has evolved from scientific and academic research in the aftermath of World War II into a comprehensive legal and policy framework to promote the development and use of outer space in a strategic manner, driven by scientific and technological advancement and international cooperation.

2.1 History of Japan's Space Development and International Cooperation

Japan's space development began as a form of academic research. Following its defeat in World War II, Japan was prohibited from conducting aerospace-related research and development until the restoration of sovereignty under the Treaty of Peace with Japan, which was signed in 1951.[8] The successful vertical launch of the Pencil Rocket at the University of Tokyo, in 1955, marked the de facto starting

[7] 佐藤雅彦・税所大輔 [Masahiko Sato and Daisuke Saisho], 「宇宙探査と宇宙資源開発の法」 ['Laws on Space Exploration and Space Resource Development'], 法学教室[*Hogaku Kyoshitsu*]No497, 有斐閣2022年, [Yuhikaku 2022] 57.

[8] 青木節子[Setsuko Aoki], 「宇宙ビジネスを支える宇宙法」 ['Space law supporting the space business'], 小塚荘一郎・佐藤雅彦編著[in Souichirou Kozuka and Masahiko Sato (eds)], 『宇宙ビジネスのための宇宙法入門』 [*Introduction to Space Law for Entrepreneurs*], 第3版 [3rd edn], 有斐閣, 2024年 [Yuhikaku 2024] 160.

point of Japan's space development. At this stage, space development was mainly regarded as part of academic research and was not considered a national strategy. In 1970, Japan successfully launched its first spacecraft, Ohsumi, aboard an L-4S rocket developed by the University of Tokyo. This made Japan the fourth country to independently launch a spacecraft, following the Union of Soviet Socialist Republics (U.S.S.R.), the USA, and France.

There was a growing need to establish a system for promoting space development as a national policy, and in 1969, the National Space Development Agency of Japan (NASDA), the predecessor of the Japan Aerospace Exploration Agency (JAXA), was founded. During parliamentary deliberation on the Law Concerning the National Space Development Agency of Japan (NASDA Act), the purpose of the agency was limited to 'peaceful purposes', and a supplementary resolution was adopted to restrict space development to 'non-military' objectives. As a result, Japan was politically constrained to focus on civilian-use spacecraft and pure scientific exploration.

In parallel, space research was also conducted by the Institute of Space and Astronautical Science (ISAS) and the National Aerospace Laboratory (NAL), each with distinct institutional cultures. ISAS originated from early rocket experiments at the University of Tokyo and, in 1981, was reorganised as an inter-university research institute open to universities across Japan. NAL, founded in 1955, added a space division in 1963 and advanced research on aerospace transport systems, supporting Japan's development of aircraft and rockets. In 1985, the Japanese government issued a 'unified interpretation' presenting the so-called 'generalisation theory', which permitted the use of spacecraft by the Self-Defence Forces (SDF) as long as such use did not involve lethal or destructive purposes.[9] This interpretation enabled the utilisation of space for security purposes. The theory was later applied to exports of rockets and components, allowing Japan to align its security use of space with the principle of peaceful purposes.

In the late 1980s, the US designated Japan as a priority watch country under Section 301 of the 1988 Omnibus Trade and Competitiveness Act, focusing on three areas of concern, including spacecraft. In response, Japan voluntarily required public entities to conduct international competitive bidding for the procurement of non-R&D spacecraft and concluded the Japan–US Spacecraft Procurement Agreement in 1990. While this led to increased market access for foreign companies, it posed long-term challenges for Japan's domestic space industry.[10]

Japan's first lunar exploration mission was independently conducted by ISAS. The spacecraft MUSES-A (renamed HITEN), launched in 1990, successfully demonstrated swing-by navigation between Earth and the Moon. In 2003, NASDA, ISAS, and NAL were merged to form JAXA. JAXA has since been positioned as the central implementation body supporting the nation's space development and utilisation from a technological perspective.

---

[9] 青木節子 [Setsuko Aoki], 『日本の宇宙戦略』 [*Japan's Space Strategy*], 初版[1st edn], 慶應義塾大学出版会, 2006年, [Keio 2006] 180–181.

[10] 鈴木一人 [Kazuto Suzuki], 『宇宙開発と国際政治』 [*Space Development and International Politics*], オンデマンド版 [on-demand version], 岩波書店, 2016年 [Iwanami 2016] 187-189.

As the first full-scale lunar exploration mission since HITEN, JAXA launched the lunar orbiter Kaguya (SELENE) in 2007. Kaguya demonstrated orbital insertion technologies for gravitational celestial bodies other than Earth and substantially updated existing knowledge of the Moon through high-precision composite observations.

In January 2024, the Smart Lander for Investigating the Moon (SLIM) achieved a high-precision lunar landing, paving the way for the establishment of practical lunar landing technologies. This marked Japan's first successful lunar landing and made it the fifth country to do so, following the Soviet Union, the US, China, and India. In the field of asteroid exploration, Hayabusa, launched in 2003, became the world's first mission to successfully return samples from an asteroid to Earth. Its successor, Hayabusa2, reached asteroid Ryugu in 2018, conducted two touchdown operations in 2019 to collect samples, and returned them safely to Earth in 2020. Moreover, international cooperation has been a cornerstone of Japan's space policy. Japan's participation in the International Space Station (ISS) project, through the Kibo module, exemplifies its contribution to multilateral projects. In 2020, Japan launched the UAE's Mars probe aboard its H-IIA rocket, reflecting broader international collaboration. In addition, JAXA is cooperating with the Indian Space Research Organisation (ISRO) on the Lunar Polar Exploration (LUPEX) mission, which investigates the presence of water and potential resource utilisation in the Moon's polar regions. JAXA is also developing the Martian Moons eXploration (MMX) mission with the National Aeronautics and Space Administration (NASA), the Centre National d'Études Spatiales (CNES), the German Aerospace Center (DLR), and the European Space Agency (ESA), to collect samples from one of the Martian moons.

Capacity-building efforts by universities and related organisations are also active. For example, UNISEC-Global, a non-profit founded in Japan in 2013, provides practical training to developing countries and promotes international cooperation through annual meetings attended by over 130 participants from approximately 33 countries.[11] Japan has long maintained a close partnership with the US. The Japan-US Space Cooperation Agreement (1969) was a foundational step. In 2020, Japan joined the Artemis Accords as an original signatory. This was followed by a Memorandum of Understanding with NASA on lunar orbital cooperation, and in 2023, a Framework Agreement on peaceful space cooperation. In April 2024, Japan and the US signed an Implementation Arrangement for Lunar Exploration, and both governments agreed that two Japanese astronauts would participate in lunar missions, marking progress in Japan's human spaceflight programme. Thus, Japan has steadily developed its technical capabilities in space resource exploration. Through continuous technological efforts, it has advanced its involvement in lunar and asteroid missions and has maintained its status as a cooperative spacefaring nation through sustained participation in international cooperation.

[11] UNISEC-Global, 'Home', <https://unisec-global.org/> accessed 21 March 2026.

2.2 Development of Japan's Space Law and Policy

Japan ratified the OST in 1967. However, no domestic legislation was immediately enacted to implement its obligations. In 1983, Japan acceded to three other UN treaties on outer space—the Agreement on the Rescue of Astronauts, the Return of Astronauts and the Return of Objects Launched into Outer Space; the Convention on International Liability for Damage Caused by Space Objects; and the Convention on Registration of Objects Launched into Outer Space—through a single legislative act. The need for implementing domestic legislation was recognised and bills were drafted, yet their submission to the Diet was ultimately postponed.

One of the reasons for this was the Japanese government's understanding at the time that private companies were not expected to engage in rocket launches, and that all spacecraft operations were conducted by national research institutions or by entities with a special relationship to the government. Since these actors were under the general supervision of the government, the government considered that there was no need to establish specific domestic legislation to implement the obligations under the space treaties. It should be noted that Japan is not a State Party to the Agreement Governing the Activities of States on the Moon and Other Celestial Bodies (Moon Agreement), as is the case other (advanced) spacefaring nations, including the US, Russia, China, the United Kingdom (UK), India, and France. On this point, many major spacefaring nations, including the US, have also not become parties to the Moon Agreement. By contrast, the US and the UK began developing national legislation concerning commercial space activities during the 1980s. In Japan, although there were cases in which private operators commissioned foreign companies to launch spacecraft, government oversight remained within the scope of the Radio Act and the Broadcast Act.

Against this backdrop, the Basic Space Act,[12] which sets forth the fundamental principles of Japan's space activities, was enacted in 2008 as a parliamentary bill with bipartisan support. This legislation was based on the recognition that Japan's space activities had progressed beyond the stage of research and development to include public services and commercial utilisation. With the enactment of the Basic Space Act as a turning point, Japan shifted its interpretation of the 'peaceful use of outer space', which had previously been understood primarily as being limited to 'non-military purposes', to the internationally recognised interpretation of 'non-aggressive use', in accordance with the purport of Article IV of the OST. Thus, the enactment of the Basic Space Act transformed Japan's space policy from a framework focused on science and technology and research and development into a comprehensive national strategy built on three pillars: science and technology, industrial promotion, and national security.

In November 2016, the SAA and the Act on Ensuring Appropriate Handling of Satellite Remote Sensing Data (Remote Sensing Data Act) were enacted, thereby achieving long-awaited legislative development. The SAA introduced a licensing

[12] Basic Space Act (Act No. 43 of May 28, 2008) <https://www.japaneselawtranslation.go.jp/en/laws/view/4194/en> accessed 21 March 2026.

system for spacecraft launches and management, a liability regime for third-party damage, and other provisions designed not only to enhance the international credibility of Japan's space activities but also to respond to the requirement under Article VI of the OST for authorisation and continuing supervision of the activities of non-governmental entities. In addition, the government's organisational structure for space policy also underwent reform, marking a transition in space governance. Japan's space activities were previously conducted within a vertically divided (stove-piped) system, under which different ministries independently carried out development projects, without integrated government oversight.[13]

Based on the Basic Space Act, the Strategic Headquarters for Space Development (Strategic Headquarters) was established (Article 25). It is chaired by the Prime Minister and its membership includes all Ministers of State (Articles 27 to 30). The Strategic Headquarters was tasked with formulating and implementing comprehensive space development policies. It has since served as the central institution responsible for drafting and revising *Basic Plan on Space Policy* (Basic Plan) of Japan, which has been formulated five times to date: in 2009 (1st), 2013 (2nd), 2016 (3rd), 2020 (4th), and 2023 (5th). These plans have strengthened interconnections between space policy and various areas such as Earth observation, positioning systems, disaster management, international cooperation, and space security. Since 2015, an *Implementation Plan of the Basic Plan on Space Policy* (roadmap) accompanying the Basic Plan has been revised annually. All relevant documents, including the Basic Plan and its associated roadmaps, are publicly available on the Cabinet Office's website.[14]

Recognising the need for specialised expertise in deliberations by the Strategic Headquarters, the Space Policy Committee was established in 2012. This committee is composed of experts and is tasked with conducting research and deliberations on key space policy matters. In April 2016, the National Space Policy Secretariat was established within the Cabinet Office as a body to support the Strategic Headquarters and the Space Policy Committee. It currently serves, in effect, as the central coordinating body for practical matters of the government's space policy. The Basic Space Act, the SAA, the Remote Sensing Data Act, and the SRA are all administered by the National Space Policy Secretariat, which is responsible for licensing and enforcement.

As a result, Japan has developed a government-led framework to advance space development in a more strategic manner. Today, the Basic Plan has evolved into a national strategic framework that encompasses broad issues such as legal systems, governance, and international responsibility. The 5th Basic Plan,[15] adopted in 2023,

[13] Hiromichi Kunitomo, 'The Office of national space policy and development of Quasi-Zenith spacecraftystem', JAXA <https://global.jaxa.jp/article/special/michibiki/kunitomo_e.html> accessed 21 March 2026.

[14] 内閣府 [Cabinet Office], 宇宙基本計画[*Basic Plan on Space Policy*], <https://www8.cao.go.jp/space/plan/keikaku.html> accessed 21 March 2026.

[15] 内閣 [Government of Japan], 「宇宙基本計画(令和5年6月13日閣議決定)」[*Basic Plan on Space Policy* (5th edn, Cabinet Decision, 13 June 2023)] <https://www8.cao.go.jp/space/plan/plan2/kaitei_fy05/honbun_fy05.pdf> accessed 21 March 2026.

sets the goal of doubling the market size of Japan's space industry from JPY 4 trillion in 2020 to JPY 8 trillion by the early 2030s. Based on the 5th Basic Plan, the Space Policy Committee formulated the *Space Technology Strategy* in March 2024,[16] identifying priority technologies for development and outlining a timeline. Notably, the 'Space Strategic Fund' was established within JAXA to support technology development by private companies and universities for up to ten years in the fields of transportation, spacecraft, and exploration.[17] The fund began operations in 2024 with a budget of JPY 300 billion and is expected to provide a total of JPY 1 trillion in support. As of February 2026, the cumulative amount of funding allocated to technology development themes under the Space Strategic Fund amounts to JPY 800 billion. Taken together, these measures illustrate how Japan has gradually established a strategic framework and legal foundation to support the expansion and diversification of its space activities.

3 Legislative Background and Policy Objectives of the SRA

This section outlines and analyses the background of Japan's policy shift concerning space resource activities and the policy objectives of the SRA. In January 2017, ispace, a Japanese startup established in 2010, attracted international attention for operating HAKUTO, one of the five finalist teams in the Google Lunar XPRIZE competition. The company proposed a private-sector business model for lunar resource exploration and transportation, and specialises in the development of lunar landers and rovers. It has established offices in Japan, Luxembourg, and the US, and positions itself as a global lunar resource development company.

Although the OST neither explicitly permits nor prohibits commercial space resource activities, its provisions have created legal uncertainty as to their permissibility. Legal clarity regarding both the legitimacy of such activities and the ownership of space resources plays an important role in facilitating access to financing and enabling the scaling of markets. In Japan, it had traditionally been understood that private ownership of space resources obtained through space exploration was not permitted.[18]However, this understanding began to evolve in the mid-2010s, influenced by the enactment of the US Commercial Space Launch Competitiveness Act

16 宇宙政策委員会 [Space Policy Committee], 「宇宙技術戦略(令和6年3月28日)」 [*Space Technology Strategy* (March 28, 2024)] <https://www8.cao.go.jp/space/gijutu/siryou.pdf> accessed 21 March 2026.

17 内閣府・総務省・文部科学省・経済産業省 [Cabinet Office, Ministry of Internal Affairs and Communications, Ministry of Education, Culture, Sports, Science and Technology, Ministry of Economy, Trade and Industry], 「宇宙戦略基金基本方針(令和6年4月26日)」 [*Basic Policy on the Space Strategy Fund* (April 26, 2024)] <https://www8.cao.go.jp/space/kikin/kihonhousin.pdf> accessed 21 March 2026; JAXA, 宇宙戦略基金 [Space Strategy Fund] <https://fund.jaxa.jp> accessed 21 March 2026.

18 原田大樹 [Hiroki Harada], 「行政法学から見た宇宙法」 ['Space law from the perspective of administrative law'], 法学教室 [*Hogaku Kyoshitsu*], No.497, 有斐閣, 2022年 [Yuhikaku 2022] 43.

in November 2015. Title IV this Act, the Space Resource Exploration and Utilization Act of 2015, granted US nationals the right to possess, own, transport, use, and sell space resources. International discussions, including the International Institute of Space Law (IISL)'s 2016 Position Paper, also encouraged a more flexible interpretation under international law.

By December 2016, arguments began to emerge in Japan had increasingly adopted the view that national legal recognition by individual countries of private ownership of space resources could be consistent with international law.[19] In response, the Cabinet Office established the Task Force on Systemic Issues for Improving the Environment to Support Space Business in 2017.[20] Under the SAA, lunar activities may be licensed as 'control of a spacecraft' pursuant to Article 20, as defined in Article 2(2), which broadly defines a spacecraft as 'an artificial object that is used by putting it into the Earth's orbit or beyond, or placing it on a celestial body other than the Earth'. For example, a lunar rover qualifies as such. Nevertheless, discussions pointed to the limitations of the existing legal framework and the need to support the development of space-related industries. However, there were few concrete private-sector activities at the time. This made it difficult to demonstrate the necessity for legislation. In Japan, an empirical or factual basis is generally required to justify the necessity of legislation. The absence of such fact patterns in space resource activities made it difficult to initiate new legislation.

At the 56th session of the LSC in March 2017, a new agenda was introduced: 'General exchange of views on potential legal models for activities in the exploration, exploitation and utilisation of space resources'. Many states expressed the view that space resource utilisation should be governed by a future international framework.[21] Against this backdrop, the Japanese government decided not to proceed with a new law on space resources during 2017–2018. Instead, the government revised the Regulations for Enforcement of SAA in 2019,[22] clarifying that space resource exploration is included as a permissible activity under Article 20(1). The revision specified that 'space science and exploration' includes 'including resource exploration' as indicated in parentheses in the 'purpose and methods of use of spacecraft'

[19] Nishimura Institute of Advanced Legal Studies, *Report by the space resource development laws study group* (Nishimura & Asahi, December 2016) 46-51 <https://www.nishimura.com/sites/default/files/images/1703_NIALS_en.pdf> accessed 21 March 2026; 中谷和弘・米谷三以・藤井康次郎・水島淳 [Kazuhiro Nakatani, Kazumochi Kometani, Kojiro Fujii, Atsushi Mizushima];「宇宙資源開発をめぐる動向と法的課題」['Trends and Legal Issues Concerning Space Resource Development'], ジュリスト [*Jurist*], No.1506, 有斐閣, 2017年 [yuhikaku 2017] 46-51.

[20] 宇宙ビジネスを支える環境整備に関する論点整理タスクフォース [The Task Force on Systemic Issues for Improving the Environment to Support Space Business] (CAO, 30 May 2018) <https://www8.cao.go.jp/space/comittee/dai70/siryou2.pdf> accessed 21 March 2026.

[21] UN COPUOS, LSC, *Report of the Legal Subcommittee on its Fifty-Sixth Session (27 March–7 April 2017)* UN Doc A/AC.105/1122 (2017) < https://www.unoosa.org/oosa/oosadoc/data/documents/2017/aac.105/aac.1051122_0.html> accessed 21 March 2026, 32-35.

[22] Regulations for Enforcement of the Act on Launching Spacecraft and Launch Vehicle and Control of Spacecraft (Cabinet Office Order No. 50 of November 15, 2017) <https://www.japaneselawtranslation.go.jp/ja/laws/view/4815/en> accessed 21 March 2026.

section in the application form.[23] Accordingly, the government's cautious approach to introducing legislation on ownership of space resources around 2017–2018 may be seen as a reasonable response to the legal and political environment at that time. Meanwhile, Luxembourg (2017) and the UAE (2019) enacted national space resource laws. The Hague Working Group also published the 'Building Blocks for the Development of an International Framework on Space Resource Activities' in 2019. These developments encouraged legislative interest in Japan. Activities conducted in Earth orbit and those involving resource extraction on celestial bodies differ significantly in their legal and technical nature. Under the existing legal framework, it was unclear which specific activities would be permissible. Accordingly, the view was put forward that a legal framework defining the procedures and conditions for authorisation was desirable. In addition, considering that the LSC, which adopts a consensus-based approach, might take many years to establish an international framework, there were growing concerns that Japanese venture enterprises might relocate their operations to other countries where legal systems recognising ownership of space resources had already been established.

In response, the Liberal Democratic Party launched the Working Team on Space Legislation and Treaties, initiating full-scale deliberations on a bill addressing the ownership of space resources. Although most laws in Japan are government-sponsored, this Act was introduced by legislators.

In October 2020, Japan signed the Artemis Accords, becoming one of the original signatories. In December 2020, ispace and its European subsidiary contracted with NASA to sell lunar regolith extracted from the Moon's surface. If this contract is carried out, it will constitute the first legal instance of space resource extraction and transfer and may also mark the beginning of the commercialisation of space resource activities.

During the legislative process, a key issue was consistency with international space law. The drafters of the Act argued that the OST does not explicitly prohibit private ownership of space resources, and that, based on the OST's principle of freedom of exploration and use, such ownership may be permissible if the activities remain within the bounds of lawful space activities.[24] The drafters also took into account alignment with the Artemis Accords when drafting the provisions.[25] In designing the framework, expert concerns about overregulation were considered. These considerations contributed to the design of a legal framework that aligns with the realities

[23] 藤井康次郎・石戸信平 [Kojiro Fujii and Shinpei Ishido], 「宇宙資源開発」 ['Space Resource Development'], 小塚荘一郎・笹岡愛美編著 [in Souichirou Kozuka and Manami Sasaoka(eds)], 『世界の宇宙ビジネス法』 [*Global space business Law*], 初版[first edn], 商事法務, 2021年 [Shojihomu 2021]258.

[24] 小林鷹之・大野敬太郎 [Takayuki Kobayashi and Keitaro Ohno], 「宇宙資源法の背景・目的・内容」 ['Background, purpose, and content of the SRA'],*NBL*, No.1203, 商事法務, 2021年 [*Shojihomu* 2021]78.

[25] 小林鷹之・大野敬太郎 [Takayuki Kobayashi, Keitaro Ohno], 『宇宙ビジネス新時代:解説「宇宙資源法」』 [*New era of space business: Commentary on the SRA*], 初版 [first edn], 第一法規, 2022年 [Daiichihohki 2022] 132-138.

of commercial activity, while also leaving several legal questions unresolved and subject to further consideration.

The SRA was enacted in June 2021. Its purpose is to promote business activities relating to exploring and developing space resources by private business operators, by providing for the acquisition of ownership of such resources under certain conditions, while ensuring the accurate and smooth implementation of conventions on developing and using outer space. The Act also signals Japan's intent to contribute to international rulemaking. It has been pointed out that this Act represents Japan's position on the interpretation of international space law concerning the ownership of space resources, and that, as a form of national practice, it has legal significance in supporting an interpretation that affirms the permissibility of such ownership and in facilitating the formation of norms in international space law.[26]

Thus, its enactment was shaped by a combination of factors, including developments in US space legal policy, the growing business activities of ispace, and, indirectly, Japan's accession to the Artemis Accords. These developments signalled a shift from the traditional focus on scientific exploration and underscored the need for a legal framework that supports private initiatives and fosters economic incentives.

4 Structure of the SRA and Its Institutional Challenges

This section examines the structure and key provisions of the SRA, and analyses institutional challenges through actual implementation cases.

4.1 Structure and Content of the SRA

The SRA adopts a concise structure consisting of only eight Articles. The Act does not require private operators to be Japanese nationals. The key parts are outlined below.

4.1.1 Definitions (Article 2)

The term 'space resources' is defined as 'water, minerals and other natural resources that exist in outer space, including the Moon and other celestial bodies' (Article 2, item (i) of the SRA). 'Natural resources' refers to tangible objects, such as solids, liquids, and gases, for example, regolith. Activities 'conducted exclusively as scientific research or for the purpose of scientific research' are excluded from the definition of 'exploring and developing space resources'. This exclusion is based on the spirit

[26] 中村仁威 [Kimitake Nakamura], 『宇宙法の形成』 [*The formation of space law*], 初版 [first edn], 信山社, 2023年 [Shinzansha 2023] 51–53.

of Article I of the OST, which ensures the freedom of scientific research in outer space, including the Moon and other celestial bodies.

The term 'exploring and developing space resources' is defined as: 'examination of the existence of space resources that contribute to the mining, extraction and other similar activities prescribed by Cabinet Office Order (referred to as 'mining, etc.' or 'mined, etc.' […])'; or 'mining, etc. of space resources and related processing, storage, and other acts prescribed by Cabinet Office Order' (Article 2, item (ii)).

Among those acts prescribed by Cabinet Office Order is the 'transportation of space resources' (Article 2 of the Regulations for Enforcement of the SRA: SRA-Reg).[27]

4.1.2 Special Provisions on Licences for the Control of Spacecraft (Article 3)

Under the SAA, a person must submit a written application specifying the name and address, the location of the spacecraft control facility, 'the purposes and methods of use of the spacecraft', the configuration of the spacecraft, the details of the measures to be taken when terminating the control of the spacecraft, and a plan stating the methods of the control of the spacecraft, pursuant to Article 20, paragraph (2). In this regard, a person intending to obtain a licence under the SAA for the control of spacecraft for the purpose of exploring and developing space resources must, in addition to the matters required under that paragraph, state in the application form, a business activity plan that specifies 'the purpose of the business activities for exploring and developing space resources', 'the duration of the business activities for exploring and developing space resources', 'the place where the exploring and developing of space resources […] is to be conducted', 'the methods of exploring and developing space resources', 'the details of the business activities for exploring and developing space resources', and 'other matters prescribed by Cabinet Office Order' (Article 3, paragraph (1) of the SRA). The 'other matters prescribed by Cabinet Office Order' include 'the financial plan and system for implementing business activities for exploring and developing space resources' (Article 3, paragraph (2) of the SRA-Reg).

In Japan, it is generally easier to amend Cabinet Office Orders than to revise statutes. Accordingly, a legislative technique has been adopted whereby the specific scope of terms such as 'exploring and developing space resources' (Article 2, item (ii) of the SRA) and 'business activity plan' (Article 3, paragraph (1) of the SRA) is left to be defined by Cabinet Office Order. This allows for flexible adaptation in response to future technological developments.

A licence for exploring and developing space resources must not be granted unless the requirements under Article 3, paragraph (2) of the SRA are satisfied, in addition to the general requirements for the licence related to the control of spacecraft set forth

[27] Regulations for Enforcement of the Act on Promoting Business Activities for Exploring and Developing Space Resources (Cabinet Office Order No. 73 of December 23, 2021) <https://www.japaneselawtranslation.go.jp/ja/laws/view/4817/en> accessed 21 March 2026.

in Article 22 of the SAA.[28] Under the SAA, 'the purposes and methods of use of the spacecraft' must comply with the basic principles of the Basic Space Act, and must not be likely to have an adverse effect on the smooth and precise implementation of the conventions on development and use of outer space and ensuring public safety. In addition, Article 3, paragraph (2) of the SRA provides that: 'the business activity plan complies with the basic principles of the Basic Space Act and is not likely to cause any adverse effects on the accurate and smooth implementation of conventions on developing and using outer space and ensuring public safety', and 'the applicant [...] has sufficient ability to execute the business activity plan'.

The examination as to whether the business activity plan for a spacecraft intended for exploring and developing space resources conforms to the conventions on outer space to which Japan is a party shall be conducted pursuant to Article 3, paragraph (2) of the SRA. This mechanism ensures that the business activity plan conforms to international treaties and helps to prevent unregulated development and maintain international credibility.

4.1.3 Public Announcement (Article 4)

When the Prime Minister grants a licence, etc. for exploring and developing space resources, the Prime Minister is to give public notice of that fact and of the following matters (if there are any changes to these matters, those matters after the changes) without delay via the Internet or by other appropriate means (main clause of Article 4, of the SRA):

(i) the name of the person that obtained the licence, etc. for exploring and developing space resources;
(ii) the matters listed in the items of paragraph (1) of the preceding Article3 (excluding item (vi), namely 'other matters prescribed by Cabinet Office Order' such as 'the financial plan and system for implementing business activities for exploring and developing space resources');
(iii) other matters prescribed by Cabinet Office Order.

The matters prescribed by Cabinet Office Order are the date and licence number of the licence under Article 20, paragraph (1) of the SAA (Article 4, paragraph (2) of the SRA-Reg).

This provision is intended to promote business activities related to exploring and developing space resources through international cooperation and to contribute to the prevention of disputes related to exploring and developing space resources. However,

[28] For example, for the purpose of preventing adverse environmental impacts, it is required that measures be implemented in the structure of satellites and other space objects to prevent the dispersion of components and parts, and that they conform to standards concerning the prevention of harmful contamination of outer space, including the Moon and other celestial bodies, as well as the prevention of harmful interference with the exploration and use of outer space(Article 22, item (ii) of the SAA).

in cases prescribed by Cabinet Office Order as those in which the interests of a person that engages in business activities related to exploring and developing space resources by obtaining the licence, etc. are likely to be unjustly harmed by the public notice, the Prime Minister may withhold some or all of the matters (proviso to Article 4 of the SRA).

The cases prescribed by Cabinet Office Order referenced in the proviso to Article 4 of the SRA are cases in which a person engaging in the business activities has submitted to the Prime Minister a document (Article 4, paragraph (1) of the SRA-Reg). This document describes matters that if publicly announced would be likely to unjustly harm interests related to business activities for exploring and developing space resources, and the reasons for that likelihood, and the reasons are considered rational and valid (Article 4, paragraph (1) of the SRA-Reg). This provision seeks to protect confidential business information, such as trade secrets. The SAA does not provide for such a public notification system.

4.1.4 Acquisition of Ownership of Space Resources (Article 5)

A person that conducts business activities related to exploring and developing space resources acquires the ownership of space resources that have been mined, etc. in accordance with the business activity plan related to the licence, etc. for exploring and developing space resources, by possessing the space resources with the intention to own them. The scope of ownership acquisition is limited to space resources that have been mined, etc. in accordance with the business activity plan related to the licence, etc. for exploring and developing space resources. This provision does not imply the recognition of territorial sovereignty over the Moon or other celestial bodies, consistent with Article II of the OST.

4.1.5 Sincere Implementation of International Agreements (Article 6); Establishment of International Systems and Securing Coordination (Article 7)

In the enforcement of the SRA, attention must be paid to not hindering the sincere implementation of treaties and other international agreements Japan has concluded (Article 6, paragraph (1)). No provision of the SRA is unjustly to harm the interests of other States exercising the freedom to explore and use outer space, including the Moon and other celestial bodies (Article 6, paragraph (2)). The national government is to endeavour to establish internationally harmonised systems for exploring and developing space resources in collaboration with foreign governments by cooperating with international organisations and other international frameworks (Article 7, paragraph (1)). It is also to take necessary measures to ensure the advancement of international information sharing, measures for international coordination, and other international alignment with regard to business activities by private businesses (Article 7, paragraph (2)). In doing so, due consideration is to be paid to the sound development of industries and the strengthening of international competitiveness related to exploring and developing space resources (Article 7, paragraph (3)). These provisions reflect Japan's commitment to international cooperation and its intention to contribute proactively to the development of global norms governing space resource activities.

4.1.6 Technical Advice (Article 8)

The national government is to provide private businesses engaged in business activities related to exploring and developing space resources with technical advice and information, and other assistance related to the business activities. This is a part of policy measures for the promotion of the development and use of outer space by private businesses as provided in Article 16 of the Basic Space Act. While self-responsibility is the basic principle, outer space activities involve high risks and uncertainties; this provision enables the government to support private businesses.

4.2 Implementation Status

By February 2026, two licences had been issued under Japan's SRA, both granted to ispace. Although neither mission succeeded, they attracted attention as pioneering private efforts towards commercial space resource development. The approved business activity plans remain publicly available on the Cabinet Office website. The first licence (Licence No. Spacecraft-22–019) pertained to ispace's Mission 1 (M1), which aimed to demonstrate the company's first lunar landing and transportation technologies using a lander (spacecraft). The licence was issued on 4 November 2022. The publicly disclosed activity plan consists of four pages, including the cover. It adopts a format that emphasises clarity and a focus on essential elements, presenting an overall picture of the business, while taking into account both transparency and confidentiality. According to the stated purpose of the business activities, ispace planned to collect lunar regolith and conduct a transaction with a customer (according to ispace's website, specifically NASA). The duration was approximately 14 days after landing. The place comprised four candidate sites, with the spacecraft scheduled to land within a radius of 4 km from each designated latitude and longitude point. The primary landing site and the three backup sites are specified with their respective latitude and longitude coordinates. Regarding the methods, the collection was deemed to consist in the presence of regolith passively deposited on the footpad of the lander by plume-induced dispersion during landing. As to the details of the business activities, a camera installed on the spacecraft was to capture images of the lunar surface, including the footpad. Upon confirming the presence of regolith deposited on the footpad, a commercial transaction for the transfer of ownership of the said regolith with the customer would be executed before implementation of termination measures. The transaction was not to involve the physical transfer of the regolith during the operation of the spacecraft. Although the launch on 26 April 2023 was successful, the lunar landing failed.

The second licence (Licence No. Spacecraft-24–025) pertained to Mission 2 (M2), submitted as a reattempt. The publicly disclosed business activity plan also consists of four pages. The purpose of M2 was to collect regolith and carry out commercial transactions on the lunar surface. It had already been confirmed that there was no previously landed spacecraft that could cause harmful interference within a 150 km radius of each candidate landing site. The licence was issued on 17 December 2024, and the launch took place on 6 June 2025; however, the lunar landing again ended in failure. ispace's efforts continue, with a view to Mission 3 scheduled for launch in 2027. Although M1 and M2 ultimately ended in failure, their commercial objectives, such as the extraction of regolith, the provision of imagery, and the commercial transaction, presented a potential model for future private space resource activities. Moreover, the licensing cases served as precedents that demonstrated the operability and transparency of Japan's legal regime. These cases may provide practical references for future application reviews under the SRA.

4.3 Institutional Challenges

4.3.1 Procedural Challenges

The SRA contains a small number of provisions, many of which employ abstract and broadly framed language. In the 5th Basic Plan (2023), the Government of Japan stated that it would appropriately promote the exploration of resources, including water resources, and the development of related infrastructure in accordance with the phased progress of lunar development. It may be understood that the Japanese government has adopted the concept of adaptive governance.

As ownership of space resources is conditionally recognised under the SRA, it is necessary to establish authorisation criteria with particular attention to the spatial scope and duration of activities. This is because activities conducted under state authorisation are deemed governmental activities under Article VI of the OST, and thus require due oversight. The setting of such criteria is also closely connected to the concept of safety zones under the Artemis Accords and may raise important legal questions in the international context.

With respect to spacecraft in general, the Cabinet Office has published several types of guidance, such as application manuals, examination standards and standard processing periods, guidelines concerning permission for the control of spacecraft, and guidelines on the prevention of spacecraft collisions.[29] However, no specific guidelines or manuals concerning licensing criteria or application procedures have

---

[29] Cabinet Office, Information and Applying for a Permission Related to the Launching of Spacecraft, etc. and License Related to the Control of Spacecraft <https://www8.cao.go.jp/space/english/activity/application.html> accessed 21 March 2026.

been developed for exploring and developing space resources. At present, the government responds carefully to consultations from the application stage. From the standpoint of foreseeability, the near-term development of guidelines and related materials would be desired.[30]

Although a licensing system has been introduced under the SRA, and general supervisory powers are provided through the application of the SAA, no specific or detailed mechanisms have been put in place for the continuous supervision or monitoring of space resource activities after the granting of licences. From the perspective of preventing overexploitation and environmental degradation, the establishment of effective monitoring and control systems is an important issue. Potential measures include a system of periodic audits and mechanisms for confirming compliance with technical standards. Furthermore, the SRA does not set forth provisions concerning the management of space resources. In other domestic sectors, such as mining or fisheries, the government appropriately maintains and manages resources and ensures their rational development by competent actors. In the context of space resources, maintaining a balance with other industries and cooperating with other countries may become future challenges. With respect to these matters, continuous deliberation is required, including consistency with the Artemis Accords.

4.3.2 Legal Issues

The SRA was enacted without any concrete legislative facts, with the aim of promoting industrial development and the formation of international rules. As such, the Act contains numerous indeterminate elements and legal issues.

First, the issue concerns the effects of ownership. The Act does not provide any provisions regarding the treatment of ownership of space resources in cases where a licence is revoked pursuant to Article 30, paragraph (1) of the SAA or where mining, etc. are conducted in violation of the business activity plan related to the licence. Therefore, it becomes unclear what happens to ownership of space resources once acquired under such circumstances. It has been noted that it is not easy to rationally explain the creation of ownership of space resources under the SRA from the perspective of administrative law.[31]

Ownership under the Act is conditionally structured: (i) a business activity plan must be submitted; (ii) a licence must be granted based on the requirements set forth in Article 3, paragraph (2) of the SRA; and (iii) the licensee must conduct mining, etc. of space resources in accordance with the business activity plan related to the licence and possess them with the intention to own them. This structure differs from the Civil Code's Article 239, paragraph (1), under which ownership of movables

[30] 小原直人・川合佑典・松本尊義 [Naoto Obara,Yusuke Kawai and Takayoshi Matsumoto], 「宇宙資源開発及び宇宙資源法」 ['Space Resource Development and The SRA'], 大久保涼・大島日向編著 [in Ryo Okubo and Hinata Oshima], 『宇宙ビジネスの法務』 [*Legal Affairs for Space Business*], 初版 [first edn], 弘文堂, 2021年, [Kobundo 2021] 116.

[31] Harada (n 18) 43.

without an owner is acquired by possessing the same with the intention to own. If the Civil Code is applied supplementarily, the counterparty in a transaction may be protected under the doctrine of good faith acquisition (Article 192 of the Civil Code).

In addition, where space resources over which ownership has been recognised under the Act are stolen or damaged by a third party within Japan, the owner can claim restitution based on such ownership, as well as compensation for damage under tort law (Civil Code, Article 709). On the other hand, it remains a legal issue whether a third party's disturbance of the 'possession' of space resources, the ownership of which is recognised under the Act, would give rise to actions for maintenance or recovery of possession, or compensation for damage, under the Civil Code (Articles 198 and 200). This is because cases are conceivable in which no physical transfer of the resource occurs.

It is essential to consider the establishment of international coordination and mutual recognition mechanisms concerning the development and ownership of space resources, alongside a multilateral supervisory regime.

Secondly, the issue of applicable law arises.[32] To apply its own law in a substantive ruling, a court must find that law applicable under its conflict-of-law rules. In Japan, the Act on General Rules for Application of Laws provides that a real right to movables is governed by the law of the place where the subject property of the right is situated (Article 13, paragraph (1)), and that acquisition or loss of a right is governed by the law of the place where the subject property of the right is situated at the time when the facts constituting the cause of the acquisition or loss were completed (Article 13, paragraph (2)).[33] However, since outer space lacks such a 'law of the place', the applicable law remains uncertain.

One view holds that, with regard to business activities authorised under the SRA, Japanese law can be considered to be the law most closely connected to the place where the activities are conducted, according to the judge's equitable discretion.[34] Nevertheless, in Japanese legal theory, there is no established consensus regarding applicable law in places without a legal regime, such as Antarctica or the high

[32] 松村啓[Hajime Matsumura], 「宇宙資源法」 ['The SRA'], 第一東京弁護士会総合法律研究所編著 [in Dai-Ichi Tokyo Bar Association Comprehensive Legal Research Institute(eds)], 『ESGと法務』 [*ESG and Legal Affairs*], 清文社, 2023年 [Seibunnsha 2023] 161–165; 武藤義行 [Yoshiyuki Muto], 「宇宙資源の所有権に関する理論的課題–宇宙資源法5条に着目した検討」 ['Theoretical Issues Concerning Ownership of Space Resources: A Study Focusing on Article 5 of the SRA'],青木節子・中谷和弘・菊地耕一編著[in Setsuko Aoki, Kazuhiro Nakatani and Kouichi Kikuchi(eds)], 『宇宙法の位相』 [*Phase of space law*], 初版 [first edn], 信山社, 2025年 [Shinzansha 2025] 77–78.

[33] 法の適用に関する通 則法 [Act on General Rules for Application of Laws (Act No. 78 of June 21, 2006)] <https://www.japaneselawtranslation.go.jp/en/laws/view/3783> accessed 21 March 2026.

[34] Kobayashi and Ohno [n 25] 227; 小塚荘一郎 [Souichiro Kozuka], 「宇宙資源法」 ['The SRA'], *Introduction to Space Law for Space Business* (n 8) 186–187.

seas. According to one scholar, the applicable law may be the law of the party's nationality,[35] whereas another suggests that it could be based on de facto control.[36]

If the nationality of the parties governs, the possibility that foreign law will apply cannot be ruled out. With respect to the criterion of de facto control, the 'safety zone' concept under the Artemis Accords may influence the applicable law. Given the unique characteristics of the SRA, it may be questionable whether courts would apply Japanese law as the law of the place most closely connected, even where the business activities are conducted pursuant to authorisation under the SRA. The governing law can be designated in private contracts between commercial entities; however, such designation does not apply to third parties and thus has limitations.

Furthermore, legal issues remain regarding the legal effect of transactions between operators from countries that recognise ownership of space resources under national legislation and those that do not, the priority in cases of competing development licences, and the risk of double registration when multinational corporations apply for licences from multiple jurisdictions. Multinational corporations may seek economically rational legal arrangements by comparing and utilising different national legal regimes, which could further complicate the legal landscape. It is conceivable that corporations may commonly obtain a launch licence from one country, conduct spacecraft control operations in another, and carry out commercial transactions in yet another. In light of Article VI of the OST, which provides that States bear international responsibility for national activities in outer space, including those conducted by non-governmental entities, States cannot remain indifferent. Accordingly, in addition to legal clarification, practical coordination and cooperation among States will be of importance.

5 Japan's Role in Space Resource Governance

In 2019, China achieved the world's first landing on the far side of the Moon with Chang'e 4, followed by a successful sample return mission with Chang'e 5 in 2020. In 2024, Chang'e 6 landed again on the far side and performed the first-ever sample return from that region. China, in collaboration with the Russian Federation, plans to begin establishing an International Lunar Research Station around 2028 and aims to conduct a crewed lunar mission in the early 2030s.

India became the fourth country to achieve a lunar landing with Chandrayaan-3 in 2023, marking the first successful landing in the lunar south polar region. The US is promoting the Artemis programme, including crewed lunar missions planned for the late 2020s. As a global trend, space science and exploration missions are

35 櫻田嘉章 [Yoshiaki Sakurada], 『国際私法』 [*Private International Law*], 第7版[7th edn], 有斐閣, 2020年 [yuhikaku2017] 204.

36 澤木敬郎・道垣内正人 [Takao Sawaki, Masato Dougauchi], 『国際私法入門』 [*Introduction to International Private Law*], 第6版 [6th edn], 有斐閣, 1990年 [Yuhikaku 1990] 268.

becoming increasingly large in scale, with private actors participating and growing international competition.[37]

The international community is currently facing urgent challenges concerning the safety and sustainability of outer space, and efforts to establish international norms have become more important than ever. There is a risk that the economic benefits of space resources may become concentrated in a few advanced countries, thereby exacerbating global inequality, and fair access and utilisation are increasingly being called into question. The responsible and sustainable use of space resources requires fair and viable rules. Particularly important issues include the prevention of overexploitation, the clarification of prohibitions to avoid celestial contamination, the establishment of rational management systems that allow the coexistence of multiple operators, and the development of mechanisms for the international recognition of resource rights.

International collaboration with foreign governments is indispensable for the development of a legal framework. Given the involvement of diverse states and stakeholders, a legal regime that relies solely on individual national approaches to ensure legal certainty is unlikely to be sufficient. Whether through bilateral or multilateral arrangements, mutual recognition and a certain level of legal interoperability are necessary to build a functional and stable system. In other sectors, such as telecommunications, the International Telecommunication Union (ITU) provides internationally agreed frameworks for spectrum allocation and technical standards, which are adopted and enforced by national regulatory authorities. In civil aviation, the International Civil Aviation Organization (ICAO) issues standards and recommended practices (SARPs), which are reflected in national regulations, while licensing and operational oversight remain the responsibility of individual states.

In the field of space resource development, a model that combines international coordination and standard-setting with national licensing and implementation mechanisms appears to be appropriate. This model is realistic, as demonstrated by the established practices in other sectors. In addition to international cooperation, Japan has steadily built international credibility through capacity-building efforts undertaken by government agencies, JAXA, universities, and NGOs. Examples of legal initiatives include institutional and financial support for the United Nations-led Global Space Law Project and the National Space Legislation Initiative (NSLI), which operates under the Asia–Pacific Regional Space Agency Forum (APRSAF) led by Japan. In the context of international rule-making, Japan possesses the capacity to make constructive contributions in both technical and legal domains.

Japan is also in a position to facilitate international dialogue on space resource activities on the Moon by formulating and presenting potential responses to concerns raised by various states in UN COPUOS, in consultation with the Artemis Accords signatories. The legal approach adopted by Japan encourages private space resource activities without impeding innovation, while ensuring the accurate and smooth implementation of its obligations arising under international space law. This approach

[37] 風木淳 [Jun Kazeki]「宇宙政策の最近の進展と宇宙法」['Recent Developments in Space Policy and Space Law'], in Aoki et al (eds), *Phase of Space Law* (n 32) 16.

is based on three fundamental principles: (i) the granting of ownership rights under domestic legislation consistent with international space law, (ii) government authorisation and transparency, and (iii) the promotion of the development of new rules for space resource activities. The SRA has the potential to function as a moderate international model.

To realise this potential, not only the assurance of transparency but also efforts to enhance the effectiveness of continuing supervision of authorised space resource activities constitute an important element. These are key elements for gaining international trust and enhancing the acceptability of the legal framework. In Japan, the public and private sectors actively collaborate in advancing space policy. Accordingly, Japan is required to undertake continuous institutional evaluation and timely improvements, establish best practices, and deepen international cooperation and capacity-building for the responsible utilisation of space resources. In doing so, it is essential to maintain a perspective that promotes the sound development of industry. The two licences granted to ispace can be regarded as practical cases of Japan's domestic authorisation system. They may also be understood as a manifestation of Japan's objective to contribute to the promotion of the formation of international norms through state practice.

Japan has steadily advanced the realisation of responsible space resource utilisation through a phased legal development of its legal framework in accordance with the progress of space resource activities. This approach enables Japan to address emerging challenges while contributing to the internationally harmonised formation of a credible governance regime.

Yumiko Takatori is a qualified attorney in Japan and runs Takatori Law Office. She graduated from the Faculty of Law, Keio University and specialises in space law, nuclear damage compensation, and regulatory law, with additional expertise in legal issues relating to artificial intelligence and crisis management. She has advised Japanese space-related companies, universities, and research institutions on the legal frameworks governing space activities, including satellite operations and rocket launches. She serves as an ad hoc member of the Cabinet Office of Japan's Space Policy Committee and as a member of its Subcommittee on Standards and Safety under the Space Activities Act and has taken part in several other governmental and municipal expert panels. She is also a member of the Study Group on Coordination Items for International On-orbit Servicing (CIOS) of the Cabinet Office.

New Zealand, the Artemis Accords and Asteroid Mining: Critiquing the Legal Challenges for Its Indigenous People

Anna Marie Brennan

Abstract On 19 October 2023, Dawn Aerospace, a space company based in New Zealand, announced the development of an in-space propulsion system for Astro-Forge, which seeks to be the first private corporation to mine an asteroid. While Article 2 of the Outer Space Treaty 1967 prohibits the national appropriation of outer space, including the moon and other celestial bodies, the Artemis Accords have considerably muddied the waters by permitting space resource utilisation. Consequently, the legal ownership of resources extracted from celestial bodies remains controversial. Several states, including Luxembourg, Japan, the United Arab Emirates and the United States, have implemented legislation permitting corporations, registered within their territorial boundaries, to retain legal ownership of resources extracted from outer space. Considering its ratification of the Artemis Accords in 2021, it is conceivable that New Zealand will also opt for a permissive regime that does not impose "unnecessary regulatory costs" that could impede the development of the space mining industry. This chapter considers the regulation of potential asteroid mining missions launched from New Zealand from the perspective of participation, partnership and protection as the core principles of New Zealand's founding document, the Treaty of Waitangi. It contends that government engagement with Māori on the development of New Zealand's space mining strategy and legislation will need to be properly informed by the concept of *kaitiakitanga* (guardianship) as a guiding framework to ensure that space, and its benefits, remain accessible for all.

Keywords New Zealand · Asteroid mining · Artemis Accords · Outer Space Treaty · Kaitiakitanga · Space resource regulation

Senior Lecturer (Above the Bar), School of Law, Politics and Philosophy, University of Waikato—New Zealand and Borrin Foundation Women Leader in Law Fellow. All errors and omissions are my own. Email: abrennan@waikato.ac.nz.

A. M. Brennan (✉)
University of Waikato, Hamilton, New Zealand
e-mail: annamarie.brennan@waikato.ac.nz

S. Bhat B and A. Variath (eds.), *Legal Responses to Space Resource Utilisation*, Space Law and Policy, https://doi.org/10.1007/978-981-95-8864-0_14

1 Introduction

As a remote nation located in the South Pacific, Aotearoa, New Zealand, is certainly not foremost in people's minds when contemplating the space race, but since its first launch in 2016, it is now the third-largest player in the world.[1] New Zealand boasts ideal launch sites, dark skies, a flexible space agency, a culture and tradition of exploration and is increasingly lauded internationally for its commitment to sustainable space exploration.[2] Globally, the space economy has grown exponentially over the last decade. Currently valued at $630 billion, Morgan Stanley has forecast it to grow to over $1.8 trillion (U.S. dollars) by 2035.[3] In 2024, the economic contribution of the space industry to New Zealand was a staggering $2.47 billion (New Zealand dollars), with several space corporations, including Zenith Tecnica, Dawn Aerospace, and RocketLab, operating on its soil.[4] Increasingly, New Zealand is widely sought after for assistance with outer space missions. For instance, to support NASA's Artemis program, New Zealand-based Rocket Lab, using an Electron rocket and Lunar Photon upper stage, sent the Cislunar Autonomous Positioning System Technology Operations and Navigation Experiment (CAPSTONE) CubeSat on an efficient transfer orbit of the Moon on 28 June 2022.[5] Widely recognised as the global leader in launch and space systems, Rocket Lab's space systems technology has enabled more than 1,700 missions from complex interplanetary scientific spacecraft such as the James Webb Space Telescope to critical communications constellations.[6] However, Rocket Lab is not the only space company operating in New Zealand aiming for the stars. Zenith Tecnica has manufactured 3D Ti6AI4V hardware in partnership with

---

[1] Mike O'Donnell, 'RocketLab and Reaching for the Stars' *The Post* (Auckland, 7 June 2025) www.thepost.co.nz/business/360713663/rocket-lab-and-reaching-stars-east-cape accessed 1 July 2025.

[2] Phil Pennington, 'NZ Plugs into Boost US Navy's Space Race with China' Radio New Zealand (Auckland, 16 April 2025) www.rnz.co.nz/news/top/558317/nz-plugs-in-to-boost-us-navy-s-space-race-with-china accessed 1 July 2025.

[3] World Economic Forum, *Space: The $1.8 Trillion Opportunity for Global Economic Growth* (2024). See also Deloitte, *Innovation for Growth: Charting the Space and Advanced Aviation Sectors* (Auckland, April 2025) 7 www.mbie.govt.nz/dmsdocument/30716-innovation-for-growth-charting-the-space-and-advanced-aviation-sectors-pdf accessed 1 July 2025.

[4] Deloitte, *Innovation for Growth: Charting the Space and Advanced Aviation Sectors* (Auckland, April 2025) 7 www.mbie.govt.nz/dmsdocument/30716-innovation-for-growth-charting-the-space-and-advanced-aviation-sectors-pdf accessed 1 July 2025.

[5] NASA, 'What is Capstone?' (Fort Lauderdale, 7 February 2025) www.nasa.gov/smallspacecraft/capstone/ accessed 1 July 2025.

[6] Aparajita Dutta, 'Rocket Lab Deploys 5 Satellites for French IoT Operator: To Buy the Stock or Let Go?' (Yahoo Finance, 11 February 2025) finance.yahoo.com/news/rocket-lab-deploys-5-satellites-200300312.html?guccounter=1&guce_referrer=aHR0cHM6Ly93d3cuZ29vZ2xlLmNvbS8&guce_referrer_sig=AQAAAICQPxZRbJgDS1aQ0SeSltIHACZ6WyfFUW862Qj2vlT1WVIoKVkM0_jVho_zzLIICFw1KLM8sLOiFq86BeCwSAOiLhgdN83jnbsy4PxuOPr0XiBdFJd4Zx3bXbEqv_-F0zwysuNI5NLQNnVng5ZtMeeoQ2bOxF7sJmyjxALIDxOt accessed 1 July 2025.

Maxar Space Systems for the NASA Psyche Mission.[7] The spacecraft, launched on 13 October 2023, is now making a 2-billion-mile journey to the asteroid Psyche.[8] Expected to arrive in approximately 2029, it will spend 26 months orbiting, mapping and studying the asteroid.[9]

Until 2017, New Zealand did not have a legislative regime to give effect to its international obligations in the outer space realm. Consequently, there was an extremely limited scope to prevent the launch of space objects from its territory. With Rocket Lab's declaration to launch from its territory from 2016, there was a growing impetus for New Zealand to develop legislation not only to give effect to its international obligations as an emerging launch state, but also to implement its Technology Safeguard Agreement with the United States.[10] Thus, the resulting *Outer Space and High Altitude Activities Act 2017* (OSHAA) was an effort not only to manage New Zealand's risk as a launching state but also the economic benefits of a promising space industry.[11] The Act provides for a permissive regime that does not impose "unnecessary regulatory costs" that could impede the development of the space industry.[12] The Act also has a deliberately broad scope, not only to ensure that New Zealand maintains sufficient oversight of high-altitude activities originating from its territory but also to provide certainty to space companies about how their activities are regulated in New Zealand.[13] Finally, by legislating for high-altitude vehicles (HAVs) New Zealand has also future-proofed in expectation of significant scientific and technological advancement.[14]

Drawing on this analysis, the chapter will examine how New Zealand will potentially regulate space resource utilisation activities launched from its territory. In particular, the *OSHAA* will be critiqued. This legislation regulates—through licences or permits—launches into outer space, launch facilities, high-altitude vehicles (HAVs) and payloads from New Zealand or by New Zealanders overseas. The sufficiency of the current legal framework will be explored to determine whether it can manage any potential or actual liability that may arise from the space resource utilisation activities and preserve New Zealand's security and national interests. Moreover,

---

[7] 'Zenith Tecnica announces manufacture of 3D Printed Ti6Al4V hardware with Maxar Space Systems for NASA Psyche Mission' (Zenith Tecnica, 5 November 2023) zenithtecnica.com/zenith-tecnica-announces-manufacture-of-3d-printed-ti6al4v-hardware-with-maxar-space-systems-for-nasa-psyche-mission/ accessed 1 July 2025.

[8] Ibid.

[9] Ibid.

[10] Madison Reidy, 'Rocket Lab Opens NZ for Space Business' *Stuff News* (Auckland, 24 January 2018) www.stuff.co.nz/business/industries/100814492/rocket-lab-opens-nz-for-space-business accessed 1 July 2025.

[11] *Outer Space and High-Altitude Activities Act* 2017 (OSHAA); Public Act 2017, No. 29, date of assent 10 July 2017.

[12] Ministry of Business, Innovation and Employment, Statutory Review of the Outer Space and High-Altitude Activities Act 2017, 5 www.mbie.govt.nz/dmsdocument/20127-report-on-the-statutory-review-of-the-oshaa-2017-pdf accessed 1 July 2025.

[13] Ibid 8.

[14] Ibid 30 and 31.

the chapter examines the regulation of the space resource utilisation under New Zealand law from the perspective of participation, partnership and protection as the core principles of Te Tiriti o Waitangi. Since a study of this kind has never been undertaken, it will be considered whether the government's engagement with Māori on the development of New Zealand's space strategy and legislation has been properly informed by the concept of *kaitiakitanga* (guardianship) as a guiding framework to ensure that space, and its benefits, remain accessible for all. Indeed, the duty to consult Māori on issues such as space resource utilisation is wide-ranging, and stems from the principles of partnership, good faith and active protection of Māori interests. While the *OSHAA* does not expressly refer to the *Treaty of Waitangi*, this chapter will examine how there is a strong Treaty-based interest in law and policy on asteroid mining, given that the Treaty was predicated on facilitating the settlement of New Zealand and preserving the right to self-determination. This chapter considers how Māori consultation and participation in the use, exploration and governance of asteroid mining can be more meaningfully supported not only at the domestic level but also internationally.

2 A Brief Overview of the Regulation of the Outer Space Industry in New Zealand

Apart from the Moon Agreement, New Zealand is a signatory party to all the main outer space treaties and conventions.[15] However, it was not until 2017 that it had a legislative regime to give effect to its international obligations in outer space. Consequently, there was limited scope to prevent the launch of space objects from New Zealand soil. With Rocket Lab's declaration to launch from New Zealand territory from 2016 onwards, there was a growing impetus to draft and enact legislation not only to give effect to its international obligations as an emerging launch state but also to implement its *Technology Safeguard Agreement* with the United States.[16] The subsequent OSHAA endeavoured to navigate New Zealand's risk as a launching state,

[15] The five treaties are: *Treaty on Principles Governing the Activities of States in the Exploration and Use of Outer Space, including the Moon and Other Celestial Bodies* (adopted 27 January 1967, entered into force 10 October 1967) 610 UNTS 205 (Outer Space Treaty); *Agreement on the Rescue of Astronauts, the Return of Astronauts and the Return of Objects Launched into Outer Space* (adopted 22 April 1968, entered into force 3 December 1969) 672 UNTS 119 (Rescue Agreement); *Convention on International Liability for Damage Caused by Space Objects* (adopted 29 March 1972, entered into force 1 September 1972) 961 UNTS 187 (Liability Convention); *Convention on Registration on Objects Launched into Outer Space* (adopted 14 January 1975, entered into force 15 September 1976) 1023 UNTS 15 (Registration Convention); See *Agreement Governing the Activities of States on the Moon and other Celestial Bodies* (adopted 18 December 1979, entered into force 11 July 1984) 1363 UNTS 3 (Moon Agreement).

[16] Rocket Lab, 'Press Release: Rocket Lab Launch Complex 1 Complete' (Auckland, 27 September 2016) www.rocketlabusa.com/updates/rocket-lab-launch-complex-1-ready-for-launches/ accessed 1 July 2025. See also: *Agreement between the Government of New Zealand and the Government of the United States of America on Technology Safeguards Associated with United States Participation*

but also the economic benefits of a promising space industry.[17] The Act provides for a permissive regime for the registration, licensing and operation of launch vehicles, payloads and high-altitude vehicles that does not impose "unnecessary regulatory costs" that could impede the development of the space industry.[18]

The legislation has a deliberately broad scope not only to ensure that New Zealand maintains sufficient oversight of high-altitude activities originating from its territory but also to provide certainty to space companies about how their activities are regulated.[19] Supplemented by the *Outer Space and High-Altitude Activities (Licences and Permits) Regulations 2017* and the *Outer Space and High-Altitude Activities (Definition of High-Altitude Vehicle) Regulations 2017,* the *OSHAA* is administered by the New Zealand Space Agency.[20] As of 30 June 2025, 129 payloads have been approved for launch by the Space Agency.[21] While some payloads have had more than one purpose, they have mainly been approved for technology demonstration, remote sensing, educational and commercial purposes.[22] With quiet airspace making safer launches possible, states are increasingly seeking to launch from New Zealand. For example, payloads owned by France, Singapore and the United States have been launched with remote sensing capabilities to track maritime traffic and gather real-time images of Earth.[23] Other payloads have conducted space weather experiments, collected radio signals from outside the earth's atmosphere, and gathered data from space "to investigate whether there is a link between … atmospheric activity and tectonic activity on Earth."[24] So far, New Zealand has mainly launched CubeSats and nanosatellites and thus, it remains to be seen whether it will develop the capacity to launch larger payloads in the future, especially for asteroid mining missions.

---

in Space Launches from New Zealand (Technology Safeguards Agreement) [2016] NZTS 14 (signed 16 June 2016, entered into force 12 December 2016).

[17] Ministry of Business, Innovation and Employment, Statutory Review of the Outer Space and High-Altitude Activities Act 2017 (Wellington, 12 May 2022) 5 www.mbie.govt.nz/assets/report-on-the-statutory-review-of-the-oshaa-2017.pdf accessed 1 July 2025.

[18] Ibid.

[19] Ibid 8.

[20] For more information about the New Zealand Space Agency see www.mbie.govt.nz/science-and-technology/space accessed 1 July 2025.

[21] Ministry for Business, Innovation and Employment, "Payloads Approved for Launch" (Wellington, 30 June 2025) www.mbie.govt.nz/science-and-technology/space/permits-and-licences-for-space-activities/payloads-approved-for-launch accessed 1 July 2025.

[22] Ibid.

[23] Ibid.

[24] Ibid.

2.1 The Sufficiency of New Zealand's Launch Licensing Regime for Future Asteroid Mining

Despite its reputation for having a permissive regime, New Zealand has arguably one of the most stringent launch licence systems of all spacefaring nations. Section 7 of the OSHAA provides the requirements that must be fulfilled for a launch licence to be granted. The provision clearly specifies that "a person must not launch a launch vehicle … unless the person has a launch licence."[25] Moreover, for the licence to be granted, the Minister must be satisfied that "all reasonable steps to manage risks to public safety have been taken" and the proposed launch is "consistent with New Zealand's international obligations."[26] Thus, any proposed launch that is not compatible with the Outer Space Treaty 1967 will not be granted a launch licence under the OHSAA. Similar to section 4(2) (c) of the Outer Space Act 1986 in the United Kingdom, a launch licence may also be declined "if the Minister is not satisfied that a proposed launch under the licence is in the national interest."[27] In determining whether there is a risk to the national interest, the "economic or other benefits to New Zealand of the proposed launch," and "any risks to national security, public safety, international relations, or other national interest" will be assessed.[28] Interestingly, there is no reference to the environmental impact of proposed launches in assessing risks. This is especially important given the concerns raised by scholars about the potential environmental harm that could be caused by space resource utilisation activities such as asteroid mining and the need for environmental impact assessments.[29] Moreover, in ascertaining the extent to which risks must be alleviated, the Minister must consult with the security Ministers under section 55 of the OSHAA.[30] Data has not yet been published on decisions to decline launch licence applications and the reasons for doing so. Thus, it is not possible to determine whether an application for a launch licence has ever been declined on the grounds that it would cause environmental harm. Applicants may only seek a review of the Minister's decision to decline a launch licence where the Minister opted to "issue a certificate that the activity or proposed activity poses a significant risk to national security."[31] This

[25] OSHAA (n 11) s 7.

[26] Ibid s 9.

[27] OSHAA, s 9(2)(a).

[28] OSHAA, s 9(3).

[29] Stephen Mustow, 'Environmental Impact Assessment (EIA) Screening and Scoping of Extraterrestrial Exploration and Development Projects' (2018) 36(6) Impact Assessment and Project Appraisal 467; William Kramer, 'Extraterrestrial Environmental Impact Assessments—A Foreseeable Prerequisite for Wise Decisions Regarding Outer Space Exploration, Research and Development' (2014) 30 Space Policy 215; William Kramer, 'In Dreams Being Responsibilities—Environmental Impact Assessment and Outer Space Development' (2017) 19(3) Environmental Practice 128; William Kramer, 'A Framework for Extraterrestrial Environmental Assessment' (2020) 53 Space Policy 101385.

[30] OSHAA (n 11) s 9(5).

[31] OSHAA, s 55(3).

review procedure involves lodging "a complaint to the Inspector-General of Intelligence and Security in relation to any advice given by an intelligence and security agency."[32] Pursuant to section 56(2), if the Inspector-General of Intelligence and Security submits a report on the matter to either one of the security Ministers, the "Prime Minister may withdraw or confirm the certificate."[33]

Under section 10, the licensee must provide intricate detail about the launch, including "the date, location, and intended trajectory" as well as "details of the intended and actual basic parameters…of any part of a launch vehicle that reaches or is intended to reach outer space."[34] This does not include a requirement for the licence applicant to provide information on whether the purpose of the launch is for research or commercial purposes, and how it will be of benefit to New Zealand. While the Space Agency does not need to be kept informed of how the proposed launch will enable the applicant to plan and develop future outer space activities and operations, section 10(1)(a)(iii) of the OSHAA does provide that the applicant must submit to the Minister "any prescribed information relating to each launch."[35] This requirement is bolstered by section 50, which empowers the Minister to request in writing any information needed to inform their decision-making on the licence application.[36] If "the applicant does not provide the information requested within a reasonable time after the requirement" the licence application can be declined.[37] Under section 19, if the decision is made to grant a licence, the date upon which it starts must be specified, as well as the expiry date.[38] Moreover, pursuant to section 27(2), licences cannot be issued for more than five years unless the applicant applies for a renewal.[39] In contrast to licensing regimes in other states, no provision is provided for the inspection of facilities and the testing of equipment to inform decision-making. While some scholars may contend that New Zealand operates a high-trust licensing model, it can be contended that there is a case for reform to include for facilities inspections and the testing of launch equipment to ensure that it will potentially harm the Earth-Space environment.[40] Section 33 provides that an applicant must submit an orbit mitigation plan but there is no elaboration upon the core components of this plan.[41]

Similar to the *Outer Space Act 1986* in the United Kingdom, there is no requirement that the person or entity conducting the launch must comply with the *UN Space*

---

32 OSHAA, s 56(1).

33 OSHAA, s 56(2).

34 OSHAA, s 10.

35 Ibid. s 10(1)(a)(iii).

36 Ibid s 50.

37 Ibid s 50(2)(a).

38 Ibid s 19 and 27.

39 Ibid s 27(2).

40 Frans von der Dunk, 'Kiwis in Space: New Zealand's Outer Space and High-Altitude Activities Act' in *Proceedings of the 60th Colloquium on the Law of Outer Space,* vol. 60 (2017) 453.

41 OSHAA (n 11) s 33.

Debris Mitigation Guidelines.[42] Neither is there a requirement to inform the Minister if the launching party is unable to adhere to the plan and measures to mitigate the orbital debris created as a result of deviating from the plan. At the same time, the Minister can declare an area to be a debris protection area. Pursuant to section 64(2) this is defined as "a site where a launch termination or an accident involving a launch vehicle has occurred; or a site on which there is a space object that has been involved in a launch termination or an accident involving a launch vehicle."[43] It is noteworthy that an area can be designated as a debris protection area following a launch termination or accident, not to facilitate an investigation into the cause of the incident or to facilitate a clean-up of the area, but instead "to comply with any international agreement relating to the protection of sensitive space technology."[44] With regard to potential asteroid mining, there is no provision for launchers to provide information or a plan on how they will endeavour to avoid harming the environment of celestial bodies. Thus, the OSHAA is very much concerned with the harmful impact of launch activities on the Earth's environment rather than the outer space environment.

2.2 New Zealand and Artemis Accords

On 31 May 2021, New Zealand became a signatory party to the Artemis Accords which seeks to provide "a practical set of principles, guidelines, and best practices in carrying out activities in outer space…intended to increase the safety of operations, reduce uncertainty, and promote the sustainable and beneficial use of space for all humankind."[45] While the Preamble affirms the "importance of compliance" with the Outer Space Treaty, it is interesting how Section 10 provides for the extraction and utilisation of space resources, but includes no reference to the term "celestial bodies."[46] Instead, it simply refers to "the surface or subsurface of the Moon, Mars, comets, or asteroids," and also affirms that extraction "does not inherently constitute national appropriation" under the Outer Space Treaty. Several countries have adopted legislation prescribing the rights and responsibilities of state and non-state actors for proposed mining activity in outer space. In particular, the United States enacted the *Spurring Private Aerospace Competitiveness and Entrepreneurship Act of 2015 (the Space Act)*. This legislation clearly states that U.S. citizens "engaged in commercial recovery of an asteroid resource or a space resource … shall be entitled

[42] *Outer Space Act 1986* (United Kingdom 1986 c 38). See also Space Debris Mitigation Guidelines of the UN Committee on the Peaceful Uses of Outer Space, endorsed by the UN General Assembly in its Resolution 62/217 www.unoosa.org/pdf/publications/st_space_49E.pdf accessed 1 July 2025.

[43] OSHAA (n 11) s 64(2).

[44] Ibid s 64(1).

[45] The Artemis Accords: Principles for Cooperation in the Civil Exploration and Use of the Moon, Mars, Comets, and Asteroids for Peaceful Purposes, Section 1 available at: https://www.nasa.gov/wp-content/uploads/2022/11/Artemis-Accords-signed-13Oct2020.pdf?emrc=68802e4a5bd2f (date accessed: 1 July 2025).

[46] Ibid The Preamble.

to any asteroid resource or space resource obtained."[47] While there is no reference to the term celestial bodies in relation to proposed extraction activities, there is instead a disclaimer of exterritorial sovereignty, whereby "the United States does not … assert sovereignty or sovereign or exclusive rights or jurisdiction over, or the ownership of, any celestial body."[48] This is also emphasised in the National Space Policy of the United States, which provides that it "will pursue the extraction and utilization of space resources in compliance with applicable law, recognising those resources as critical for sustainable exploration, scientific discovery, and commercial operations."[49]

As discussed above, New Zealand has yet to legislate for space resource utilisation activities on celestial bodies. Indeed, the only reference to celestial bodies in the OHSAA is contained in section 3 which provides that the purpose of the Act is to ensure that New Zealand fulfils its international obligations relating to outer space use and exploration.[50] This includes obligations not to place weapons or instal military bases on celestial bodies.[51] With no reference to space resource extraction activities in the OHSAA, New Zealand has arguably lost a valuable opportunity to future-proof its outer space regulatory system. Pursuant to Section 86 of the *OHSAA* a review of the legislation was conducted by the Ministry of Business, Innovation and Employment (MBIE) from mid-August until the end of September 2021. Constructive feedback was sought from several international regulatory agencies especially in Australia, Europe and the United States, as well as industry stakeholders such as space companies. On 12 May 2022, a report of the review's findings was tabled at Parliament, but there was no discussion of space resource utilisation or the potential for such activities to be launched from New Zealand territory.[52] There was also no discussion of the need to amend the legislation to give effect to its obligations under the Artemis Accords. Indeed, no detailed analysis has been published by MBIE on the legal implications of the Accords for the space industry. With some of the technology to enable future asteroid mining operations being designed and manufactured by Dawn Aerospace in New Zealand, questions are beginning to emerge about how and the extent to which compliance with the Artemis Accords will be ensured.

The principal concern is the legal protection of the outer space environment from the impacts of asteroid mining, which has been described by Alessandra Marino and Thomas Cheney as deficient.[53] Arguably, Article IX of the Outer Space Treaty

---

[47] *Spurring Private Aerospace Competitiveness and Entrepreneurship Act of 2015 (the Space Act)*, Section 51303.

[48] Ibid s 403.

[49] National Space Policy of the United States of America (9 December 2020) page 3 available at: https://trumpwhitehouse.archives.gov/wp-content/uploads/2020/12/National-Space-Policy.pdf (date accessed: 1 July 2025).

[50] OHSAA (n 11) s 3.

[51] Ibid.

[52] Statutory Review (n 17).

[53] Alessandra Marino and Thomas Cheney, 'Centring Environmentalism in Space Governance: Interrogating Dominance and Authority Through a Critical Legal Geography of Outer Space' (2023) 63 Space Policy 101521.

provides some degree of protection by necessitating that state parties "shall pursue studies of outer space, including the Moon and other celestial bodies, and conduct exploration of them so as to avoid their harmful contamination and also adverse changes in the environment of the Earth, ... and where necessary, shall adopt appropriate measures for this purpose."[54] In addition, Article VI of the Outer Space Treaty affirms that states "shall bear international responsibility for national activities in outer space, including the Moon and other celestial bodies, whether such activities are carried on by governmental agencies or non-governmental entities."[55] The provision goes on to state that "activities of non-governmental entities in outer space ... [also] require authorization and continuing supervision by the appropriate State Party to the Treaty."[56] In view of the fact that the phrase "authorization and continuing supervision" is not defined in the Outer Space Treaty, Epstein opines that it creates an indeterminate "liability relationship between the state and private actors, and does not imply that the responsible nation exercises ownership, possession or sovereignty over anything affected or undertaken by any private party."[57] Thus, the extent to which New Zealand will need to supervise the activities of space companies engaged in asteroid mining remains unclear.[58] Apart from launching activities, no clear guidance on what other types of activities need to be supervised is outlined in the OHSAA. Space companies such as Karman+ and AstroForge will be required to apply for permission to launch asteroid mining ventures from New Zealand soil pursuant to section 7 of the OHSAA.[59] If permission is granted and space companies succeed in launching asteroid mining ventures, Article VI provides that New Zealand would have a duty to continuously monitor their activity.[60] With the country increasingly viewed as an attractive location to launch space operations, it is unclear why it has not yet published guidelines on how it will seek to fulfil these supervision obligations. With technological developments gathering pace, New Zealand will need to examine how it will seek to supervise asteroid mining ventures launched from its territory with the potential to last decades. Until the OHSAA is amended to include a duty to protect the outer space environment, especially celestial bodies, it is unlikely that space companies will shift towards "stakeholder capitalism" and

---

[54] Article IX of the Outer Space Treaty (n 15) Article IX.

[55] Outer Space Treaty (n 15) Article VI.

[56] Ibid.

[57] David Epstein, 'Protecting the Cosmos: Defining Celestial bodies in the Outer Space Treaty' (2024) 15 (1) Space and Defense 35, 41.

[58] William Kramer, 'Extraterrestrial Environmental Impact Assessments—A Foreseeable Prerequisite for Wise Decisions Regarding Outer Space Exploration Research and Development' (2014) 30 Space Policy 215, 217. See also Melissa de Zwart, 'Authorization and Continuing Supervision (Article VI)' in Mahulena Hofmann and PJ Blount (eds), *Elgar Concise Encyclopedia of Space Law* (Edward Elgar 2025) 17, 18.

[59] OSHAA (n 11) s 7.

[60] Melissa de Zwart, Stacey Henderson and Michelle Neumann, 'Space Resource Activities and the Evolution of International Space Law' (2023) 211 *Acta Astronautica* 155, 157.

abandon the "shareholder bias of prevailing business models associated with profit maximisation, externalisation, and corporate short-termism."[61]

3 Asteroid Mining from the Perspective of The Treaty of Waitangi: A Critical Analysis

A crucial issue which has not been widely considered in the scholarly literature on the regulation of asteroid mining is its potential impact on Indigenous Peoples. A Western Worldview continues to dictate outer space law and policy and indeed, the key treaties and conventions that regulate outer space activity were drafted and negotiated without their input or consultation.[62] As noted by scholars such as Varaghese, Henry and Morris, developing a sustainable future for outer space use and exploration necessitates building respectful relationships with Indigenous communities who continue to face a myriad of challenges due to colonisation.[63] Thus far, plans to conduct asteroid mining have embraced a Western view of sustainability which is inherently anthropocentric by permitting the exploitation of celestial bodies for economic gain for spacefaring nations.[64] These plans have not considered the potential impact of future asteroid mining on the sovereignty of Indigenous Peoples, especially their access to dark skies and the use of their land resources for launching space missions and operations, which have often been confiscated and thus displaced them from their culture, history and way of life.[65]

In New Zealand, despite considerable efforts to foster a regulatory framework that "encourages the growth of a safe responsible and sustainable space industry that creates economic development," a review of the OSHAA undertaken in 2021 raised considerable concerns about the government's engagement and partnership with Māori, the Indigenous Peoples of New Zealand on the development of the country's space strategy.[66] The main concern emerging from the review was whether the OSHAA was compatible with the Treaty of Waitangi and "informed by the concept of *kaitiakitanga* (guardianship) as a guiding framework to ensure that space,

[61] Daniel, Augenstein, 'Towards a New Legal Consensus on Business and Human Rights: A 10th Anniversary' (2022) 40(1) Netherlands Quarterly of Human Rights 3, 42. See also: John Ruggie, Caroline Rees and Rachel Davis, 'Ten Years After: From the UN Guiding Principles to Multi-Fiduciary Obligations' (2021) 6 Business and Human Rights Journal 179.

[62] Carolle Varughese, Lenaq Henry, Adam Morris *et al.*, 'The Intersection of Space and Sustainability: The Need for a Transdisciplinary and Bi-Cultural Approach' (2023) 211 *Acta Astronautica* 684.

[63] Ibid.

[64] Ushana Jayasuriya and Krushil Watene, 'Just Transitions as Relationship-Building' (2024) 20(2) Journal of Global ethics 171.

[65] Ibid.

[66] Phil Twyford, 'New Space Payload Principles Protect the National interest' (New Zealand Parliament, 17 December 2019) www.beehive.govt.nz/release/new-space-payload-principles-protect-national-interest accessed 1 July 2025.

and its benefits, remain accessible for all," including Māori.[67] The Treaty of Waitangi is New Zealand's founding document. Signed at Waitangi in the Bay of Islands on 6 February 1840, it: "is a broad statement of principles on which the British and Māori made a political compact to found a nation state and build a government in New Zealand. The document has three articles. In the English version, Māori cede the sovereignty of New Zealand to Britain; Māori give the Crown an exclusive right to buy lands they wish to sell, and, in return, are guaranteed full rights of ownership of their lands, forests, fisheries and other possessions; and Māori are given the rights and privileges of British subjects."[68]

Pursuant to the Treaty, there is a comprehensive duty to consult Māori, and this is derived from principles of partnership, good faith and protection of Māori interests.[69] Importantly, the duty is not restricted to *taonga* (treasure) under Article 2 of the Treaty such as natural resources or land.[70] Instead, consultation is required where the New Zealand Government is formulating general policy (such as space policy) that may engage Treaty principles or affect Māori interests.[71] Even though the *OSHAA* does not directly refer to the Treaty of Waitangi, the results of the consultation published in 2022 led to the publication of the first National Space Policy by the Minister for Economic Development on 31 May 2023. This policy sought to be a "reference to consistently communicate New Zealand's interests both on and offshore, to inform the creation of future space policies, and our overall direction in … [the] sector …"[72]

Several values formed the cornerstone of the 2023 policy, which sought to "promote through New Zealand's space activities, engagements and the use of space technologies."[73] In particular, it emphasised that there is a strong Treaty-based interest in space policy, given that the Treaty was predicated on facilitating the settlement of New Zealand and preserving the right of Māori to self-determination and self-governance.[74] Indeed, the broad language of certain provisions in the OSHAA imports consideration of the Treaty of Waitangi and mandates a high level of consultation and engagement with Māori.[75] Core values outlined in the National Space Policy to steer future policy-making included stewardship in respect of preserving space and Earth environments for future generations; innovation to develop outer space activity and drive economic growth; and responsibility for compliance with domestic

---

[67] Ministry for Business, Innovation and Employment, 'Summary of Feedback: New Zealand Space Policy Review Consultation' (Wellington, 14 December 2022) 12 www.mbie.govt.nz/assets/new-zealand-space-policy-review-summary-of-submissions-report.pdf accessed 1 July 2025.

[68] 'The Treaty in Brief' /nzhistory.govt.nz/politics/treaty/the-treaty-in-brief accessed 1 July 2025.

[69] MBIE (n 67).

[70] Ibid 49.

[71] Ibid 55.

[72] New Zealand Space Agency, National Space Policy (Wellington, May 2023) 1 www.mbie.govt.nz/dmsdocument/26656-national-space-policy accessed 1 July 2025.

[73] Ibid 3.

[74] Ibid 2.

[75] Ibid 4.

and international space laws.[76] These values were a stepping stone towards guiding "New Zealand government officials to articulate New Zealand's interests in space on and offshore and to inform the development of future space policies."[77] Furthermore, the policy concluded that "[t]he government is committed to the ongoing engagement and involvement of Māori in the development and implementation of space policies."[78]

Following public consultation on the National Space Policy, it was decided in a Cabinet Paper that various workstreams would be developed to address the concerns raised.[79] In particular, the Cabinet Paper noted that "permitting any payloads with national security applications would breach New Zealand's international obligations, domestic laws and policies and ... [the Treaty of Waitangi] and therefore against New Zealand's national interests."[80] Consequently, the government sought guidance "as to what is—and is not—in New Zealand's national interests" and whether it included "payloads with national security applications" and "weapons systems."[81] Based on guidance received from officials, cabinet affirmed that there would be "no blanket ban on permitting payloads with national security applications" under the *OSHAA* and instead payloads would "continue to be assessed on a case-by-case basis."[82] While Cabinet did concede that New Zealand had launched payloads with national security applications none of these had been deemed incompatible with its national interests.[83] Moreover, it was affirmed that any payload "inconsistent with ... domestic laws and policies, as well as international obligations, [would] ... not be progressed."[84] Cabinet also committed to initiating a workstream that fulfils the National Space Policy aim of "clarifying that space activities are inconsistent with New Zealand's national interest—including through applying the National Space Policy value and objectives to existing regulatory processes."[85]

Published in July 2023, the Aotearoa New Zealand Aerospace Strategy 2023–2030 sought to "establish a distinct New Zealand approach to developing the aerospace sector, by building on ... national strengths, while managing national security risks."[86] It outlined several goals for the state to accomplish, including being at the pinnacle of sustainable space exploration. An action plan comprising three phases "to

---

[76] Ibid.

[77] Ibid.

[78] Ibid.

[79] Ibid.

[80] Office of the Minister of Economic Development, Space Policy Review: National Space Policy (Wellington, 31 May 2023) para. 27 www.mbie.govt.nz/dmsdocument/26664-space-policy-review-national-space-policy-proactiverelease-pdf accessed 1 July 2025.

[81] Ibid.

[82] Ibid para. 28.

[83] Ibid.

[84] Ibid para. 29.

[85] Ibid 11.

[86] Ministry of Business, Innovation and Employment, *Aotearoa New Zealand Aerospace Strategy 2023-2030* (Wellington, July 2023) 3 www.mbie.govt.nz/assets/te-rautaki-atea-a-rangi-o-aotearoa-2023-2030-aotearoa-new-zealand-aerospace-strategy.pdf accessed 1 July 2025.

deliver on this ambitious vision"[87] by 2030 was endorsed in the Strategy with a "focus on cross-cutting enablers that support development of the sector and ... close collaboration between government, the aerospace sector and other interested groups."[88] While the meaning of the phrase "interested groups" was not elaborated upon in the Strategy, presumably it refers to all stakeholders with an interest or concern in the development of New Zealand's aerospace sector and the use and exploration of outer space more broadly. Plans to support the aerospace sector in accordance with the Treaty of Waitangi were not expanded upon in detail in the Strategy. While there was an undertaking to conduct "analysis of current Māori participation and aspirations for the sector, as well as work on specific economic opportunities for Māori in aerospace" this has not come to fruition.[89] The Aerospace Strategy appeared to be achieving some of these goals with MBIE inviting proposals in April 2023 under the Catalyst Fund for collaborative Earth observation feasibility studies between New Zealand-based researchers and NASA.[90]

However, the adoption of a revised Space and Advanced Aviation Strategy in September 2024 has led to considerable consternation.[91] There is no mention of the necessity to consult with Māori on outer space use and exploration, and to protect their interests in accordance with the Treaty of Waitangi. In order to "[e]stablish a world-leading regulatory environment for space and advanced aviation" it instead seeks to "[p]artner internationally on advanced research world-leading research institutes ... [and] [e]xplore innovative ways for the government to apply aerospace technologies to address scientific, societal and national security challenges."[92] In view of the concerns raised in the published review of the OHSAA in 2022 about the need for in-depth consultation and meaningful partnership with Māori on outer space use and exploration, the fact that there is no reference to the Treaty of Waitangi, the founding document of New Zealand, in the Strategy without any explanation is disconcerting. Launch facilities built on Indigenous lands by RocketLab on the Māhia Peninsula have sparked controversy and conflicts with the values of the local Māori community, especially access restrictions to ancestral lands before and during payload launches.[93]

---

[87] Ibid.

[88] Ibid 25.

[89] Ibid 9.

[90] For further information on this Fund please see: www.mbie.govt.nz/science-and-technology/science-and-innovation/funding-information-and-opportunities/investment-funds/catalyst-fund accessed 1 July 2025.

[91] Ministry of Business, Innovation and Employment, New Space and Advanced Aviation Strategy 2024-2030 (Wellington September 2024) www.mbie.govt.nz/assets/new-zealand-space-and-advanced-aviation-strategy-2024-2030.pdf accessed 1 July 2025.

[92] Ibid 7 and 8.

[93] 'RocketLab Scope Creep Angers Mahia Mana Whenua' *waateanews.com* (Wellington, 22 April 2021) waateanews.com/2021/04/22/rocket-lab-scope-creep-angers-mahia-mana-whenua/ accessed 1 July 2025.

This is just one example of a "failure to fully understand Indigenous concerns, as legitimate concerns."[94] Moving forward, how will it be possible to develop democratisation and meaningful inclusion of Māori in outer space use and exploration, especially asteroid mining, that is not based in "institutional and political inequalities that date back to colonial times."[95] As evidenced by the revised Space and Advanced Aviation Strategy published by MBIE in 2024, the failure to include any references to Māori is ultimately exclusionary, and the use of ambiguous language in the OSHAA has enabled discrimination and marginalisation to continue. In New Zealand and several other countries, outer space exploration has had a profound impact on Indigenous Peoples, especially their access to dark skies to gather food, facilitate navigation and practice their cultural and religious beliefs. Asteroid mining will inevitably increase space traffic and light pollution but thus far, meaningful solutions to counteract these future challenges have not been put forward. As a result, Article 1 of the Outer Space Treaty which states that "[o]uter space … shall be free for exploration and use by all States without discrimination of any kind" is rather tokenistic, since it is left to states like New Zealand to determine how Indigenous Peoples and other side-lined communities are included in outer space use and exploration. The future distribution of space resources as a result of asteroid mining, to foster self-determination continues to be largely ignored by the international community.[96]

4 Conclusion

As stated in the introduction, outer space law and policy in New Zealand has so far had extraordinarily little to do with the regulation of proposed asteroid mining, and this chapter has considered a fundamental consequence of that actuality with reference to the rights and interests of Māori, the Indigenous Peoples of New Zealand. It has also explored the potential impact of the continued lack of regulation for asteroid mining activity and made proposals for how the potential social and environmental impacts of asteroid mining on Māori could be monitored. This chapter also highlights how New Zealand's outer space law and policy do not currently fully capture the dynamics of the space industry, despite corporations such as RocketLab and Dawn Aerospace having quickly emerged over the last two decades as the industry's foremost players. The discussion has highlighted the need for a bespoke legislative regime to be developed to regulate asteroid mining operations launched from New Zealand territory as a solution to this dilemma. While some of the major spacefaring nations have already adopted legislation to regulate space resource utilisation activity, New Zealand will need to carefully explore a range of policy and regulatory approaches in light of

[94] Tony Milligan, 'From the Sky to the Ground: Indigenous Peoples in an Age of Space Expansion' (2023) 63 Space Policy 101520.

[95] Ibid.

[96] Perpetua Adar, 'Space and the Future of Humanity: A TWAIL Critique of International Space Law and Space Discourse' (2022) 3 Third World Approaches to International Law Review 92.

the Treaty of Waitangi. It will also need to safeguard consultation and engagement with Māori on the impacts that corporate asteroid mining will have, not only on their access to dark skies for navigation and cultural reasons, but also on their lands. Unfortunately, the international human rights framework still provides no guidance on how the international community can monitor corporate human rights due diligence in outer space and ensure that space ventures designed for profit maximisation do not impede or ignore the interests and rights of Indigenous Peoples. An additional challenge is that many of the space companies operating in New Zealand are also registered in the United States.

While the United States is a signatory party to the International Covenant on Civil and Political Rights (but not the International Covenant on Economic, Social and Cultural Rights), it has unfailingly contested its extraterritorial application. It has also rather abruptly withdrawn from negotiations on the UN Draft Treaty on Business and Human Rights for the same reason. Future research needs to consider how companies based in New Zealand can be encouraged to comply with human rights law and safeguard Māori rights and interests. Space companies should be required to consult with the Indigenous community to determine how their activities and operations in outer space could impact them, especially their access to dark skies. If consultations reveal that their asteroid mining operations will have a detrimental impact, space companies should be mandated to take steps to resolve these concerns.

Asteroid mining is largely unregulated in New Zealand apart from launches. Efforts to develop a legislative regime for asteroid mining will also likely be challenging at best. This chapter could thus be described as disappointing: like a lot of scholarly research in outer space law and policy, it seeks to examine why a particular challenge has arisen and why any attempts to resolve it are likely to be difficult. Nonetheless, the consequences for New Zealand of not regulating asteroid mining operations launched from its territory are likely to be far-reaching. Without regulation, grave environmental damage in outer space is likely to occur and the interests and rights of Māori will likely continue to be marginalised in the quest for corporate profit. Space companies in New Zealand need to be able to demonstrate that they can conduct asteroid mining safely and with integrity. Their risk assessment, mitigation and remediation processes will also need to fully consider the potential risks of their operations not only to their own business but also to impacted stakeholders, especially Māori. With a proactive Space Agency and dedicated Space Minister, time will tell whether asteroid mining will become a legislative priority.

Dr. Anna Marie Brennan is a Senior Lecturer (Above the Bar) in Law and a Borrin Foundation Women Leader in Law Fellow at the University of Waikato. She serves on the International Institute for Space Law's Working Group on the Moon Agreement and chairs the New Zealand chapter of the AI and Space Law Society. Dr Brennan has held academic posts at the University of Liverpool and University College Cork, and her research spans international criminal law, counter terrorism law, and outer space law. She is the author of Transnational Terrorist Groups and International Criminal Law (2018) and editor of The Future of Outer Space Law (2025), published by Routledge.

The UAE's Approach to Space Resource Governance: International Cooperation or National Interests?

Chelsea Francek

Abstract This chapter explores the United Arab Emirates' (UAE) evolving legal and policy framework for space resource utilisation, with particular emphasis on the 2023 Space Resources Regulation and the Emirates Mission to the Asteroid Belt. The mission, aligned with the UAE's 2030 National Space Strategy, reflects both scientific ambitions, expanding knowledge of solar system formation, and economic aspirations, including the potential for future asteroid mining. Against this backdrop, the chapter critically examines the UAE's attempt to position itself as a responsible leader in space governance while simultaneously advancing national and private-sector interests. It interrogates whether the 2023 Regulation establishes clear and enforceable guidelines, consistent with the principles of the Outer Space Treaty and related instruments, or whether it remains broad, anthropocentric, and ultimately tilted towards sovereign and commercial benefit. Furthermore, the analysis considers the extent to which UAE regulations, while domestically oriented, contribute to international governance efforts by promoting sustainable, peaceful, and equitable access to outer space resources. In doing so, the chapter situates the UAE's initiatives within broader debates on the balance between national regulation, private enterprise, and global frameworks for space resource governance.

Keywords United Arab Emirates · Space resource utilisation · Space Resources Regulation 2023 · Asteroid Belt Mission · Outer Space Treaty · Space governance

C. Francek (✉)
University of Waikato, Hamilton, New Zealand
e-mail: cf189@students.waikato.ac.nz

S. Bhat B and A. Variath (eds.), *Legal Responses to Space Resource Utilisation*, Space Law and Policy, https://doi.org/10.1007/978-981-95-8864-0_15

1 Introduction

In 2023, the United Arab Emirates ("UAE") Space Agency introduced its latest innovative space-focused project, the Emirates Mission to the Asteroid Belt.[1] This venture is an exploratory scientific mission that is meant to increase the understanding of the formation and composition of the entire solar system, showcasing the nation's determined initiatives and advanced technologies. It is also the start of a significant economic opportunity for the UAE, its private sector, as well as its global partners. The scientific objectives of the mission, as well as those within the broader 2030 National Space Strategy,[2] signify a desire for future asteroid mining and space resource utilisation. The UAE is attempting to position itself as an ambitious yet responsible leader in the space sector, reflected in its recent Space Resources Regulation,[3] which renews its commitment to international legal obligations and the sustainable use of outer space. This chapter will examine the UAE's full scope of legal obligations, regulations, and policies related to outer space activities, focusing on space resource utilisation. It will consider whether the 2023 Space Resource Regulation is sufficient to establish clear guidelines and enforcement mechanisms that, combined with the existing principles found within international treaties such as the Outer Space Treaty,[4] will truly ensure equitable access and use of outer space resources, or whether it is another broad, anthropocentric, and ambiguous document that fails to efficiently regulate, resulting in a sovereign benefit for UAE private industry shareholders and partners. Lastly, this chapter will assess the extent to which these regulations, while primarily applicable to activities originating within the UAE, contribute to broader international efforts by examining if and how the UAE is promoting a global regulatory framework, one that works to safeguard and preserve the peaceful and equitable use of outer space.

---

[1] 'Emirates Mission to the Asteroid Belt|Space Exploration |…' https://space.gov.ae/en/projects-and-initiatives/space-exploration/emirates-mission-to-the-asteroid-belt accessed 19 March 2025.

[2] 'National Space Strategy 2030|The Official Portal of the UAE Government' https://u.ae/en/about-the-uae/strategies-initiatives-and-awards/strategies-plans-and-visions/industry-science-and-technology/national-space-strategy-2030 accessed 12 March 2025.

[3] UAE Space Agency, 'Space Resources Regulation. Regulatory Framework on Space Activities of the United Arab Emirates.' <space.gov.ae>.

[4] UN General Assembly, '2222 (XXI). Treaty on Principles Governing the Activities of States in the Exploration and Use of Outer Space, Including the Moon and Other Celestial Bodies' (Outer Space Treaty). Available online at https://www.unoosa.org/oosa/en/ourwork/spacelaw/treaties/outerspacetreaty.html. The text of the treaty is reproduced in 33J. AIR L. 8r COM. 132 (1967).

1.1 Historical Foundation of the UAE's Space Ambitions

Guided by Zayed's vision, we eagerly anticipate scripting new achievements and raising our flag amidst the stars.[5]—H.H. Sheikh Hamdan bin Mohammed Al Maktoum

The United Arab Emirates (UAE) journey to the stars began only five years after the country's formal establishment.[6] In 1976, the "Father of the Nation", founder and former President Sheikh Zayed bin Sultan Al Nahyan, hosted a delegation from the National Aeronautics and Space Administration (NASA) to discuss his vision for the UAE, one that saw his fledgling state becoming a competitive spacefaring nation.[7] This was a move that acted not only in the interest of diplomatic curiosity, but as a bold statement of the UAE's long-term national vision. Even from the earliest days, the UAE positioned itself as a nation looking beyond terrestrial borders.

Over the following decades, these visions began to form into tangible institutions and programs. In 1997, the UAE secured a majority investment of Thuraya Telecommunications Company (Thuraya), a provider of mobile satellite communications, that signalled a commitment to private sector space capabilities.[8] By 2006, the government established the Emirates Institution for Advanced Science and Technology (EIAST) to bolster research and development in space. EIAST would later be merged with the Mohammed bin Rashid Space Centre (MBRSC), anchoring the UAE's civil space activities in a single institution.[9] As stated by the current Director General of MBRSC, H.E. Salem Humaid AlMarri: "MBRSC stands as a testament to the UAE's role as a beacon of science, innovation, and hope, solidifying our legacy in space exploration and inspiring the next generation of dreamers and innovators".[10] These words illustrate how the MBRSC and its space initiatives function as a strategic tool for both national development and international positioning.

Building on its space ambitions, the UAE now required a more formal and centralised governance structure. A few years later, the UAE Space Agency (Agency) was established by Federal Decree No. 1 of 2014 (Decree).[11] The Agency holds

---

[5] 'MBRSC' https://www.mbrsc.ae/ accessed 20 March 2025.

[6] 'History | UAE Embassy' (History | UAE Embassy) https://www.uae-embassy.org/discover-uae/history accessed 19 March 2025; see also Unwalla K, Avakyan S and Jeong V, 'Global Outer Space Guide: United Arab Emirates (UAE)' https://www.nortonrosefulbright.com/en/knowledge/publications/e2a4dae0/global-outer-space-guide-uae accessed 19 March 2025.

[7] 'Founder of UAE—Sheikh Zayed Bin Sultan Al Nahyan' https://www.mofa.gov.ae/en/the-uae/the-founder accessed 19 March 2025.

[8] Tair HA, 'Securing UAE's Independent Space Industry: Building, Launching, Operating'.

[9] MBRSC functions like a science and technology hub and incubator for the UAE Space Agency. The UAE Space Agency, 'History of UAE Space Sector' (UAE Space Agency) https://space.gov.ae/en/about-us/history-of-uae-space-sector accessed 20 March 2025; see also Abashidze A, Solntsev A and Mirzaee S, 'The United Arab Emirates Approach towards International Space Law: Divergence or Convergence?' (2023) 213 Acta Astronautica 81.

[10] 'About—MBRSC' https://www.mbrsc.ae/about/ accessed 26 May 2025.

[11] UAE Space Agency n 9.

the responsibility to establish partnerships, support academia, advance space exploration throughout the Middle East, and manage the UAE's investment in space.[12] More specifically, the Decree states that the Agency must develop the National Space Framework, which consists of four components: Space Policy, Space Strategy, Space Law, and Space Regulations.[13] The establishment of this future-focused governance framework was a turning point in the UAE's approach to space. The creation of the Agency, supported by the Decree, illustrated the nation's serious commitment to space exploration, as well as its awareness of the responsibilities that come with this level of ambition. As previous legal scholars have noted, emerging space powers often shift strategy from traditional powers towards more inclusive participation.[14] This carefully crafted blend of responsibility and ambition signified the UAE's desire to compete as a leading spacefaring nation not just for the Middle East, but for the world.

It has been over a decade since the establishment of the UAE Space Agency, and the nation has forged ahead with its spacefaring ambitions. The 2021 Emirates Mars Mission, *Amal* (also known as "Hope probe"), was an internationally recognised milestone that made the UAE the fifth country or organisation to reach Mars.[15] Beginning in 2006, the *Amal* mission was curated to develop the UAE's scientific community, as well as further awareness and education of the nation's space program and activities.[16] Building on its success, in 2023, the UAE announced its next ambitious endeavour, Unveiling the Secrets of the Asteroid Belt: The Emirates Mission (EMA). The EMA is "not just about scientific discovery, but also a catalyst for the UAE's space industry".[17] Through this mission, the UAE hopes to contribute to the knowledge of the origins and composition of the entire solar system and search for the "fundamental building blocks of life within the asteroid belt". Simultaneously, the nation intends to evaluate the possibility of resource extraction within the belt, eyeing asteroid mining as a future industry that could lead to substantial economic growth and increased global partnerships.[18] Another project that the UAE has established with the aim of global collaboration is the Emirates Mars Mission. This, in

[12] UAE Space Agency n 9.

[13] Abashidze n 9.; see also Al Rashedi N, Al Shamsi F and Al Hosani H, 'UAE Approach to Space and Security' in Kai-Uwe Schrogl (ed), Handbook of Space Security: Policies, Applications and Programs (Springer International Publishing 2020) 621–652.

[14] Cross MKD and Pekkanen SM, 'Introduction. Space Diplomacy: The Final Frontier of Theory and Practice' https://brill.com/view/journals/hjd/18/2-3/article-p193_1.xml accessed 27 May 2025; see also Jakhu RS, National Regulation of Space Activities (Springer Science & Business Media 2010); see also Beard JM, 'Soft Law's Failure on the Horizon: The International Code of Conduct for Outer Space Activities' (2016) 38 University of Pennsylvania Journal of International Law 335.

[15] The US, Russia, India, and the European Space Agency had all previously reached Mars. Clark S, 'Spacewatch: UAE's Hope Mars Mission Hits First Anniversary' The Guardian (23 July 2021) https://www.theguardian.com/science/2021/jul/23/spacewatch-uaes-hope-mars-mission-hits-first-anniversary accessed 27 May 2025.

[16] Clark n 15.

[17] 'Emirates Mission to the Asteroid Belt' (UAE Space Agency) https://space.gov.ae/en/projects-and-initiatives/space-exploration/emirates-mission-to-the-asteroid-belt accessed 12 March 2025.

[18] 'Emirates Mission to the Asteroid Belt' n 17.

addition to related projects such as Mars 2117, has been created to understand the Martian climate and to determine how humans can establish a long-term presence on the planet within the next one hundred years.[19] As these missions and projects grow in scale and complexity, so does the need for a strategic, transparent, and adaptable regulatory framework. The UAE has begun to address this need through its 2030 Space Strategy.

1.2 The 2030 UAE Space Strategy

Following the formation of the Agency by Decree, the next formal establishment within the UAE's space sector was the release of the National Space Strategy 2030 (Strategy). As mentioned, the Strategy is one of the four components of the UAE's space regulatory framework, in addition to the National Space Policy, Space Law, and Regulations. In order to achieve policy objectives, those which set forth the nation's principles, ambitions, and directions in the space domain, the Strategy states its vision and then specifically defines several national initiatives and programs.[20]

The core vision of the Strategy is for the UAE to "be one of the most pioneering and advanced countries in the field of space".[21] In order to realise this vision, the Strategy has outlined six strategic goals: (1) provision of competitive and leading space services; (2) development of advanced local capacities in space technology manufacturing and R&D; (3) launching inspiring space scientific and exploration missions; (4) creating space culture and expertise; (5) promote effective local and global partnerships and investments in the space industry; and (6) ensure a supporting legislative framework and infrastructure to match the future developments in the sector.[22] The Strategy includes eighteen programs and seventy-one associated initiatives to illustrate how the nation will practically implement these goals, including the launch of targeted educational programs, develop the manufacturing facilities for space technologies, create the Emirati astronauts program, design a roadmap for the stages of the Mars 2117 program, focus on partnerships and agreements with leading International Space organisations such as NASA, the Japan Aerospace Exploration Agency (JAXA), Roscosmos, and the Indian Space Research Organization (ISRO), and attract the private sector into space exploration activities (i.e. space resource utilisation).[23]

---

[19] 'Space Exploration' (UAE Space Agency) https://space.gov.ae/en/projects-and-initiatives/space-exploration accessed 28 May 2025.

[20] 'National Space Strategy 2030 | The Official Portal of the UAE Government' https://u.ae/en/about-the-uae/strategies-initiatives-and-awards/strategies-plans-and-visions/industry-science-and-technology/national-space-strategy-2030 accessed 12 March 2025 (National Space Strategy 2023).

[21] National Space Strategy 2023 n 20.

[22] National Space Strategy 2023 n 20.

[23] National Space Strategy 2023 n 20.

This Strategy is not simply a domestic, small-scale proposal. It is a carefully executed strategic plan with a nod to the nation's global aspirations, drafted with the support of public and private partners, as well as international collaborators. The Strategy explicitly affirms its commitment to align with international policies and laws, such as the Long Term Sustainability Guidelines approved by the United Nations Committee on the Peaceful Use of Outer Space (COPOUS) and the Treaty on the Peaceful Use of Outer Space (OST).[24] There is also a section of the Strategy which delineates seven main global directions in the space sector: (1) increasing number of players, (2) growth in the space economy, (3) bigger international cooperation, (4) a leap in space security and military utilisation, (5) higher quality technologies, (6) new activities, and (7) crowded space environment.[25] This is another example of the UAE positioning itself as an ambitious yet responsible spacefaring nation, a move which could either be a show of global leadership, or an attempt to be first-in-time and an assertion of national dominion.

2 The 2023 Space Resources Regulation

As mentioned, the Strategy outlines seven main global directions. The "New Activities" direction includes specifics such as sub-orbital flights, launched from space, space tourism, and space resources extraction.[26] Space resource utilisation is also mentioned as a main research area within the space sector, and as an initiative under the program titled "Sustainability and expansion of the national space exploration program scope".[27] The goal within this program is to "attract the private sector into the space exploration activities, including space resources utilization".[28] As noted by scholars, space exploration and utilisation not only increase human knowledge of the universe, but it can also improve human life here on Earth through diversified communication, increased observation, and navigation abilities.[29] Further, a space resource utilisation undertaking like off-Earth mining is predicted to become a highly profitable industry and an indicator of economic growth for both public and private actors.[30]

It is important to define what "space resource utilisation" (SRU) is, or at least the generally accepted understanding. In the United States, a space resource is defined

[24] National Space Strategy 2023 n 20.

[25] National Space Strategy 2023 n 20.

[26] National Space Strategy 2023 n 20.

[27] National Space Strategy 2023 n 20.

[28] National Space Strategy 2023 n 20.

[29] Al Rashedi n 13.

[30] Chernykh I and Gugunskiy D, 'Space Resource Utilization as Sustainable Economic Growth Perspectives: Legal Aspects' in Agnessa O Inshakova and Elena I Inshakova (eds), New Technology for Inclusive and Sustainable Growth: Technological Support, Standards and Commercial Turnover (Springer 2022) https://doi.org/10.1007/978-981-16-9808-8_18 accessed 28 May 2025.

as "an abiotic resource in situ in outer space".[31] Similarly, Japan's 2021 Act on the Promotion of Business Activities for Exploring and Developing Space Resources states that space resources "means water, minerals, and other natural resources that exist in outer space, including the Moon and other celestial bodies".[32] Additionally, according to NASA, the "use of space resources, commonly referred to as In Situ *Utilization* (ISRU), involves the processes and operations to harness and utilise resources in space (both natural and discarded) to create products for subsequent use".[33] Examples of potential resources include water, natural gases (i.e. hydrogen, carbon, nitrogen), the various metals and minerals found in extraterrestrial soils, solar energy, regions of permanent light and darkness, the vacuum and zero-gravity provided by space itself, waste produced by space travellers or settlers, and discarded or damaged hardware (i.e. defunct satellites).[34]

SRU is much more than simple resource extraction via machinery. The process and execution involved are cross-disciplinary, as it typically integrates space systems, robotics, materials handling and beneficiation, chemical process engineering, as well as a required knowledge of planetary geology.[35] While seemingly complex, SRU could provide a critical return and aid space exploration and commercialisation efforts by increasing the understanding of the resources themselves, allowing a characterisation and mapping process, producing consumables, facilitating the construction of the necessary space infrastructure, producing and storing energy, as well as other manufacturing needs.[36]

As with any great venture, SRU must be regulated to ensure the process and result are safe, transparent, and equitable. The UAE has already taken its own initiative to show how it will regulate its space resource utilisation activities. In 2023, the UAE issued four new decisions under its Federal Law No. (12) of 2019 (the Law) on the Regulation of the Space Sector, all intended to set rules of conduct for space-related activities.[37] One of these decisions was the approval of the Space Resources Regulation (2023 Regulation). The 2023 Regulation has jurisdiction over the following:

---

[31] 'Definition: Space Resource from 51 USC § 51,301(2) | LII/Legal Information Institute' https://www.law.cornell.edu/definitions/uscode accessed 29 May 2025; see also 'The Space Review: US Space Resources Law Needs Clarification by Congress' https://www.thespacereview.com/article/4945/1 accessed 29 May 2025.

[32] Act on the Promotion of Business Activities for Exploring and Developing Space Resources 2021. Japan Ministry of Justice. (Act No 83) Available at https://www8.cao.go.jp/space/english/resource/documents/act83_2021.pdf.

[33] Sanders GB, 'SPACE RESOURCE UTILIZATION: TECHNOLOGIES AND POTENTIAL SYNERGIAM WITH TERRESTRIAL MINING' https://ntrs.nasa.gov/api/citations/20150003499/downloads/20150003499.pdf accessed 29 May 2025.

[34] Sanders n 33.

[35] Hadler K and others, 'A Universal Framework for Space Resource Utilisation (SRU)' (2020) 182 Planetary and Space Science 104811.

[36] Sanders n 33.

[37] Abreo S, 'Key Features of the New United Arab Emirates Regulations on Space Activities' (Watson Farley & Williams, 10 August 2023) https://www.wfw.com/articles/key-features-of-the-new-united-arab-emirates-regulations-on-space-activities/ accessed 29 May 2025.

> a Person who carries out Space Activities, Space-Support Flight activities, High-altitude Activities, Space date management and distribution activities, or any other activities related to the space sector subject to The Law.[38]

This jurisdiction extends over activities that are carried out by UAE nationals, private organisations that have a main headquarters in the UAE, and even foreign entities with a subsidiary in the UAE.[39] Additionally, the activities to be regulated are those that target the "Identified Area", which is essentially the UAE's definition of outer space: "an area eighty kilometres or more above the average level of sea level".[40] Further, the UAE also provides its own definition of space resources in the 2023 Regulation as "any non-living resources present in outer space, including minerals and water".[41] While there is no explicit acknowledgement of preceding definitions from other nations, it has been noted that the UAE's understanding closely resembles the United States' reference to "abiotic", or non-living, resources.[42] As for an understanding of the space resources activities themselves, the Regulation stipulates the following:

> Activities relating to the exploration, exploitation and use of space resources, whether for commercial, scientific or other purposes, including the extraction, recovery, refining, processing, utilization, ownership, purchase, sale, trade, transport or storage of space resources, as well as logistics related activities conducted in the identified area, such as transporting, storing or supplying space resources.[43]

The specific conditions for those wanting to participate in space resource activities within the UAE jurisdiction, as well as the Agency's Board of Directors themselves, are outlined in Article (4) and Article (5). These conditions include safety and compliance requirements such as the need for operators to take all necessary measures to prevent or mitigate risks to other persons, damage to property, harm or contamination to the environment, creation of space debris, or harmful interference with other space activities.[44] The Regulation continues to set conditions to ensure that any proposed activities are in compliance with international legal obligations, not just those that prohibit any interference with lawful activities of other States, but also those that protect intellectual property rights and commercially sensitive data.[45]

[38] UAE Space Agency, 'Space Resources Regulation. Regulatory Framework on Space Activities of the United Arab Emirates.' <space.gov.ae> (2023 Regulation).

[39] 2023 Regulation n 38.

[40] Interesting to note that the UAE defined parameter for outer space at eighty kilometers is a slight deviation from the universally accepted parameter of outer space, one which is known as the "Karman Line" and usually set at one hundred kilometers above sea level McDowell JC, 'The Edge of Space: Revisiting the Karman Line' (2018) 151 Acta Astronautica 668; see also 2023 Regulation n 38.

[41] 2023 Regulation n 38.

[42] Janikowski A, 'Bold Steps Forward: The Investment Impact of Enacting Space Resources Legislation' [2025] Space Policy 101675.

[43] 2023 Regulation n 38.

[44] 2023 Regulation n 38.

[45] 2023 Regulation n 38.

To align with the broader strategic goals of the Agency, particularly the objective to increase international collaboration, the Regulation also sets conditions that promote the sharing of scientific information pertaining to these space resources activities. Although a strong sentiment, there is no mandate to share information. Rather, there is a simple encouragement that the Agency share "to the greatest extent feasible and practicable, on a good-faith basis, and in accordance with the applicable national laws".[46] In actuality, none of these conditions has any mandated component to them, but are simply considerations that the Agency must take prior to granting any official authorisation to the potential operator.

It is important to critically consider whether this self-regulation may simply be a tactic to ensure that stronger international regulation efforts do not occupy the field and dictate the potentially highly profitable operations for the UAE. But, just as it is important to critically examine the UAE's self-regulation intent, it is equally important to examine the immense effort the UAE has put into international legal obligations and compliance. The alternative perspective is that this could be a genuine attempt for the UAE to secure a global leadership position, using the field of space resource utilisation to do so.

3 International Legal Obligations and Compliance

The United Arab Emirates has developed not only its national strength but, strategically, its place amongst global partners. From its inception, the UAE has been explicitly clear about how important international cooperation is to its national space program, ensuring that diplomacy and multilateral participation have been incorporated into its national space strategy. This section examines these global partnerships, the UAE's role in soft power diplomacy, and its position with international obligations and norms related to space resources.

The UAE has demonstrated an overt acceptance and support of international legal obligations and space governance by signing foundational outer space treaties. The core instruments drafted by members of the United Nations include the following: the Outer Space Treaty,[47] the Registration Convention,[48] the Liability Convention,[49] and the Rescue and Return Agreement.[50] These treaties, although all primarily drafted decades ago in the 1960s and 1970s, represent modern outer space law. They also

[46] 2023 Regulation n 38.

[47] Outer Space Treaty n 4.

[48] UN General Assembly, '3235 (XXIX). Convention on Registration of Objects Launched into Outer Space.' (1974). Available online at https://www.unoosa.org/oosa/en/ourwork/spacelaw/treaties/registration-convention.html (Registration Convention).

[49] UN General Assembly, '2777 (XXVI). Convention on International Liability for Damage Caused by Space Objects'. Available online at https://www.unoosa.org/oosa/en/ourwork/spacelaw/treaties/liability-convention.html (Liability Convention).

[50] UN General Assembly, '2345 (XXII) Agreement on the Rescue of Astronauts, the Return of Astronauts and the Return of Objects Launched into Outer Space' (1967). Available online

emphasise critical principles that spacefaring nations are intended to uphold, such as the peaceful use of outer space, the non-appropriation of celestial bodies, and the obligation of signatory member States to be responsible for their own activities within and related to outer space.[51] By signing these treaties, the UAE has formally accepted these international norms and obligations.

In addition to its participation in international treaties, the UAE claims to demonstrate its commitment to international collaboration through its use of partnerships and bilateral agreements to successfully achieve its space objectives. For example, the UAE has formed agreements with established spacefaring nations such as the United States, France, Japan, and Russia, all of which have facilitated the development of its national technical and regulatory capacities. Specifically, the UAE's "Hope probe", discussed earlier as the spacecraft that would monitor the atmosphere of Mars, was launched in collaboration with the University of Colorado Boulder (CU Boulder) and NASA.[52] In 2014, CU Boulder partnered with the UAE as the leading U.S. scientific academic partner on the Emirates Mars Mission (EMM).[53] Over the next six years, researchers, scientists, and engineers at the Laboratory for Atmospheric and Space Physics (LASP) at CU Boulder worked on a mission led by the Mohammed Bin Rashid Space Centre (MBRSC) in the UAE to create the "Hope probe".[54] This key partnership illustrates how international collaboration has enabled the UAE, an emerging space power, to execute an interplanetary mission within six years of establishing its space agency.

Further, to perform the EMM after research and development, the UAE partnered with the Japanese organisation iSpace Inc.,[55] utilising a Falcon 9 rocket from SpaceX.[56] This demonstrates an integration of both public and private partners, spanning jurisdictions, coming together to achieve objectives that support those involved individually, as well as collectively. It embeds the UAE within a network

---

at https://www.unoosa.org/oosa/en/ourwork/spacelaw/treaties/introrescueagreement.html (Rescue and Return Agreement).

[51] Outer Space Treaty n 4; see also Registration Convention n 48; see also Liability Convention n 49; see also Rescue and Return Agreement n 50.

[52] Dearnley G, 'Everything You Need to Know About CU's Involvement in the Emirates Mars Mission' (University of Colorado Boulder, 31 March 2021) https://www.colorado.edu/coloradan/2021/03/31/everything-you-need-know-about-cus-involvement-emirates-mars-mission accessed 3 June 2025.

[53] Dearnley n 52.

[54] Dearnley n 52.

[55] iSpace Inc. is a "lunar exploration company with a vision to extend human presence into outer space", wanting to utilize the Moon's water resources to develop a space-based economy. "Imagine the Moon supporting construction, energy, steel procurement, communications, transportation, agriculture, medicine, and tourism… We believe that by 2040 the Moon will support a population of 1,000, with 10,000 people visiting every year. ispace will be instrumental in supporting life on Earth through space-based infrastructure." 'Ispace' (ispace) https://ispace-inc.com/ accessed 3 June 2025.

[56] 'UAE Partners with Japan's Ispace to Send Rover to the Moon in 2022' Reuters (14 April 2021) https://www.reuters.com/world/middle-east/uae-partners-with-japans-ispace-send-rover-moon-2022-2021-04-14/ accessed 3 June 2025.

of shared space governance norms, many of which include memoranda of understanding (MOUs) in bilateral agreements, such as the 2016 Framework Agreement signed by the UAESA and NASA.[57] These agreements are immensely important because, although the legal enforceability may be a grey area, they show an alignment with universal best practices and a commitment to international cooperation. As mentioned by H.E. Dr Ahmad Belhoul Al Falasi, Minister of State for Higher Education and Advanced Skills and Chairman of the UAE Space Agency:

> This significant agreement builds on our bilateral relationship with NASA and the US space sector. It also reflects the deep economic, cultural and diplomatic ties shared between the United States of America and the United Arab Emirates. The global space sector is built on international collaboration, and we welcome these substantial opportunities to work closer than ever before with our counterparts at NASA.[58]

In addition to bilateral agreements, the UAE is an active participant in the United Nations Committee on the Peaceful Use of Outer Space (COPUOS), established by the UN General Assembly in 1959 to "govern the exploration and use of space for the benefit of all humanity: for peace, security and development".[59] This shows a strong desire for the UAE to be a part of direct efforts that shape global space governance. In a statement to COPUOS, the UAE again stressed the importance of partnerships to critical space programs, such as the EMM as well as the Emirates Mission to the Asteroid Belt (EMA): "With the EMA mission we expanded strategic partnerships with public, private, and academic entities locally …. In addition, international partners such as the Italian Space Agency, the University of Colorado Boulder, Arizona State University, and Northern Arizona University". An effort was made to note that the EMA was created to explore the asteroid belt "with scientific objectives including understanding the origins and evolution of water-rich asteroids and evaluating their potential as resources for future space missions".[60] The UAE went on to describe

---

[57] In 2016, the UAESA and NASA signed a Framework Agreement (FA) that "established a framework for areas of cooperation in ground-based research; sub-orbital research; research and flight activities in low-Earth orbit (LEO); and human and robotic exploration in the vicinity of the moon, on the lunar surface, and beyond". Later in 2018, the two parties signed an Implementing Agreement, one which falls under the FA, that "outlines cooperation across a range of areas related to space exploration and human spaceflight". This was signed at the International Astronautical Congress held in Bremen, Germany. UAE Space Agency, 'NASA and UAE Space Agency Sign Historic Implementing Arrangement for Cooperation in Human Spaceflight' (UAE Space Agency, 2 October 2018) https://space.gov.ae/en/media-center/news/2/10/2018/nasa-and-uae-space-agency-sign-historic-implementing-arrangement-for-cooperation-in-human accessed 3 June 2025.

[58] UAE Space Agency, 'NASA and UAE Space Agency Sign Historic Implementing Arrangement for Cooperation in Human Spaceflight' n 52.

[59] The UAE chaired the COPUOS in 2022. 'COPUOS' (Committee on the Peaceful Uses of Outer Space) https://www.unoosa.org/oosa/en/ourwork/copuos/index.html accessed 3 June 2025; see also 'UAE to Chair Committee on The Peaceful Uses of Outer Space' (2 June 2022) https://www.mofa.gov.ae/en/mediahub/news/2022/6/2/02-06-2022-uae-outer-space accessed 3 June 2025.

[60] 'Agenda item 15: Space exploration and innovation. Statement by the United Arab Emirates'. 67th Committee on the Peaceful Uses of Outer Space (COPUOS). https://www.unoosa.org/documents/pdf/copuos/2024/statements/15_UAE.pdf accessed 30 May 2025.

their commitment to develop national capabilities, such as its "Space Means Business" workshop, to engage space companies and highlight opportunities within space exploration, as well as the continued initiatives that will develop the next generation of scientists and engineers.[61] Through this participation, the UAE acknowledges the importance of global partnerships, illustrates its contribution while discussing its aspirations, and reaffirms its commitment to international policies and principles, such as those found within the OST.

Another demonstrated agreement to international norms related to space exploration and utilisation is the signing of the Artemis Accords (Accords). In 2020, the Accords were established by NASA, in coordination with the U.S. Department of State, with the assistance and support of seven other initial signatory nations, including the UAE.[62] Her Excellency Sarah bint Yousef Al Amiri, Minister of State for Advanced Technologies and Chairwoman of the UAE Space Agency, signed on behalf of the UAE on October 13th, 2020.[63] The Accords, which have fifty-five signatories as of May 2025, acknowledge the complexities presented by the many countries and companies conducting missions around the Moon and were developed to "provide a common set of principles to enhance the governance of the civil exploration and use of outer space".[64] Further, this coordination effort reinforces the commitment to previously established legal obligations and international principles, such as those delineated in the OST, the Registration Convention,[65] the Rescue and Return Agreement,[66] as well as other "best practices and norms of responsible behaviour for civil space exploration and use".[67]

The Accords make an explicit reference to space resources, reminding the signatories that any mining that takes place on celestial bodies, including the Moon, Mars, comets, and asteroids, needs to be executed in a manner that complies with previously established international obligations, such as the OST, and must support safe and sustainable space activities.[68] Additionally, the signatories must understand that any resource extraction "does not inherently constitute national appropriation under Article II of the Outer Space Treaty" and that any other agreements related to resource

[61] 'Agenda item 15: Space exploration and innovation. Statement by the United Arab Emirates' n 55.

[62] 'Artemis Accords—NASA' https://www.nasa.gov/artemis-accords/ accessed 3 June 2025.

[63] 'The Artemis Accords. Principles for cooperation in the civil exploration and use of the Moon, Mars, comets, and asteroids for peaceful purposes' (2020). Available online at https://www.nasa.gov/wp-content/uploads/2022/11/Artemis-Accords-signed-13Oct2020.pdf?emrc=683e48c1e4c79.

[64] 'Artemis Accords—NASA' n 63.

[65] UN General Assembly, '3235 (XXIX). Convention on Registration of Objects Launched into Outer Space.' (1974) Available online at https://www.unoosa.org/oosa/en/ourwork/spacelaw/treaties/registration-convention.html.

[66] UN General Assembly, '2345 (XXII) Agreement on the Rescue of Astronauts, the Return of Astronauts and the Return of Objects Launched into Outer Space' (1967). Available online at https://www.unoosa.org/oosa/en/ourwork/spacelaw/treaties/introrescueagreement.html.

[67] 'Artemis Accords—NASA' n 63.

[68] 'Artemis Accords—NASA' n 63.

extraction should be consistent with the OST.[69] Seemingly, this is intended to say that countries can extract resources from celestial bodies, but that use of that area does not equate to a proprietary ownership right. Instead, it suggests the use of Safety Zones to facilitate the resource extraction.[70] It is important to note that there has been ample criticism of this approach, as it is seen to avoid the development of a multilateral governance framework, one that is acceptable to the entire international community as well as the commercial sector, it lacks key definitions and parameters, and that its reach extends beyond what has been agreed to within international treaties.[71] Nonetheless, the UAE's involvement in both COPUOS and the Artemis Accords can be interpreted as its keen awareness of the need for multilateral partners on a global scale in order to achieve its nationally leaning initiatives, including the appropriation of any space resources.

4 Soft Power and Diplomatic Efforts

The UAE's international space strategy is part of its broader ambitions in diplomatic efforts. For example, the UAE Centennial 2017, a long-term plan that extends over five decades, focuses on initiatives curated to "fortify the country's reputation and its soft power".[72] In the context of space exploration, security, and general outer space affairs, soft power is defined as the "ability to promote the global agenda for space exploration and security, to develop and to maintain international space cooperation ties, and to increase one's own international prestige through demonstration of ambitious and impressive achievements in space activity".[73] Through its high-profile missions, such as the Emirates Mars Mission and the Emirates Mission to the Asteroid Belt, the UAE is clearly attempting to increase its international prestige and reputation amongst spacefaring nations. Further, its participation in international governance efforts, such as the Artemis Accords and COPUOS, demonstrates the

[69] 'Artemis Accords—NASA' n 63.

[70] 'Artemis Accords—NASA' n 63.

[71] McKeown B, Dempster AG and Saydam S, 'Artemis Accords: Are Safety Zones Practical for Long Term Commercial Lunar Resource Utilisation?' (2022) 62 Space Policy 101504; see also de Zwart M, 'The Impact of the Artemis Accords on Resource Extraction' (John Wiley & Sons, Ltd 2022) In-Space Manufacturing and Resources, 351–368; see also Gilbert AQ, 'Implementing Safety Zones for Lunar Activities under the Artemis Accords' (2023) 10 Journal of Space Safety Engineering 103.

[72] The core four pillars of this plan include a future-focused government, excellent education, a happy and cohesive society, and a diversified knowledge economy, all of which included a specific mention of focus on the space sector. 'UAE Centennial 2071 | The Official Portal of the UAE Government' https://u.ae/en/about-the-uae/strategies-initiatives-and-awards/strategies-plans-and-visions/innovation-and-future-shaping/uae-centennial-2071 accessed 30 May 2025.

[73] Pekkanen SM and Blount PJ, The Oxford Handbook of Space Security (Oxford University Press 2024).

UAE's ability and willingness to promote space security and sustainability. Additionally, on a regional level, the UAE is showcasing its preparedness to take responsibility and leadership within the space sector, such as its promotion of the Arab Space Cooperation Group (ASCG). The ASCG, currently sitting at fourteen members, has a vision for "cooperation amongst the Arab region and its leadership in the field of space", created with the intent to assist these nations in establishing Arab space agencies and sustainable space programs, supporting and building space capacities for ASCG members, stimulating research and development, inspiring education and developing capabilities, as well as an "effective presence on the international scene and enhanced participation in global forums".[74]

These diplomatic efforts have lasting implications for not just the UAE but for every global citizen. By positioning itself as a responsible and inclusive space actor, the UAE is building a reputation that others will rely upon to promote consensus-based approaches to space resource governance. The UAE functions as a bridge between the North, South, East, and West, giving it substantial influence in future debates over legal and regulatory frameworks for space resource utilisation.

5 Compliance, Coherence, or Convenience: The UAE Regulation Versus International Space Law

While the UAE has signalled strong support for international space law through its ratification of treaties, participation in global governance efforts, and diplomatic duties, the question remains as to whether its national regulatory framework for space resource utilisation, the 2023 Space Resource Regulation, is fully aligned with the intent and obligations of foundational instruments. National implementation of international legal principles is obviously a convincing tactic to demonstrate a commitment to global governance. But inserting a written line item that simply states that one complies is not comparable to reading the national regulation as a whole and determining that its practical implementation, as well as intent, is truly coherent and in compliance with international principles. This section will discuss a few key concerns with the UAE Regulation and use these points to analyse if the Regulation falls within the guardrails of international space law, or whether it reflects a more pragmatic approach that prioritises national and commercial convenience.

Article 7 of the Regulation defines the ownership rights over space resources, allowing operators the "right of ownership, purchase, sale, trade, transportation, storage, use, or dispose any of Space Resources extracted…".[75] Yet, the OST's Article II states that "Outer space, including the moon and other celestial bodies, is not subject to national appropriation by claim of sovereignty, by means of use or

[74] 'Vision. Arab Space Cooperation Group.' https://www.arabspacegroup.org/vision.html accessed 3 June 2025.

[75] Article 7. 2023 Regulation n 38.

occupation, or by any other means".[76] If we think about the practical implication of the Regulation's extended ownership right, it is allowing an operator to occupy an area and claim full ownership over any resources in that authorised area. This scale of an operation will not be short-term, nor non-invasive. The amount of capital, time, and resources that companies and nations will be forced to invest in these mining operations will make certain that these are lengthy, long-term, and invasive missions. That level of disruption will surely make it difficult for others to utilise the space as well, which defeats the purposeful intent that outer space is to be a shared space. It is not too drastic to presume that the operators will leave only when they feel they have exhausted the available resources. If there are no enforcement mechanisms that state the operators must leave the area as they discovered it, does that not destroy the area for future use and enjoyment? This conundrum has been discussed by previous legal scholars: by allowing property rights over extracted resources, nations risk creating de facto appropriation rights over the celestial bodies themselves.[77]

Additionally, it is also imperative to remember the broader intent behind the OST found in Article I: "The exploration and use of outer space, including the moon and other celestial bodies, shall be carried out for the benefit and in the interests of all countries, irrespective of their degree of economic or scientific development, and shall be the province of all mankind".[78] If the UAE Regulation requires potential operators to first be approved by the Agency, the UAE now has selective oversight over who can conduct activities that fall within the UAE's jurisdiction. This could open the door to the nation, selectively allowing only those who have created substantially beneficial agreements with the UAE or its representatives, a situation which does not seem to represent "all of mankind". Further, the operators, those that have the economic and technological capabilities to reach these celestial bodies and conduct the required operations, have complete ownership rights over the extracted space resources. There is no mention of how the activities and use will "be carried out for the benefit and in the interests of all countries, irrespective of their degree of economic or scientific development".[79] Many have argued that outer space needs to be governed as a commons area, as this would ensure shared access, use, and distribution of resources.[80] The fact that these national regulatory frameworks, like the UAE's Regulation, have failed to include any provision that would ensure an

---

[76] Article II. Outer Space Treaty n 4.

[77] Tronchetti F, 'The Exploitation of Natural Resources of the Moon and Other Celestial Bodies' [2009] The Exploitation of Natural Resources of the Moon and Other Celestial Bodies https://www.academia.edu/67862642/The_Exploitation_of_Natural_Resources_of_the_Moon_and_Other_Celestial_Bodies accessed 4 June 2025; see also Su J, 'Legal Status of Abiotic Resources in Outer Space: Appropriability, Ownership, and Access' (2022) 35 Leiden Journal of International Law 825.

[78] Article I. Outer Space Treaty n 4.

[79] Outer Space Treaty n 4.

[80] Janssen MA and Yap X-S, 'Governing Outer Space as a Commons Is Critical for Addressing Commons on Earth' (2024) 18 International Journal of the Commons https://thecommonsjournal.org/articles/10.5334/ijc.1378 accessed 6 June 2025; see also Pic P, Evoy P and Morin J-F, 'Outer Space as a Global Commons: An Empirical Study of Space Arrangements' (2023) 17 International Journal of the Commons https://thecommonsjournal.org/articles/10.5334/ijc.1271 accessed 6 June

equitable distribution of resources is arguably a violation of international space law principles. But, until there is further clarification within space law and regulation in general, it is difficult to claim conclusively.

6 Conclusion and Future Directions

Despite the UAE's ambitions and increasing responsibility in the space sector, there are clearly still regulatory and strategic gaps to be addressed, particularly within the context of space resources utilisation. As discussed in this chapter, the 2023 Space Resources Regulation is upheld as a milestone national initiative for the UAE, but there is a question about the practical implementation and true intent. As current regulation stands, space activities are to be governed by international standards and policies, such as those set within the Outer Space Treaty (OST) that attempt to secure the peaceful use and equitable access of outer space.[81] The UAE's Regulation states that the operators conducting space resource activities that fall within the jurisdiction of the Regulation "shall not carry out any activities…which jeopardise or risk jeopardising the ability of the State to comply with any international legal obligations".[82] This seems like a reasonable condition, but, other than requiring a few risk assessments, what are the enforcement mechanisms that the UAE is implementing to ensure that these operators are doing all that is necessary in order for the UAE, as a state actor, to stay in compliance with international law? And if international law itself does not have the appropriate regulation in place that can hold both private and public actors accountable, in addition to clearly establishing parameters, definitions, and enforcement mechanisms, how much legal and regulatory weight does this compliance language carry? Without management systems, clear procedural guidelines, active oversight, or binding international norms to determine if and when a violation may occur, in addition to the enforcement mechanisms to impose actual consequences, this Regulation's compliance language risks being nothing more than a symbolic gesture. In the absence of a universally accepted regulatory framework for space resources, or even for space in general, national or regional level legislation remains susceptible to diverse interpretations, a varying

2025; see also 'Space Is a Great Commons. It's Time to Treat It as Such.' (Carnegie Endowment for International Peace) https://carnegieendowment.org/research/2021/03/space-is-a-great-commons-its-time-to-treat-it-as-such?lang=en accessed 6 June 2025; see also Schauer WH, 'OUTER SPACE: The Boundless Commons?' (1977) 31 Journal of International Affairs 67; see also Cross MKD, 'Outer Space and the Idea of the Global Commons' (2021) 35 International Relations 384.

[81] The Annex of the Outer Space Treaty affirms the following: "Recognizing the common interest of all mankind in the progress of the exploration and use of outer space for peaceful purposes… Believing that the exploration and use of outer space should be carried on for the benefit of all peoples irrespective of the degree of their economic or scientific development". UN General Assembly, '2222 (XXI). Treaty on Principles Governing the Activities of States in the Exploration and Use of Outer Space, Including the Moon and Other Celestial Bodies'. Available online at https://www.unoosa.org/oosa/en/ourwork/spacelaw/treaties/outerspacetreaty.html.

[82] 2023 Regulation n 38.

understanding that could effectively undermine any current or future international efforts. In essence, these national regulatory frameworks, such as the Space Resources Regulation implemented by the UAE, could be doing more harm than good.

Although a critical perspective, it is possible that these national space resource utilisation regulatory frameworks are in fact carefully crafted strategic tools to ensure that the State positions itself as a private investor's oasis, rather than being the responsible multilateral partner that these nations are portraying. Previous scholars have discussed a trend found amongst emerging space powers where it seems as if national regulatory frameworks are being prioritised over a global governance regime and standards that would be binding on participating members of the international community.[83] It is not a question that the UAE has a clear intent to use the growing space industry as a means to bolster its own economy, as can been seen in the objectives of its previously discussed Space Strategy, those which include the following: promote effective local and global partnerships and investments in the space industry, and ensure a supporting legislative framework and infrastructure to match the future developments in the sector.[84] The nation has even previously declared that it wishes to "provide a competitive and regulatory environment that will attract the private sector and investments in the field of space", causing some to wonder if all the posturing is more about economic competitiveness than responsibility.[85] The question then becomes whether a spacefaring nation can be driven by national pride and interest, but act on and maintain its commitment to global and collective responsibility.

Ultimately, the UAE's dual position, as a future-thinking national actor and an advocate of international cooperation, highlights a defining paradox in contemporary space governance. The UAE's strategic initiatives, diplomatic efforts, and active participation in global governance indicate a commitment to multilateral values, while its national regulatory framework may, either intentionally or deliberately, surpass the questionable scope of what is currently permissible under space law. This serves as a prime example for why further global space resource utilisation regulation is needed, a system which is universally accepted, with parameters and terms that are clearly defined, reinforced by enforcement mechanisms, and containing the necessary principles to truly secure the peaceful use and access to outer space. In this context, the regulation of space resource utilisation reflects how well the international community is adapting to technological innovation and societal shifts. It is important for the international community to evaluate the UAE's actions, as it presents an opportunity to analyse whether an emerging spacefaring nation can equitably balance national interest and economic ambition with its actionable commitment to the shared governance of outer space.

[83] Tronchetti n 77.

[84] National Space Strategy 2023 n 20.

[85] Anih S and others, Space Resource Utilization: A View from an Emerging Space Faring Nation (Annette Froehlich ed, Springer International Publishing 2018).

Chelsea Francek is a lawyer and Ph.D. candidate in Law at the University of Waikato, New Zealand, specialising in the regulation of biotechnology and emerging technologies for outer space settlement. Her doctoral research examines how biotechnological adaptations will be critical for human survival beyond Earth, while also posing significant ethical, legal, and existential risks if left unregulated. She approaches space law through the lens of existential risk and species preservation, seeking to integrate ecofeminist and post-humanist approaches into a framework for equitable and anticipatory space governance. Chelsea's professional background includes senior legal, regulatory, and compliance roles in the corporate, international development, and non-profit sectors, with expertise in governance, security, and international regulatory systems.

Afterword

Mining the Sky: Reflections, Challenges, and Some Ending Thoughts

Karl Grossman

Abstract There has been a push to extract valuable minerals from space, boosted especially by the publication in 1997 of the book *Mining the Sky: Untold Riches From the Asteroids, Comets, and Planets* by Dr. John S. Lewis, along with the University of Arizona Lunar and Planetary Laboratory. Many nations are interested in space mining, including the United States, China, Japan and Russia. Companies have been formed to pursue it. "Asteroid Mining Could Solve Rare Metal Shortage" was the headline of a typical article encouraging it that ran on manufacturing.net in 2020. The sub-head: "Some near-earth asteroids could be worth billions. Now we just have to get there and back—safely and cost-effectively." But how realistic and necessary is this? " An article on Bloomberg Businessweek in 2023, typical of the critical viewpoint on space mining, related that "there's widespread scepticism that space mining can move beyond science fiction in the foreseeable future." It said "the outlay for extracting galactic resources remains astronomical. Prices of metals such as platinum or palladium would need to skyrocket for the ventures to pay off, and there's little to indicate that asteroid miners could extract enough metal to supplant Earth-based mining in any meaningful way." Meanwhile, there are the military and nuclear power links. Bruce Gagnon, coordinator of the Global Network Against Weapons and Nuclear Power in Space (of which the author was a co-founder) says "a key reason for the militarization of space is resource extraction on the planetary bodies." And, says Gagnon, "The aerospace industry says it needs nuclear power in space for mining the sky." He speaks about "nuclear reactors on rockets" being "needed to give them the kind of lift capability needed."

Keywords Space mining · Asteroid resources · Economic feasibility · Militarisation of space · Nuclear power in space · Resource extraction

K. Grossman (✉)
State University of New York at Old Westbury, Old Westbury, NY, USA
e-mail: grossmank@oldwestbury.edu

S. Bhat B and A. Variath (eds.), *Legal Responses to Space Resource Utilisation*, Space Law and Policy, https://doi.org/10.1007/978-981-95-8864-0_16

1 Early Visions of Space Mining

There has been a push to extract valuable minerals and other resources from space boosted especially by the book *Mining the Sky: Untold Riches from the Asteroids, Comets, and Planets*[1] by Dr. John S. Lewis, published in 1996. Lewis "has given us a glimpse of the future as it will be," states Hans Mark, a former deputy administrator of the National Aeronautics and Space Administration, on the book jacket. (NASA was established by the United States in 1958 and declares[2] as a mission: "Exploring the secrets of the universe for the benefit of all. NASA investigates the unknown in air and space, innovates for the benefit of humanity, and inspires the world through discovery." Under the words of Mark are those of Gordon Pettingill, professor at the Centre for Space Research at the Massachusetts Institute of Technology: "John Lewis forces us to accept the outrageously improbable as a logical part of our future."

Then there is this description of the book on its jacket: "While we worry over the depletion of the earth's natural resources, the pollution of our planet, and the challenges presented by the Earth's growing population, billions of dollars worth of metals, fuels, and life-sustaining substances await us in nearby space. In this visionary book, noted planet scientist John S. Lewis explains how we can mine these precious metals from the asteroids, comets and planets in our own solar system for use in space construction projects. And this is just one of the possibilities. Join John S. Lewis as he contemplates milking the moons of Mars for water and hollowing out asteroids for space-bound homesteaders—all while demonstrating the economic and technical feasibility of plans that were once considered pure fiction."

When he wrote the book, Lewis was a professor of planetary sciences at the University of Arizona's Lunar and Planetary Laboratory. He is a former member of the board of directors of the American Rocket Company. Now, a professor emeritus, he is chief scientist at Deep Space Industries. Among his chapters in *Mining the Sky* is one titled "The Asteroid Belt: Treasure Beyond Count." In it, Lewis writes: "The wealth of materials available in the belt asteroids, revealed by the study of meteorites, staggers the imagination. There is enough raw material in the asteroids to support a truly vast population….It appears that sharing the belt's wealth among five billion people leads to a shameless level of affluence."[3]

Lewis concludes *Mining the Sky* by declaring in its last chapter, "Civilization in Space," that "Scientifically and technically improvements in launch vehicles will make departure from Earth easy and inexpensive. Once we have a foothold in space, the mass of the asteroid belt will be at our disposal, permitting us to provide for the material needs of a million times as many people as Earth can hold….Using less than 1 percent of the helium-3 energy resources of Uranus and Neptune for fusion propulsion, we could send a billion interstellar arks, each containing a billion

[1] John S Lewis, *Mining the Sky: Untold Riches from the Asteroids, Comets, and Planets* (Basic Books 1996).

[2] NASA, 'NASA Missions' https://www.nasa.gov/nasa-missions/ accessed 21 August 2025.

[3] John S Lewis, *Mining the Sky: Untold Riches from the Asteroids, Comets, and Planets* (Basic Books 1996), 193,196.

people, to other stars. There are about a billion Sun-like stars in our galaxy. *We have the resources to colonise the entire Milky Way.* (Original italics) But there is no need to wait several billion years before beginning the cosmic diaspora; we could begin within a few decades."[4]

Other books since the publication of *Mining the Sky* include *When the Heavens Went on Sale: The Misfits and Geniuses Racing to Put Space Within Reach*[5] by Ashlee Vance, published in 2023. Vance's earlier books include *Elon Musk: Tesla, Space X, and the Quest for a Fantastic Future.* In *When the Heavens Went on Sale,* Vance writes: "Plenty of people question the assumptions behind this gamble. They're quite sure that the commercial space bubble will eventually pop with little to show for all the excitement. But while there will be some painful moments along the way, I'm confident that the next chapter in our technological evolution will continue and result in profound changes in how our world functions. This is the nature of technology and of the human spirit when offered a new playground."[6] And there is *The Space Economy: Capitalize on the Greatest Business Opportunity of Our Lifetime* by Chad Anderson also published in 2023.[7] Anderson writes: "Whether you work in satellite manufacturing or software development, rocket design or real estate, what's happening in space will become increasingly relevant to your work and industry over the next decade. The Space Economy is on track to become the dominant source of growth across the global economy. No one can ignore or downplay this phenomenon. Financial analysts, venture capitalists, and investment bankers aren't the only ones who must understand the investment side if they hope to navigate the next generation of digital infrastructure, the 'invisible backbone' already powering the world's largest industries, you could say we're all space professionals now."[8]

And there is *Space Enterprise*[9] by David Dietzler, published in 2024, in which, in a chapter titled "Interplanetary Trade," he writes: "Every nation on Earth trades with other nations. It seems that no nation can produce everything its people need and want, be it raw materials or finished products, or both. Some nations are rich with natural resources, and other nations have lots of cheap labour. A similar situation is likely to exist in space. The Moon has lots of raw materials, but it lacks many needed elements. Substitutes will be found or these elements will be imported from Earth, Mars or asteroids. Asteroids might supply Earth, the Moon and Mars with platinum group metals. The Moon and orbital space might import hydrocarbons from asteroids. Orbital space industries will sell energy, real estate, vacations and

---

[4] John S Lewis, *Mining the Sky: Untold Riches from the Asteroids, Comets, and Planets* (Basic Books 1996) 255.

[5] Ashley Vance, *When the Heavens Went on Sale: The Misfits and Geniuses Racing to Put Space Within Reach* (Ecco/HarperCollins 2023).

[6] Ashley Vance, *When the Heavens Went on Sale: The Misfits and Geniuses Racing to Put Space Within Reach* (Ecco/HarperCollins 2023) 284.

[7] Chad Anderson, *The Space Economy: Capitalize on the Greatest Business Opportunity of Our Lifetime* (John Wiley & Sons 2023).

[8] Chad Anderson, *The Space Economy: Capitalize on the Greatest Business Opportunity of Our Lifetime* (John Wiley & Sons 2023).133.

[9] David Dietzler, *Space Enterprise* (self-published, Middletown, Delaware 2024).

spaceship transportation. The Moon will sell vacations and travel, also. Mars presents a problem. What can Mars sell to the Moon that can't be obtained locally or from asteroids? What can the Moon sell to Mars?" Dietzler goes on: "The first order of business will be the establishment of a base on the Moon that bootstraps itself up to a point at which millions of tons of raw material are launched into space every year. This will cost ten or twenty billion dollars based on estimates."[10]

This book follows Dietzler's *Mining The Moon, Bootstrapping Space Industry*,[11] published in 2020. In it, he writes: "With lunar materials, large fleets of ships with thousands of settlers could be sent to Mars and the planet could be terraformed or, if you prefer, reformed or rejuvenated. What we learn on Mars will be applied to a more distant future when we travel to other star systems and find Mars-like planets. There could be other worlds where life has never evolved by but the right conditions exist for the planting of life on those worlds. The things we learn about mining asteroids and building space systems—veritable cities in space, will be applied to other star systems, which most probably have asteroids and comets also…Let's not forget that filmmakers will have so many opportunities that several cable channels could be filled with shows about what's going on up on the High Frontier and that these new frontiers will inspire untold generations of new artists, musicians, poets, writers, sculptors and other creative types. There could be televised athletic competitors in low-G, weightless dancers, circuses, and more broadcast live…Hope for a long-lasting future for the human race and no more end-of-the-world scenarios. The development of trade between the Moon, Mars, asteroids, space colonies, settlements on other worlds and subsequently a larger economy that offers new business opportunities and a chance for entrepreneurs to tap new markets."[12]

And there is *Space Mining and Manufacturing, Off-World Resources and Revolutionary Engineering Techniques* published in the United Kingdom in 2019.[13] The author, Davide Sivolella, an aerospace engineer, wrote as a preface: "I have been a life-long advocate for the development of space, but I must admit that since an early age I have always felt a pang of distress at seeing how space exploration seemed so far removed from assisting society in overcoming the predicaments that are increasingly threatening our future on this planet. Out of such dissatisfaction. I started to embrace the notion of moving our manufacturing industries into orbiting factories that draw their resources from the Moon or asteroids. Many studies dating back to the 1970s and 1980s have discussed off-world mining and manufacturing. Today they fall under the banner of 'living off the land' and envisage permanent outpost on the Moon or Mars. This is a worthy notion, but it is a restricting way of thinking because there is so much scope for exploiting such research to the benefit of humankind here

[10] David Dietzler, *Space Enterprise* (self-published, Middletown, Delaware 2024) 39.

[11] David Dietzler, *Mining the Moon: Bootstrapping Space Industry* (self-published, Middletown, Delaware 2020).

[12] David Dietzler, *Mining the Moon: Bootstrapping Space Industry* (self-published, Middletown, Delaware 2020) 11.

[13] Davide Sivolella, *Space Mining and Manufacturing: Off-World Resources and Revolutionary Engineering Techniques* (Praxis Publishing 2019).

on Earth. Mining and manufacturing activities can produce waste products of such kinds and in such amounts that they cannot be metabolized and neutralized by the air, water, and soil in the environment. In exceeding the capability of the environment to absorb what we throw at it, we face long-term pollution, Factored across the whole globe, we have now created the conditions for climate change, endangering of animal species, and loss of human lives. Our frantic way of living is also exhausting the resources of the planet. Many conflicts between peoples arise from efforts to exert control over dwindling resources. "There is no danger of environmental pollution in space, since there is no biosphere," he continues. "Water and minerals are available on the Moon and countless mountain-sized asteroids. We have grown familiar with such ideas in science fiction stories, but rest assured it is not beyond of the realm of real engineering….Rest assured, however, that unlike many fellow space advocates I am not preaching the development of space as a panacea that will enable us to write off environmental degradation, resource scarcity, and the like. Instead, I do believe that it has the potential to contribute to the mix of strategies available for tackling those issues. For the last 60 years of space exploration, we have been guilty of seeing space as a place for discovery and exploration. Now it is time to consider it also as a resource, and set a new space exploration and exploitation path that will genuinely benefit humankind."

2 Expanding the Space Economy

Among the very many articles about "mining the sky" was this piece on the website *manufacturing.net* in 2020 headed "Asteroid Mining Could Solve Rare Metal Shortage" with a sub-head, "Some near-earth asteroids could be worth billions. Now we just have to get there and back—safely and cost effectively." The article begins: "This may sound like science fiction, but since at least the 1970s, organizations like NASA have been considering the possible advantages of asteroid mining for resources. Now, in the midst of a new privatized space race and a global rare metal shortage, companies are revisiting the possibility of sources materials from outer space. A single asteroid could contain trillions of dollars' worth of previous metals, and sourcing materials from asteroids could enable large-scale construction in space….So, will the world economy soon be flooded by a vast new supply of previously-scarce precious metals? Probably not. At present, no company is quite ready to risk billions of dollars on a venture that requires technologies which have not been fully developed and techniques that are not proven." But it said "it may only be a matter of time until an asteroid with enough previous metals passes near the Earth and someone decides it's worth the risk to try to catch it."[14]

[14] 'Asteroid Mining Could Solve Rare Metal Shortage' (Manufacturing.net, 2019) https://www.manufacturing.net/technology/blog/21113380/asteroid-mining-could-solve-rare-metal-shortage accessed 21 August 2025.

An article in Bloomberg *Businessweek* in 2023 on space mining cast doubt about the "Asteroid Gold," and said its sub-head: "Despite entrepreneurs' hopes, the technical challenges and huge costs mean no one will be retrieving ore from the heavens anytime soon." The piece said "there's widespread skepticism that space mining can move beyond science fiction beyond the foreseeable future. While the price tag of space exploration has declined drastically—NASA says the cost of launching rockets fell from 92% from 2000 to 2018—the outlay for extracting galactic resources remains astronomical. Prices of metals such as platinum or palladium would need to skyrocket for the ventures to pay off, and there's little to indicate that asteroid miners could extract enough metal to supplant Earth-based mining in any meaningful way….Still, the skepticism hasn't dissuaded many space executives, who point to rapid advancements in technology and innovations from companies such as Elon Musk's Space Exploration Technologies Corp. as evidence of what's possible."[15]

Then there are the military and nuclear power links. Bruce Gagnon, coordinator of the Global Network Against Weapons & Nuclear Power in Space, puts exploiting space in a historical perspective with a military angle citing a 1997 report by the U.S. Space Command titled *Vision for 2020.*[16] He quotes passages of this 1997 U.S. military plan: "Historically, military forces have evolved to protect national interests and investments—both military and economic. During the rise of sea commerce, nations built navies to protect and enhance their commercial interests. During the westward expansion of the continental United States, military outposts and the cavalry emerged to protect our wagon trains, settlements, and railroads….During the early portion of the 21st century, space power will also evolve into a separate and equal medium of warfare. Likewise, space forces will emerge to protect military and commercial national interests and investment in the space medium due to their increasing importance." Says[17] Gagnon: "This is essentially a declaration of war as the document later states the U.S. will 'control and dominate space and deny other nations access to space.'"

"In a similar publication," Gagnon said, "this one called *Military Space Forces: The Next 50 Years*[18] published by the U.S. Congress in 1989," the militarists were already planning to project power in and through space in order to control the pathway on and off our planet." This publication, Gagnon notes, says, "'Military space forces at the bottom of the Earth's so-called gravity well are poorly positioned to accomplish offensive/defensive/deterrent missions, because great energy is needed to overcome gravity during launch. Forces at the top, on a space counterpart of 'high ground', could initiate action and detect, identify, track, intercept, or otherwise respond more rapidly to attacks.' Put simply, it takes less energy to drop objects down a well than

[15] Lizette Chapman, 'Space Mining Will Likely Draw More Venture Capital than Asteroid Metals' *Bloomberg* (6 December 2023) https://www.bloomberg.com/news/articles/2023-12-06/space-mining-will-likely-draw-more-venture-capital-money-than-asteroid-metals accessed 21 August 2025.

[16] Vision 2020 Report https://thecommunity.com/wp-content/uploads/2018/08/Vision2020.pdf accessed 21 August 2025.

[17] Personal Interview of Gagnon by author (March 2024).

[18] *Military Space Forces: The Next 50 Years* (Biblio.com) https://www.biblio.com/book/military-space-forces-next-50-years/d/1138544814 accessed 21 August 2025.

to cast them out." "The report concludes," Gagnon continues, "with control of the Moon and with armed space stations at the L4 and L5 quadrants on either side of the Moon, it says 'Armed forces might lie in wait at that location to hijack rival shipments on return.' That's called piracy—something the Western colonial powers have perfected during their more than 500-year reign over our globe."

"Fast forward to today, and we find two key developments to this story," he says. "First, the U.S. has created the Artemis program, an American-led effort to return to the Moon by 2026, with the ultimate goal of expanding space exploration to Mars and beyond. As of early 2024, 35 nations have signed the accords. The U.S today has more than $35 trillion of national debt and can't afford to pay for these expensive space dreams by itself. So, Washington is signing up the allies to help cover the costs of these grand missions. But the US intends to stay in charge of the tip of the spear." "Not to be left out, Russia and China are considering joint construction of a nuclear power plant on the lunar surface between 2033 and 2035," says Gagnon, whose organization is based in the U.S. state of Maine, goes on: "A nuclear reaction could provide the robust energy supply needed for lunar operations a Russian spokesman has said."

3 Nuclear Power as the Enabler of Off-World Extraction

Returning to *Military Space Forces*, we find this important claim by the U.S. Congress in *Military Space Forces: The Next 50 Years*, Nuclear reactors thus remain the only known long-lived, compact source able to supply military space forces with electric power needed for 'space-based lasers, neutral particle beams, mass drivers and rail guns. Nuclear reactors must support major bases on the moon until better options, yet to be identified, become available." "So, the race to the moon and beyond is underway," says Gagnon, "And the U.S. at least has decided that space weaponization and nuclear power will be essential in their mission to 'control and dominate' the heavens. Russia and China are not far behind as they view nuclear power as a necessary component to their plans, as they too are rounding up allies to be with them in this coming period of likely conflict." "The war system, that has been deeply woven into the fabric of life on Earth, is now being taken into the heavens," says Gagnon. "I call it the 'bad seed' of war, greed and environmental degradation." "Thus, it is ever the more crucial," says Gagnon, "that a global dialogue must begin asking those paying the freight for these expensive space operations if they wish to see the competitive global war system move above our heads. And if not, what should be put into place to prevent such an insanity from ever happening? This has been the work of the Global Network Against Weapons & Nuclear Power in Space for the past 32 years." The organization was founded in 1992. And, Gagnon says, "a key reason for the militarization of space is resource extraction on the planetary bodies" and nuclear power is seen as a space energy source for that and for weapons in space but also for power for spacecraft involved in mining the sky. "The aerospace industry says it needs nuclear power in space for mining the sky," says Gagnon. He speaks about "nuclear "nuclear reactors on rockets" being considered "required to give them the kind of lift capability needed."

Indeed, an extensive paper on this was published in 2013 titled "Nuclear Power and Associated Environmental Issues in the Transition of Exploration and Mining on Earth to the Development of Off-World Natural Resources in the 21st Century."[19] Its six authors were scientists and engineers for various corporations. The abstract of their paper declared: "Once humans landed on the Moon on July 20, 1969, the goal of space exploration envisioned by United States President John F. Kennedy in 1961 was already being realized. Achievement of this goal depended on the development of technologies to turn his vision into reality. One technology that was critical to the success of this goal was the harnessing of nuclear power to run these new systems. Nuclear systems provide power for satellite and deep space exploratory missions. In the future, they will provide propulsion for spacecraft and drive planet-based power systems. The maturation of technologies that underlie these systems ran parallel to an evolving rationale regarding the need to explore our own solar system and beyond. Since the Space Race, forward-looking analysis of our situation on Earth reveals that space exploration will one day provide natural resources that will enable further exploration and will provide new sources for our dwindling resources and offset their increasing prices or scarcity on Earth. "Mining is anticipated on the Moon for increasingly valuable commodities," the abstract continued, such as thorium and samarium and on selected asteroids or other moons as a demonstration of technology at scales never before imagined. In addition, the discovery of helium03 on the Moon may provide an abundant power source on the Moon and on Earth through nuclear fusion technologies. However, until the physics of fusion is solved, that resource will remain on the shelf and may even be stockpiled on the Moon until needed.

"It is clear," the abstract of the paper went on, "that nuclear power will provide the means necessary to realize these goals while advances in other areas will provide enhanced environmental safeguards in using nuclear power in innovative ways, such as a space elevator or by a ramjet space plane to deliver materials to and from the Earth's surface and personnel and equipment into space and a space gravity tractor to nudge errant asteroids and other bodies out of collision orbits. Nuclear systems will enable humankind to expand beyond the boundaries of Earth, provide new frontiers for exploration, ensure our protection, and renew critical natural resources while advancing spin-off technology on Earth. "During the past ten years, China, Japan, India, and other countries have mounted serious missions to explore the Moon and elsewhere," it said. "Recent exploration discoveries by Japan on the Moon may mark the beginning of a new race to the Moon and into space to explore for and develop natural resources including water (from dark craters to make hydrogen for fuel and oxygen, etc.), nuclear materials (uranium, thorium, and helium3), rare- earth minerals, and other industrial commodities needed for use in space and on Earth in the decades ahead."

[19] 'The History and Path Forward of the Human Species…' (Academia.edu, 2023) https://www.academia.edu/108893790/ accessed 21 August 2025.

The use of nuclear power in space and accidents involving it have been a focus of this writer. My work has included authoring the book *The Wrong Stuff: The Space Program's Nuclear Threat to Our Planet.*[20] In it, for example, I relate the accident on April 21, 1964 involving a United States navigational satellite powered by a SNAP-9A plutonium-fueled radioisotope thermoelectric generator (RTG). The satellite failed to achieve orbit and fell from the sky, breaking up along with the RTG as it hit the atmosphere. The SNAP-9A was fueled with 2.1 pounds of plutonium. I quote from a publication titled *Emergency Preparedness for Nuclear-Powered Satellites,* a report of Europe's Organization for Economic Cooperation and the Swedish National Institute for Radiation Protection, which stated that the satellite with its plutonium system broke up "over the West Indian Ocean north of Madagascar" and a "worldwide soil sampling program carried out in 1970 showed radioactive SNAP-9A "debris to be present at all continents and at all latitudes." I quoted Dr. John Gofman, a former professor of molecular and cell biology at the University of California at Berkeley, an M.D. and Ph.D., who developed the first methods for isolating plutonium for the Manhattan Project and co-discovered several radioisotopes, including Uranium-233, as saying that the dispersal of plutonium from the SNAP-9A accident added to global lung cancer rates.

The Wrong Stuff starts with the story of another nuclear-fueled space probe, the Russian Mars 96 space probe—with a half-pound of plutonium on board—falling back to Earth in November 1996. It was expected to break up over east-central Australia as then U.S. President Bill Clinton was to pay a visit there. It ended up, however, disintegrating over Chile and Bolivia. And *The Wrong Stuff* tells of how the next mission of the ill-fated Challenger space shuttle, slated for May 1986, involved it lofting the Ulysses space probe with an RTG with 24.2 pounds of plutonium. If the Challenger had exploded on that next mission rather than on January 28, 1986, it would not have only been six astronauts and a "teacher in space," Christa McAuliffe, losing their lives, but, if the plutonium was released as dust, many more people would have.

David Dietzler in his book *Mining the Moon,* declares under a heading "Nuclear Power on the Moon," that: "On the Moon, there is no life and no air or water to contaminate. If nuclear waste did escape into the environment, it would not dissolve into groundwater and spread because there is no groundwater on the Moon. Permafrost has been detected all over Mars, and we must wonder about nuclear waste giving off heat and melting the frost and spreading. Nuclear power could be used on the Moon safely. The environment there is already so highly radioactive due to cosmic rays that some uranium would just be a drop of water in a swimming pool. Our Earth's geomagnetic field deflects many cosmic rays, and the thick atmosphere blocks the rest of them from reaching the surface with the exception of some rare ultra-high energy rays. Inhabitants of the Moon will need to cover their shelters with about nine

[20] Karl Grossman, *The Wrong Stuff: The Space Program's Nuclear Threat to Our Planet* (Common Courage Press 1997).

tons of regolith per square meter or a layer about six meters (20 feet) thick. They might also locate bases inside lava tubes."[21]

"The Moon miners could go into business selling nuclear fuels to interplanetary spaceship owners," he writes. As for nuclear-powered rockets,[22] "Nuclear electric propulsion (NEP) would use far less reaction mass than chemical rockets and it could deliver higher speeds to short transit times to other planets. While solar electric propulsion (SEP) might be considered for ships in the inner solar system....NEP will be essential for travel into the outer solar system. Time must also be considered. A minimum energy orbit from Earth to Jupiter could be attained by chemical rockets, but it will take 2.7 years and for Saturn, 6 years. Radiation exposure, psychological stress, life support and supply problems would make this impractical for human travelers. A six-month journey to Mars could be shortened to 39 days with fast NEP using VASIMIR and vapor core reactors generating 200 megawatts of electric power. Voyages to Jupiter and Saturn or their moons, more properly could be completed in less than a year with powerful enough reactors and lunar nuclear fuels." Dietzler adds: "The fear of nuclear power has something to do with confusion with nuclear weapons. It also has something to do with the fear of cancer and the fear of genetic mutation. Nobody wants deformed offspring due to radiation exposure and a curse on all subsequent generations. Medicine might ultimately cure all cancers, and genetic engineering might rid humanity of all inherited diseases and deformities without turning to eugenics programs that breed people like livestock. Radiation won't be so universally feared when this is achieved. Even so, nobody wants a nuclear reactor to come crashing in from space. No businesspersons want to be sued because one of their nuclear vessels crashed into up burned up in the air over a large city."[23] Beyond the dangers with using nuclear power in space, there are the perils of space itself—questioning how realistic is "mining the sky" and other schemes about the heavens having gone "on sale."

Sarah Novak wrote an article[24] in 2023 in *Discover Magazine* titled "Road Trip to the Red Planet" in which she spoke about "what it would be like to stay or live on Mars." In it, she said: "Mars is an arid, inhospitable desert, with temperatures reaching minus 81 degrees Fahrenheit regularly. What's more, the Red Planet would not be habitable without spacesuits and a completely enclosed environment, because the planet's air is about 95 percent carbon dioxide." She continued: "Colonists on Mars would face other challenges, too. For starters, it would be difficult to grow plants in Mars' regolith, or soil, which contains poisonous compounds of chlorine in

---

[21] Karl Grossman, *The Wrong Stuff: The Space Program's Nuclear Threat to Our Planet* (Common Courage Press 1997) 118.

[22] Karl Grossman, *The Wrong Stuff: The Space Program's Nuclear Threat to Our Planet* (Common Courage Press 1997) 120.

[23] Karl Grossman, *The Wrong Stuff: The Space Program's Nuclear Threat to Our Planet* (Common Courage Press 1997) 124.

[24] Eric Betz, 'What Would a Trip to Mars Look Like for a Tourist?' *Discover Magazine* (2023) https://www.discovermagazine.com/the-sciences/what-would-a-trip-to-mars-look-like-for-a-tourist accessed 21 August 2025.

molecules called perchlorates." She writes: "All the elements that we take for granted on Earth would not exist on Mars."

Or consider the book published in 2023, *A City on Mars: Can We Settle Space, Should We Settle Space, and Have We Really Thought This Through?*[25] In it, Kelly and Zach Weinersmith write: "The truth is that settling other worlds, in the sense of creating self-sustaining societies somewhere away from Earth, is not only quite unlikely anytime soon, it won't deliver on the benefits touted by advocates. No vast riches, no new independent nations, no second home for humanity, not even a safety bunker for the ultra elites. Yet, we find ourselves in a world where space agencies, huge corporations, and media-savvy billionaires are promising something else. According to them, settlements are coming, perhaps as soon as 2050 or so..." The Weinersmiths continue: "Jeff Bezos likely got his theory of space settlement from Dr. Gerard K. O'Neill, a professor at Princeton whose lectures Bezos attended as a young student. O'Neill's philosophy for space oriented around large solar-powered stations as the way to save Earth's economy and ecology. This argument may have been plausible circa 1970, when it was widely believed that space would keep getting cheaper and that energy and food crises would result in unprecedented worldwide famines by the 1980s. Today, you do a much better job of saving Earth's biosphere with Earth-based solar and wind power. Even if we thought space settlements would take pressure off of Earth's seas and lands, they will absolutely not arrive in time to thwart an environmental calamity."

They close with: "Having devoted all this time to trying to see a more realistic vision of space settlement, one of the few things we can say with certainty is that we have more appreciation of this grass by our (original italics) home. In fact, not just the grass, but the home itself, and the people inside it...Earth isn't perfect, but as planets go, it's a pretty good one. We aren't saying you should give up on the hope for a life off-world—that's too pretty a dream to part with. What we are saying is that if you *do* (original italics) want that dream, you have to see the challenges as they are—real, profound, and present at every level from molecules to sociology." In 2024 on *Jacobin,* Peter Howson wrote a piece[26] titled "Get These Rich People Off The Moon." He wrote how in January of 2024 a "U.S. private venture crashed back to Earth. Astrobotic's Peregrine lander had been supposed to dispose of at least seventy dead rich people (and one rich dog) on the lunar surface. Spending billions of dollars dumping odd things in space has become a tradition among the lunar classes. Elon Musk famously sent a Tesla Roadster as the dummy payload for the 2018 Falcon Heavy test flight. Driven by a mannequin in a spacesuit dubbed 'Starman,' the car is now an enduring satellite of the sun."

NASA paid Pittsburgh-based Astrobotic $108 million toward the Peregrine mission," related Howson. Its mishap, described by *The New York Times* as a "propulsion malfunction" led to it being aimed back at Earth. "American Company's Moon

[25] Kelly Weinersmith and Zach Weinersmith, *A City on Mars: Can We Settle Space, Should We Settle Space, And Have We Really Thought This Through?* (Penguin Press 2023).

[26] Luke Savage, 'Space Billionaires' *Jacobin* (February 2024) https://jacobin.com/2024/02/space-billionaires-intuitive-machines-musk accessed 21 August 2025.

Lander Disintegrates in Earth's Atmosphere" was the headline of a piece[27] on this in *The Times* by Kenneth Chang who, the newspaper related, "has reported on four failed lander missions, and three successful landings, since 2019." Howson referenced how earlier a "Texas start-up Intuitive Machines…achieved the first moon landing by a private firm. It's dumping rich people's detritus on the lunar surface—a grim sign of how the superrich plan to plant their flag beyond our planet." "Other than allowing billionaires and private companies to benefit from taxpayer-funded pipe dreams and advertising, the value of going to the moon for all mankind is not at all clear," he wrote. Further, said Howson, "It all sounds very exciting. But realizing these fantasies have costs for us. According to Atrium, a big insurer for rocket-makers, early space-faring outfits should typically expect 30 percent of their launches to end in catastrophic failure."

4 Conclusion

Space "is one of the most extreme environments imaginable," the European Space Agency emphasizes on its website.[28] The website states: "The first extreme condition a spacecraft has to deal with is that of launch. The rocket that places the spacecraft into orbit will also shake it violently and batter it with extremely loud sound waves. Either of these phenomena can shatter delicate pieces of equipment….Temperatures in space can range from the extremely cold, hundreds of degrees below freezing, to many hundreds of degrees above especially if spacecraft ventures close to the Sun. Although there is no air in space, energy is carried by radiation, usually coming from the Sun, that causes heating when it is absorbed by spacecraft, planets or other celestial bodies." "Radiation in space can be split into 'trapped' and 'transient' types. The trapped particles are the subatomic particles, mainly protons and electrons, trapped by Earth's magnetic field which creates the so-called Van Allen radiation belts around our planet….The transient radiation is mainly composed of protons and cosmic rays that constantly stream through space and are enhanced during the magnetic storms on the Sun known as 'solar flares'. When this radiation collides with electronic circuits, they can change the contents of memory cells, cause spurious currents to flow around the craft or even burn out computer chips," the agency continues. "Meteor showers can also damage spacecraft."

"The Problems with Space Mining No One Is Talking About" was the heading of a 2022 article on the Impakter website by Andrew S. Rivkin, a planetary astronomer at Johns Hopkins University in Baltimore, Maryland. Its sub-head: "Asteroid mining could unlock untold riches and thorny ethical problems. Are we ready as we race to

[27] Kenneth Chang, 'American Company's Moon Lander Disintegrates in Earth's Atmosphere' *New York Times* (18 January 2024) https://www.nytimes.com/2024/01/18/science/moon-lander-peregrine-nasa.html accessed 21 August 2025.

[28] European Space Agency, 'Surviving Extreme Conditions in Space' https://www.esa.int/Science_Exploration/Space_Science/Extreme_space/Surviving_extreme_conditions_in_space accessed 21 August 2025.

the launch pad?"[29] "Asteroid mining is coming sooner than people realise," wrote Rivkin. "Several asteroid mining companies have developed serious business cases. Demonstration missions could occur within a few years for interested, patient and well-funded investors. The most enthusiastic advocates of asteroid mining suggest it could unlock trillions of dollars of wealth; more sober analysts: tens of billions." He continued: "Space resources are occasionally compared to those of the sea. But the barrier to entry for the ocean is a fishing pole or net or the ability to dive. Asteroid mining, by contrast, requires advanced technology and large amounts of starting capital." Further, "The level of wealth required to pursue an asteroid mining venture is concentrated in the hands of a very small number of people. Large disparities exist between those able to take advantage of the resources and those most at risk of harm by exploitation.

It also seems possible, if not likely, that the earliest successes in asteroid mining will be the only successes. Competition with established companies will be an additional barrier, and a monopoly or cartel may develop." He cites Daniel Pilchman, a legal philosopher, as saying, "asteroid mining is likely to increase inequality on Earth. He argues it will therefore be an unethical practice, unless it can be regulated to bring benefits to all. James Schwartz, also a philosopher, says mining asteroid resources is unlikely to 'significantly improve the well-being of average human beings,' and by extension, would be unethical." A "legacy of colonialist decision-making harming Indigenous people throughout history has left a stain on the profession of mining—a legacy space miners would do well to avoid," writes Rivkin. "Asteroid mining may not harm humans in a way that's comparable to terrestrial mining, but disruption and dust from mine operations is still possible. Physicist Paul Wiegert studied the spread of dust when NASA and the European Space Agency tried nudging the asteroid Didymos to test an asteroid impact prevention system. He concluded that while the released dust and rubble posed no threat to Earth, mining operations could plausibly generate lots of such debris."

"Asteroid resources are unlikely to harbour life; meanwhile, the only planetary body with known life in the Solar System, Earth, continues to be exploited," Rivkin points out. "The United Nations takes the view that space exploration should be done for the benefit of all." "Space science is often lauded for its ability to inspire future generations. That inspiration can cut both ways: how humans act in taking these steps into the cosmos will set precedents that subsequent generations will either follow or have to undo. The question is not only how to make technical progress but whether we should." Indeed, the question is "whether we should." We reside on this exquisite blue marble in space that sustains life, and we so need to be stewards caring for the Earth, not indulge in dangerous, ultra-expensive schemes of getting rich by mining the heavens and dubious fantasies of escaping from Earth.

[29] 'The Problems with Space Mining No One is Talking About' *Impakter* (2024) https://impakter.com/the-problems-with-space-mining-no-one-is-talking-about/ accessed 21 August 2025.

Prof. Karl Grossman is a full professor of journalism at the State University of New York at Old Westbury, which is in the United States. He has written extensively and done television programs about weapons and nuclear power in space and has given presentations about this all over the world. Books Professor Grossman has written include *The Wrong Stuff: The Space Program's Nuclear Threat to Our Planet* and *Weapons in Space.* He wrote and presented the television documentaries *Nukes in Space: The Nuclearization and Weaponization of the Heavens* and *Star Wars Returns.* Grossman has given presentations about weapons and nuclear power in space at the United Nations in New York and Geneva, and twice addressed members of the British Parliament on this. For 20 years, he was a member of the Commission on Disarmament Education, Conflict Resolution and Peace of the United Nations and International Association of University Presidents. Honors he has received for journalism include the George Polk, James Aronson, Generoso Pope, Leo Goodman and John Peter Zenger Awards.

manufacturer's authorised representative in the EU is Springer
ature Customer Service Centre GmbH, Europaplatz 3, 69115 Heidelberg,
Germany. If you have any concerns regarding our products, please
contact ProductSafety@springernature.com

Printed and bound by CPI Group (UK) Ltd, Croydon, CR0 4YY
07/10/2026
02229357-0011